The Silences of Science

Over the last half century scholars from a range of disciplines have attempted to theorise silence. Naively we tend to think of silence negatively, as a lack, an emptiness. Yet silence studies shows that silence is more than mere absence. All speech incorporates silence, not only in the gaps between words or the pauses that facilitate turn taking, but in the omissions that result from the necessary selectivity of communicative acts. Thus silence is significant in and of itself; it is a sign that has socially constructed (albeit context-dependent and ambiguous) meanings.

To date, studies of science communication have focussed on what is said rather than what is not said. They have highlighted the content of communication rather than its form, and have largely ignored the gaps, pauses and lacunae that are an essential, and meaningful, part of any communicative act. Both the sociology of science and the history of science have also failed to highlight the varied functions of silence in the practice of science, despite interests in tacit knowledge and cultures of secrecy. Through a range of case studies from historical and contemporary situations, this volume draws attention to the significance of silence, its different qualities and uses, and the nature, function and meaning of silence for science and technology studies.

Felicity Mellor is Senior Lecturer in Science Communication at Imperial College London, UK. Her research examines the representation of science in the media and the ideological dimensions of scientists' public discourse. She is editor, along with Alice Bell and Sarah Davies, of *Science and Its Publics* (Cambridge Scholars Publishing, 2009).

Stephen Webster is Director of the Science Communication Unit at Imperial College London, UK. His books include *Charles Darwin* (Stroud: The History Press, 2015) and *Thinking about Biology* (Cambridge: Cambridge University Press, 2003).

The Silences of Science
Gaps and pauses in the communication of science

Edited by
Felicity Mellor and Stephen Webster

LONDON AND NEW YORK

First published 2017
by Routledge
2 Park Square, Milton Park, Abingdon, Oxon OX14 4RN

and by Routledge
711 Third Avenue, New York, NY 10017

Routledge is an imprint of the Taylor & Francis Group, an informa business

© 2017 Felicity Mellor and Stephen Webster

The right of the editor to be identified as the author of the editorial material, and of the authors for their individual chapters, has been asserted in accordance with sections 77 and 78 of the Copyright, Designs and Patents Act 1988.

All rights reserved. No part of this book may be reprinted or reproduced or utilised in any form or by any electronic, mechanical, or other means, now known or hereafter invented, including photocopying and recording, or in any information storage or retrieval system, without permission in writing from the publishers.

Trademark notice: Product or corporate names may be trademarks or registered trademarks, and are used only for identification and explanation without intent to infringe.

British Library Cataloguing in Publication Data
A catalogue record for this book is available from the British Library

Library of Congress Cataloging in Publication Data
Names: Mellor, Felicity, editor. | Webster, Stephen, 1957- editor.
Title: The silences of science : gaps and pauses in the communication of science / edited by Felicity Mellor and Stephen Webster.
Description: Abingdon, Oxon ; New York, NY : Routledge, [2016] | Includes bibliographical references and index.
Identifiers: LCCN 2016006616| ISBN 9781472459978 (hardback) | ISBN 9781315609102 (ebook)
Subjects: LCSH: Communication in science. | Science--Methodology--Social aspects.
Classification: LCC Q223 .S544 2016 | DDC 501.4--dc23
LC record available at http://lccn.loc.gov/2016006616

ISBN: 978-1-4724-5997-8 (hbk)
ISBN: 978-1-315-60910-2 (ebk)

Typeset in Times New Roman
by Taylor & Francis Books

Contents

List of illustrations — vii
List of contributors — viii

Introduction: The communicative functions of silence in science — 1
FELICITY MELLOR

PART I
Choosing silence — 29

1 'He didn't go round the conference circuit talking about it': Oral histories of Joseph Farman and the ozone hole — 31
PAUL MERCHANT

2 Darwin's silence: An anatomy of quietude — 48
STEPHEN WEBSTER

3 'Tired with this subject ...': Isaac Newton on publishing and the ideal natural philosopher — 65
CORNELIS J. SCHILT

4 Engineers at the patient's bedside: The case of silence in inter-institutional educational innovation — 89
NICK W. VEROUDEN, MAARTEN C. A. VAN DER SANDEN AND NOELLE M. N. C. AARTS

PART II
Cultures of silence — 113

5 Talking about secrets: The Hanford nuclear facility and news reporting of silence, 1945–1989 — 115
DANIELE MACUGLIA

6 Silence and selection: The 'trick cyclist' at the War Office
Selection Boards 135
ALICE WHITE

7 The silenced subject: Oral history and the experience of cancer
research 152
CATRIONA GILMOUR HAMILTON

8 Reconstructing ancient thought: The case of Ancient
Egyptian mathematics 172
ELIZABETH HIND

9 Meditations on silence: The (non-)conveying of the experiential in
scientific accounts of Buddhist meditation 193
BRIAN RAPPERT, CATELIJNE COOPMANS AND GIOVANNA COLOMBETTI

PART III
Silences in the public sphere 219

10 The silent introduction of synthetic dyestuffs into
nineteenth-century food 221
CAROLYN COBBOLD

11 Having it all: Ownership in open science 241
ANN GRAND

12 Shocking silences: The management and distribution of
silences around TASER™ 253
ABI DYMOND

13 'An outcry of silences': Charles Hoy Fort and the uncanny
voices of science 274
CHARLOTTE SLEIGH

Index 296

Illustrations

Figure

1.1 Joseph Farman with a Dobson Spectrophotometer at British Antarctic Survey headquarters, Cambridge, 2011 — 32

Tables

0.1 A typology of silence — 8
8.1 The relative frequency of multiplication techniques in the Rhind Mathematical Papyrus — 181
8.2 Words used for geometric dimensions in the Rhind Mathematical Papyrus — 186

Contributors

Noelle Aarts is Professor of Strategic Communication at the University of Amsterdam and Professor of Communication and Change in Life Science Contexts at Wageningen University, the Netherlands. Focusing on conversations between people, she studies inter-human processes and communication for creating space for change. She has published on topics such as communication of organizations with their environment, conflict and negotiation in life science domains, dealing with ambivalence, network-building and self-organization.

Carolyn Cobbold investigated the use of synthetic dyes in food production in the nineteenth century as part of her Ph.D. studies at Cambridge University, having completed a Master's in the History of Science at Imperial College and UCL. As a historian, she is particularly interested in the relationship between science, society and risk management, having worked as a risk management journalist for three decades investigating and reporting on technical, human and natural risks faced by companies and organisations throughout the world. Carolyn is also project leader for a partnership of local communities and statutory authorities which pioneered the UK's largest coastal realignment scheme and a new approach to integrated planning on the south coast of England.

Giovanna Colombetti is Associate Professor in the Department of Sociology, Philosophy and Anthropology at the University of Exeter. She works primarily in the philosophy of cognitive science, and is particularly interested in debates on 'embodied' and 'enactive' cognition, on emotion and affectivity and on the relationship between natural science and phenomenological philosophy. Her book *The Feeling Body: Affective Science Meets the Enactive Mind* (2014) proposes a reconceptualization of various affective phenomena from an 'enactive' perspective. More recently she has worked on the idea that the mind, including affectivity, can 'extend' beyond the boundaries of the living organism, to include parts of the material world.

Catelijne Coopmans is a Research Fellow at the Asia Research Institute at the National University of Singapore and a Fellow and Director of Studies at

Tembusu College at the same university. Much of her research draws on ethnography and discourse analysis to examine the arrangements in and through which (new) forms of visual evidence are used and valued. Recurrent themes include revelation and concealment, expertise and accountability. Catelijne is co-editor (with Janet Vertesi, Michael Lynch and Steve Woolgar) of *Representation in Scientific Practice Revisited*, published with MIT Press in 2014.

Abi Dymond is an ESRC-funded Ph.D. candidate in the Department of Sociology, Philosophy and Anthropology at the University of Exeter and the University of Bristol Law School. Amongst other roles, she works part-time for the UK NGO the Omega Research Foundation and was previously a member of the Metropolitan Police's TASER Reference Group. All views expressed are her own.

Catriona Gilmour Hamilton completed her Ph.D. in 2016 at the Centre for Medical Humanities at Oxford Brookes University. Her Ph.D. thesis, *A Cohort of One: Oral History and Cancer Research in Britain, 1970–2010*, uses oral history to challenge and augment the historiography of cancer research. She completed an M.A. in History of Medicine in 2011, with a dissertation on the subject of changing doctor–patient communication about cancer in late twentieth-century Britain, and the roles of doctors and cancer support organisations in facilitating that change. Her research interests include: oral history; illness narrative; cancer cultures; research ethics; medical ethics. She has a background in oncology nursing and, before taking up full-time postgraduate study, worked in the voluntary sector for a national cancer support organisation.

Ann Grand is Lecturer in Science Communication at the University of Western Australia. Her research interests lie in the use of online and digital platforms to engage the public with science. She is particularly interested in the ways in which the emerging ideas and practice of open science can be a medium for public access to, and involvement in, the process of science and an innovative method for real-time science communication. She has been a research fellow in the Science Communication Unit at the University of the West of England, Bristol and a research associate with the Open University Catalyst for Public Engagement with Research.

Elizabeth Hind completed research into Egyptian mathematics in the Science Communication Unit at the University of Liverpool. She is interested in the cultural dimensions of science and mathematics and uses this interest to inform her work in STEM education. She has worked for organisations such as the British Science Association and STEMNET. She currently combines this work with independent research.

Daniele Macuglia is a Ph.D. candidate at the University of Chicago's Fishbein Center for the History of Science and Medicine. He earned a Laurea

Magistrale in physics from the University of Pavia in Italy and a Master's degree in the history of science from the University of Chicago. He won the 13th Italian National Contest for Young Scientists and received a European Union Contest for Young Scientists Special Prize.

Felicity Mellor is Senior Lecturer in Science Communication at Imperial College London, where she runs the M.Sc. in Science Communication. Her research interests concern the representation of science in the media, with a particular focus on the ideological dimensions of scientists' public discourse. Her publications include analyses of public discourse about asteroid impacts, the narrative structure of recent cosmology and science journalists' reliance on a set of 'non-news values'. In 2010 she led the content analysis for the BBC Trust's review of the impartiality of the BBC's science output and she is co-editor with Alice R. Bell and Sarah R. Davies of *Science and Its Publics* (2008).

Paul Merchant is an oral history interviewer for 'An Oral History of British Science', led by National Life Stories at the British Library. He is also a Research Associate on the project 'Science and Religion: Exploring the Spectrum' at Newman University, Birmingham and York University, Toronto. His background is in cultural geography and oral histories of the production of knowledge.

Brian Rappert is a Professor of Science, Technology and Public Affairs at the University of Exeter. His long term interest has been the examination of the strategic management of information; particularly in the relation to armed conflict. His books include *Controlling the Weapons of War: Politics, Persuasion, and the Prohibition of Inhumanity; Biotechnology, Security and the Search for Limits*; and *Education and Ethics in the Life Sciences*. More recently he has been interested in the social, ethical and political issues associated with researching and writing about secrets, as in his book *Experimental Secrets* (2009) and *How to Look Good in a War* (2012).

Cornelis J. (Kees-Jan) Schilt was educated at Utrecht University and the University of Sussex, and is currently finishing his D.Phil. in History of Science at Linacre College, Oxford. He is also Transcription Manager for the online Newton Project, which since 1998 has been transcribing and publishing all of Isaac Newton's writings. Schilt specializes in early modern history, history of science, and in particular Isaac Newton's life and works. During the summer of 2015 he was a Dibner Research Fellow at the Huntington Library in San Marino, CA. He has published on Newton's optics, his alchemy, and his chronological works. He regularly blogs about his research on www.corpusnewtonicum.wordpress.com.

Charlotte Sleigh is Reader in History of Science at the University of Kent. Her research has ranged across diverse areas in the science humanities including animal history, literature and science, and science communication. Her most recent book is *Scientific Governance in Britain, 1914–1979*, co-edited with Don Leggett (Manchester University Press, 2016). She is editor of the *British Journal for the History of Science* and is currently writing a book about science and science fiction in interwar Britain.

Maarten C. A. van der Sanden is an Assistant Professor of Science Communication in the Department of Science Education and Communication at the Delft University of Technology, the Netherlands. He specializes in the social design of science communication processes and its supporting tools for scientists and science communication practitioners, from a social systems perspective. He teaches social systems and social design courses in both M.Sc. and B.Sc. programmes.

Nick Verouden is currently finishing his Ph.D. in the Science Communication Department at the Delft University of Technology, the Netherlands. His research focuses on the functions and consequences of silence in interdisciplinary research collaboration. Previously he has worked on the topic of silence in the context of health communication. Other research interests include water-related human–environmental interactions and the social and cultural dynamics of borders and borderlands. He holds an M.A. in anthropology.

Stephen Webster is Director of the Science Communication Unit at Imperial College London. His academic background is in zoology and the philosophy of biology. He was a school science teacher before moving into science writing and broadcasting. He is the author of *Thinking About Biology* (2003) and *Charles Darwin* (2014).

Alice White is completing a Ph.D. at the University of Kent, examining efforts to develop a science of human relations in Second World War and post-war Britain. She is the editor of the history of science magazine, *Viewpoint*. Her published work includes 'Governing the Science of Selection: The Psychological Sciences, 1921–1945' in *Scientific Governance in Britain, 1914–79*, edited by Don Leggett and Charlotte Sleigh.

Introduction

The communicative functions of silence in science

Felicity Mellor

Discussions about the communication of science often rest on an unquestioned assumption that open and efficient channels of communication are always of greatest benefit to both science and society. Increasingly, scientists are urged to maximise their communications by joining large collaborations, increasing their publication rates, sharing their data online, taking part in public outreach events, advising policymakers and talking to the media. Journalists and other media producers, for their part, are often told that they should cover more science, more often and at greater length. Yet too much communication can sometimes become a barrier to effective communication. For instance, a recent overview of efforts to engage the public with science suggests that unreflexive public engagement can close down debate rather than open it up. Despite some successful and productive engagement initiatives, the authors reflect that they now find themselves advising scientists and policymakers 'how and when *not* to engage' (Stilgoe et al., 2014: p. 11). Sometimes it might be best – for the scientific community and for wider society – if scientists stay silent.

This book emerges out of a series of AHRC-funded workshops which aimed to draw attention to the role that silence can play in the communication of science. Contrary to the ideal of science as an open enterprise, scientific innovation relies as much on discontinuities – on barriers and lacunae – as it does on the free flow of information. For instance, the fear of competing research groups stealing ideas can restrict scientists' willingness to discuss their work openly. Journalists, too, may hold back on a story if publishing could provoke litigation suits or compromise future access to sources.

Such closing down of communication can be understood as the production of silence. Crucially, these silences can communicate in their own right. The silence of a scientist who shuts herself away in her laboratory, for example, could be variously interpreted as signifying hard work, an imminent breakthrough, an uncollegiate attitude or disengagement from society; whatever the interpretation, her silence has carried meaning for those who attend to it. Not all silences have communicative value, but those silences that do communicate have the potential to complement and enhance, rather than just limit, communication through verbal language.

Silence, then, is more than simply the absence of noise; silence signifies. This is the fundamental message of scholars who have studied the nature of silence. Scholars working in a range of disciplines – including linguistics, rhetoric, literary studies, feminist studies, political theory, organisation studies, theology and philosophy – have examined the nature and meanings of silence in different contexts. Even as early as 1973, Thomas Bruneau was able to draw up a bibliography of some 250 academic studies of silence (1973: p. 42). One aim of this book is to show that science is another context where silence is worthy of academic attention.

As well as adding to the silence studies literature, the contributions in this book also complement and extend recent work in the emerging field of ignorance studies. There, scholars have drawn together interests in uncertainty, risk and absent knowledge to examine the social construction of nonknowledge. One branch of work in this field, known as 'agnotology' (Proctor and Schiebinger, 2008), takes nonknowledge to be either the outcome of cultural biases that suppress knowledge production or the product of a deliberate strategy to obstruct the dissemination of knowledge. Focusing on issues where nonknowledge claims have been deployed to prevent or delay policies that would be in the public interest – such as tobacco legislation or climate action – agnotology draws attention to the construction of ignorance as the outcome of political struggles.

This approach has been criticised by some for its normative tendency to identify the strategic use of ignorance claims with anti-scientific interest groups and 'bad' science (Gross and McGoey, 2015; Pinto, 2015). By contrast, other work in ignorance studies takes nonknowledge to be an inherent feature of the knowledge society: knowledge and nonknowledge of scientific issues are, by this account, co-produced (e.g. Gross, 2010). Areas of interest highlighted by such work include the varying degrees of intentionality regarding nonknowledge on the part of social actors, the existence of differing epistemic cultures of nonknowing within science, and the strategic uses to which nonknowledge claims are put in public debates (Böschen et al., 2010). For silence studies, too, individuals' intentions regarding their silences, the cultures that produce silence, and the circulation and distribution of silences in public fora, are all of interest. This book is therefore divided into three parts – 'Choosing silence', 'Cultures of silence', and 'Silences in the public sphere' – which consider each of these aspects in turn.

Like ignorance studies, the study of silence draws attention to the ways in which absences are constructed and the ends to which such absences can be put. But where ignorance studies highlights epistemological issues arising from nonknowledge claims, the study of silence emphasises the communicative value of absent speech – something that, crucially, is both relative and context-dependent. Since, as I discuss below, silence carries positive as well as negative connotations, a focus on silence also reminds us that leaving some things unsaid may aid knowledge production rather than being necessarily obstructive.

The case studies presented in this book examine the varied meanings that silence takes in the production and communication of scientific knowledge. This Introduction develops a theoretical framework in which these case studies can be placed. In what follows, I first present an overview of some of the key insights of the silence studies literature and then propose a typology of silence which I use to explore some of the contours of silence in modern science.

Signifying silence

Whilst the aim here is to move beyond a view of silence as absence, it is worth first considering how even absence implies that silence is more than nothing. What is it that is absent and to what degree? Absence is relative, implying a potential presence through the very act of negating it. In his composition 4'33" – in which a musician sits at an instrument for 4 minutes and 33 seconds without playing anything – John Cage famously drew attention to the environmental noises that constitute a situation that would otherwise be perceived as silent. Cage's piece, highlighting the impossibility of absolute silence, was inspired by his experiences in an anechoic chamber. Even in this specially engineered sound-insulated chamber, he had found that he could still hear noises, which the engineers later explained were the pumping of his heart and the hum of his nervous system (Cage, 1957). Cage concluded that an objective demarcation between sound and silence was untenable; whether or not a specific situation is silent is a subjective judgement and the qualities of that silence are dependent on context and expectations (Cage, 1955).

At the first of the series of workshops that led to the current volume, pianist Rolf Hind performed Cage's piece. As the workshop participants listened to the silent performance, we heard many things, including the creaks and groans of the building as preparations were made for some renovation works. Those who worked in the building could locate and identify these noises, conjuring a familiar architectural and social space. For the rest of us, the noises signified the unknown and anonymous space that surrounded us that day. Not only did the performance resignify silence, but the meanings the silence carried depended on the listener.

Since absolute silence is never possible, the absences that characterise silence will vary, depending on the situation. Linguist Michal Ephratt (2008) distinguishes stillness, the antonym of noise, from silence, the antonym of speech (or, we might add, of verbalisation more generally). Stillness describes a state that is external to the act of communication. By contrast, silence that is defined as the absence of speech is located within the communicative act. It is these silences that have been of most interest to silence studies scholars.

Silence is both a necessary accompaniment to speech, as with the gaps that separate words and sentences, and can also be identified with that which is excluded from speech. As the rhetorician Robert L. Scott put it:

4 Introduction

> Every decision to say something is a decision not to say something else, that is, if the utterance is a choice. In speaking we remain silent.
> And in remaining silent, we speak.
>
> (Scott, 1972: p. 146)

Silence is thus a constant presence in speech, as much as it is an absence of speech. The conception of silence as absence fails to attend to the complexities implied by this co-existence of speech and silence. For instance, although silence can be thought of as corresponding to a gap in communication, speech too can generate a communication deficit, as the writer W. R. Espy highlighted when he joked: 'I have nothing whatever to communicate, and words are the best means of non-communication I know' (cited in Sobkowiak, 1997: p. 55).

Conversely, silence can lead to a proliferation of speech. Maggie MacLure and co-authors (2010) give the example of a five-year-old child called Hannah. At the start of each school day, a register is taken and each child responds by saying 'good morning' to the teacher – each child, that is, except Hannah, who remains silent when her name is called. Paradoxically, Hannah's silence calls forth an excess of speech in response – of meetings and charts and analyses – as teachers, parents, researchers and even the other children all try to work out what Hannah's silence means and how they can make it stop. Hannah's silence is a presence which others try, unsuccessfully, to undo through an onslaught of language.

As this example illustrates, speech and silence are closely entwined, each capable of generating the other. Adam Jaworski thus recommends that they should be treated 'as fuzzy, complementary categories, and not as discrete and opposite ones' (1993: p. 48). Similarly, Kris Acheson argues that: 'silence and speech, paradoxically, are parallel communicative events in addition to opposite poles of a binary' (2008: p. 543). For Acheson, not only can silence function as a zero sign – an empty signifier that replaces missing words – but it can also function as a sign in its own right:

> silence and speech (like stillness and action) gain the capacity to live in a paradox of simultaneous opposition and correspondence because they both constitute signs in a semiotic system. Because, together, they comprise language, they are oppositional in their relationship with each other and corresponding in the relationship they share with language.
>
> (Acheson, 2008: p. 543)

Central to this conception of silence as a sign is a distinction between communicative and non-communicative silences. Not all silences signify; for instance, Jaworski identifies muteness as a non-communicative silence (1993: p. 4). Communicative silences, by contrast, enfold some measure of intent – they have meaning because they are meant. This implies that such silences are an active construction in contrast to the typical assumption that silence is a passive state. As Scott argues, the production of silence requires great effort:

Introduction   5

'Silence is the withholding of speech, it is desire conquering desire, it is not simply not saying. Not saying is failure. Silence is to be achieved' (1993: p. 6).

As well as a degree of intention on the part of the producer of the silence, communicative silences also rely on the expectations of the receivers of the silence. If no communication is expected, a silence will not be interpreted as communicative. Thus communicative silence is co-created by speaker and listener (Scott, 1993). It was Hannah's teachers, confronted with the child's silence at a moment when they expected speech, who construed the silence as meaningful, even though what precisely it meant – fear? anxiety? excitement? – remained unclear to them (MacLure et al., 2010).

The role of the listener also implies that silences can be directed at specific addressees. Giving someone 'the silent treatment', for instance, sends a message to that particular person. In an analysis of silence as political strategy, Barry Brummett notes that a silence can also be 'overheard' and that this may be intentional on the part of the producer of the silence (1980: p. 295). He gives the example of China being expected to take note if the United States stops speaking to Taiwan. A critical analysis of silence therefore needs to take account of both the targeted audience and the indirect audience of the silence.

One prominent feature of silence is its potential for ambiguity. This again relates to expectation. The meanings of conventional silences – such as a minute's silence to mourn a death or the silence of a quiet space like a library – are highly constrained, denoting, in these particular examples, respect and concentration respectively. By contrast, freely chosen silences, such as Hannah's silent response to the calling of the register, can be highly ambiguous. For instance, a silence during a conversation could convey any of the following contradictory meanings among many others (adapted from Johannesen, 1974):

The silence expresses agreement.
The silence expresses disagreement.

The silent person is bored.
The silent person is attentive.

The silence marks anger.
The silence marks empathy.

The silence signals disrespect for the other.
The silence signals respect for the other.

The silent person feels ill at ease.
The silent person feels entirely at ease.

The degree of ambiguity in the meaning of a specific instance of silence depends on the culture in which the silence occurs. In his landmark study of the extensive use of silence in Western Apache society, anthropologist Keith Basso (1970) found that his subjects understood acts of silence to have tightly constrained functions as appropriate responses to socially ambiguous and unpredictable situations. To Basso's subjects, these silences had well-defined meanings. Onlookers from outside the culture, however, may misread such silences, as happened with the stereotyping of Native Americans as sullen and uncooperative or as dignified stoics. Since communicative silences are part of the language system, it follows that their meanings must be learnt; to learn a language thoroughly entails learning how it uses silence.

Whilst the meanings of culturally learnt silences are circumscribed by their conventional uses in specific contexts, other silences escape this semiotic delimitation. Indeed, the inherent ambiguity of silence in contexts where speech would fix meaning is one of its most powerful communicative functions. Jaworski argues that one aspect of the semiotic openness of silence is that it can serve to keep communication channels open in situations where speech would close them down. Giving the example of an angry exchange in which a silent response can avoid the long-term damage inflicted by words said in anger, Jaworski suggests that: 'It is easier to undo silence than to undo words' (1993: p. 25).

As this implies, like speech, silence can be used to positive ends as well as negative. In addition to the negative silences of, say, a sulk or anxiety, there are the positive silences of intimacy or respect. This again moves beyond the assumptions embedded in the view of silence as absence, where silence is most readily construed as negative, as a failure to speak. Rather, whilst communicative silence can be suppressive, refusing to acknowledge that which is unsaid, it can also be generative, opening up multiple interpretations and allowing for new understanding.

So far, this overview has focussed on silence as a feature of the communicative system that is freely deployed by the person who produces it. However, silence is not always chosen, as the verb form of the word reminds us: to be *silenced* is to have silence imposed. Silencing enacts power relations, entailing submission to whatever authority enforces the silence. Indeed, Scott (1993) argues that silence can mark out hierarchies of authority even more effectively than speech can, since assent verbalised at least makes possible the idea of non-assent in a way that the tacit agreement of silence does not.

Silence studies scholars have paid particular attention to the historically gendered significance of silencing; in Western societies, male speech and female silence has dominated the public sphere, whilst female speech and male silence has characterised the domestic sphere. Silence in the public sphere – for instance, from a political leader at a time of crisis – is typically interpreted as a sign of weakness, whilst the interpretation of silence in the domestic sphere is more varied (Brummett, 1980). The values attached to silence thus both mirror and sustain power relations. Those in positions of power can choose to use silence strategically; for instance, managers can use silence as a form of punishment or as a way of distancing themselves from others, whilst the silences of those in subordinate positions are more likely to indicate that they are prevented from speaking out or feel unable to do so (Glenn, 2004).

However, as Thiesmeyer (2003) notes, silencing is not necessarily coercive. Furthermore, it can be most effective when concealed, the silences masked by other discourse. For instance, John Keane (2012) gives the example of complex large-scale engineering projects that sidestep democratic accountability by masking corporate silences regarding safety and governance with extensive media relations and fabricated positive stories.

Although silencing enacts power relations, silence can also be used as a form of resistance to power. The sulking teenager, the accused who exercises the right to silence, the victim of torture who refuses to yield, and the activists who protest in silence as a response to an oppressive regime – all use silence in their confrontation with the agents of power. Silence, then, should not automatically be read as submission.

Similarly, silence does not necessarily indicate political passivity. Sean Gray argues that the equation of silence with disempowerment betrays the speech-centric foundation of democratic theory, which he characterises as the 'speech clause':

> The speech clause is premised on the idea that citizens rule when democracies empower public forms of self-expression – enabling citizens to voice opinions, interests, beliefs, arguments, reasons, and, above all, judgements over what to support and whom to hold to account. The speech clause carries the underlying logic of this idea to its ultimate conclusion: if practices of democratic citizenship are artefacts of language, then citizens' capacity for self-rule is contingent upon their ability to speak.
> (Gray, 2012: p. 3)

Attempts to enhance democracy, such as exercises in deliberative democracy regarding new technologies, thus strive to facilitate civilian speech as a counterbalance to expert speech. The speaking citizen is valorised as the ideal type whilst the silent citizen is interpreted as a 'deficit in democracy' (ibid.: p. 7). Such a view, Gray argues, fails to acknowledge forms of silence that are deeply embedded within the functioning of democracy. Although silence can be used to suppress dissent – as when certain subjects are circumscribed as out of bounds for citizen input in favour of, say, expert-led decision making – it can also mark tacit approval, enable consensus building or signal dissatisfaction with the choices available. In democratic processes, as in other forms of communication, silence should not be reduced to a deficit.

A typology of silence

In summary, then, silence is an element of discourse that can both suppress and enhance communication and that can be both actively chosen from within the discourse community or imposed from without. This suggests a typology that can help highlight the agency of communicative silences by characterising silence along a producer-oriented dimension of choice (is the silence internal, chosen by the party that is silent, or is it external, imposed by another party?) and an audience-oriented dimension of the actual or intended communicative effect (is the silence suppressive or generative?). This allows us to distinguish between four types of silence that help draw attention to both the similarities and differences between different manifestations of silence (see Table 0.1). For instance, both censorship imposed by the state (an externally imposed silence) and an individual's choice to withhold information in order to protect their privacy (an internally produced silence) have in common that the use of silence suppresses communication. By contrast, the chosen silence of an intimate relationship and the imposed silence of a remembrance service both serve to enhance the meaning of the communication.

It is worth emphasising that neither the dimension of choice nor that of communicative effect correspond with positive or negative attributes. An externally imposed, suppressive silence can have positive effects – as in legislation that prevents racial abuse. Equally, a chosen, generative silence can have negative effects – if, for instance, it results in a failure to share information that others may need.

The distinction between silences that are chosen by those who are silent and those that are imposed externally is not clear cut, since individuals can sometimes choose whether or not to comply with that which is imposed on them. However, foregrounding the dimension of choice helps remind us of the difference between being *silent* and being *silenced*, and thus draws attention to the power hierarchies that silence can reproduce. Distinguishing between suppressive and generative forms of silence also has its limitations, ignoring as it does the fact that enhancing communication in one respect frequently entails degrading it in other respects. Nonetheless, the focus on the communicative effect of silence helps draw attention to the ends to which silence works within the wider communication context. Focussing on the dimensions of choice and effect thus ensures that we consider both the production context and the reception context for specific instances of silence. My aim in

Table 0.1 A typology of silence

	Suppressive (minimising communicative efficacy)	*Generative (maximising communicative efficacy)*
External (being silenced)	Censorship	Libraries
	Defamation and anti-hate laws	Memorial silences
	Official secrets legislation	Media embargoes
Internal (being silent)	Conflict avoidance	Retreats
	Protecting privacy	Intimacy
	Self-censorship	

proposing this typology is therefore not to reify these four different types of silence, but rather the opposite: to better understand the struggles that are entailed in both the production and interpretation of silence by highlighting how silences migrate across these blurred boundaries.

In what follows, I use the above typology as an organising device to draw out some of the ways in which silence is enacted in scientific contexts. I discuss exemplars of each type of silence, paying particular attention to the dynamics of the production of silence as different parties struggle to control communication. These examples all focus on silence as enacted. However, silence also exists in the analytical plane when researchers encounter gaps in the historical record, areas of missing knowledge or the ineffability of aspects of human experience – silences that may have had no communicative value for the actors involved, or may not even have been present at the time, but that must nevertheless be interpreted by the scholar. I finish with a discussion of these epistemological silences.

External, suppressive silences

The secrecy of state-sponsored security regimes is perhaps the most notable mechanism through which imposed, suppressive silences are produced within science. A secret consists of a silence produced with the deliberate intent to conceal information from those who might wish to be informed. Secrecy can be freely chosen by individuals, but regimes of secrecy – which by their nature entail an institutionalised, collective system – typically involve an element of imposition and coercion.

In his famous 1942 essay outlining the institutional imperatives comprising the ethos of science, Robert K. Merton characterised science as opposed to secrecy. For Merton, science was an open enterprise entailing full disclosure. His norm of communalism required that all scientific knowledge be shared collectively: 'Secrecy is the antithesis of this norm; full and open communication its enactment' (Merton, 1973: p. 274). Open science remains a widely accepted ideal in science (see Grand, Chapter 11), yet, in practice, scientific research has also frequently depended on secrecy.

Merton's idealisation of science was born out of the wartime context in which he was writing, a riposte to the conditions for science in Nazi Germany (Kellogg, 2006). It is ironic, but not a coincidence, that just as Merton was praising the openness of democratic science, many US scientists were being recruited into a regime of institutionalised secrecy that would continue long after the end of the war. The Manhattan Project serves to illustrate how suppressive silences can be imposed through a tightly controlled security regime. It also shows how this imposition can involve an element of choice and reveals the partial nature of the silences that constitute a state of secrecy.

When General Leslie Groves took control of the Manhattan Project in 1942, he instituted a strict policy of 'compartmentalization' across the research laboratories and production facilities that were to build the first atom bombs. Information about the project was circulated on a strict need-to-know basis managed through a system of classification. In addition, letters were censored, access to the five sites was controlled by the military, media outlets were asked to refrain from reporting stories containing certain key words, and scientists were forbidden from talking to scientists at other sites without the permission of their manager.

Through such means, silences were imposed on all aspects of the Manhattan Project with the aim of suppressing its communication. Compartmentalisation not only prevented information about nuclear research reaching the Germans, but also suppressed communication between those working on the project. However, to the knowledgeable listener, an unmasked silence could itself communicate that which it was intended to suppress. As early as 1941, the Russian physicist Georgii Flerov had interpreted the absence of publications by US nuclear scientists as evidence that they were working on an atom bomb: 'a stamp of silence has been laid on this question, and this is the best sign of what kind of burning work is going on right now' (Vermeir and Margócsy, 2012: p. 10). Robert Oppenheimer, director of the Los Alamos Laboratory, would later claim that US physicists not involved in the project had also known about its work but had kept silent for the sake of national security (Quist, 2002: p. 85).

The strict secrecy regime was resisted by some, both through individual actions and through more organised means. For instance, one formal initiative often interpreted as a challenge to compartmentalisation came when Oppenheimer instituted a weekly seminar open to all the Manhattan scientists (Dennis, 1999). The scientists thus continued to appeal to the ideal of the free exchange of ideas within the scientific community; this was despite a version of compartmentalisation having been first introduced by physicist Gregory Briet, before Groves took control of the project, in an attempt to persuade the military authorities that they could work with the scientific community (Goldberg, 1995; Quist, 2002).

Groves himself also ignored compartmentalisation when he thought it would impede progress, and on occasion he would allow a scientist to communicate with another site (Goldberg, 1995). Indeed, the creation of the centralised Los Alamos site was itself a result of the difficulties scientists experienced working within a compartmentalised system when they were scattered across the country (Hales, 1997). By gathering many of the scientists in one place, and thus containing their talk geographically, an external silence could more easily be maintained whilst still enabling some level of communication between scientists.

The scientists perceived the silences imposed by compartmentalisation as restrictive, even if to some degree necessary. By contrast, General Groves argued that it aided the production of knowledge by preventing the scientists from being distracted by the many interesting, but not relevant, problems that came up in the course of the research (Quist, 2002: p. 85). As this suggests, whether the communicative effect of a silence is seen as suppressive or generative depends on the point of view. Those who silence others will see benefits where those who are silenced do not.

The secrecy of the Manhattan Project was achieved not only by prohibiting communication but also by generating alternative communication. Code words replaced sensitive terms such as 'uranium' or 'bomb' and, in the later years of the war, the media were fed a steady stream of misleading information to cover up the true purpose of the production sites. Despite these measures, information did leak out. By 1944, Groves had noted 104 published references to the project over the previous five years (Washburn, 1990). By then, the existence of military establishments at the large production sites was common knowledge, newspapers had made passing reference to the production of a weapon that could end the war, and Cleveland Press reporter Jack Raper had identified the Los Alamos site and had even commented on its policy of compartmentalisation (Jones, 1985; Wellerstein, 2013). Yet, despite the partial nature of the silences surrounding it, the project failed to impinge on public consciousness until the Little Boy bomb was dropped on Hiroshima.

In Chapter 5 of this volume, Daniele Macuglia examines how, as the nuclear sites began to come under scrutiny from concerned citizens in the decades after the war, silences about serious problems were maintained through the selective release of information about lesser incidents. Macuglia argues that local residents near the Hanford nuclear site existed in a midway state between knowing and not-knowing. Decoy stories successfully masked Hanford's silences for some years, but eventually – as those living downwind of the site started to record instances of animal deformities, human miscarriages and high rates of cancer – the bodily manifestations of the site's history drew attention to the silences. At this point, local journalists, who initially had helped circulate the partial silences emanating from Hanford, transformed their role and joined with activists in attempting to break the silences.

The history of nuclear silences shows that even those silences that are imposed through an elaborate culture of secrecy are dynamic and partial in nature. That the silences of the Manhattan Project and later nuclear weapons development co-existed with, and were in part produced by, carefully chosen language and that they could dominate even as information leaked out, reveals both the relativity of silence and the effort that has to be invested in its maintenance. Security regimes have constructed silences around biological and chemical weapons research through similar means. Suppressive silences can also be found in technoscientific contexts other than weapons development. In current-day Russia, all scientific research that could be used to develop 'new products' is treated as potentially classifiable and researchers must apply for security clearances before publication of their work in journals or as conference talks (Schiermeier, 2015).

Science and Technology Studies scholars who have examined the dynamics of secrecy in contexts such as these argue against an opposition between, on the one hand, open communication associated with the production of knowledge and, on the other hand, suppressive regulation associated with the construction of ignorance. Vermeir and Margócsy (2012), for instance, argue that secrecy entails a dynamic process of veiling and revealing in which different actors are privy to varying degrees of knowing, as seen in the above example of nuclear silences. Similarly, Balmer (2012) challenges the assumption that the object of secrecy is unperturbed by the silences that surround it. Secret science, he suggests, is not just open science done behind closed doors. Rather, secrecy dictates specific geography, cultural practices and social structures; by so doing, it actively constructs new forms of knowledge as well as generating ignorance.

Not all suppressive silences are created through the deliberate implementation of a secrecy regime. They can also arise through less direct means, as the result of a complex interplay of multiple social and cultural factors. Carolyn Cobbold (Chapter 10) examines the case of the introduction of synthetic dyes into food in the second half of the nineteenth century. The use of these new additives, which in some cases were later found to be highly toxic, went largely unremarked in the media, despite widespread concerns about other aspects of food adulteration. Cobbold finds that a combination of factors – including a dominant media discourse celebrating the new chemistry, the vulnerabilities of the newly emerging profession of public analysis and consumer expectations about what certain foodstuffs should look like – all contributed to the imposition of a public silence around the use of the new dyes in food.

The opposition between silence and transparent open communication is further challenged if we consider the ways in which openness itself can engender silences. Catriona Gilmour Hamilton (Chapter 7) shows how a culture of openness can serve to generate new forms of silence. She traces the evolution of informed consent in cancer research in the UK over the past four decades. The substitution of the imposed silences of the paternalistic health care system of the 1970s (which assumed that patients should not be troubled with the details of their diagnosis and treatment options) with today's culture of openness (which envisages patients as research partners capable of giving their informed consent) has not eliminated silences from the consent process. Rather, Gilmour Hamilton finds, preoccupations with the evidence hierarchy have resulted in the silencing of individual, subjective experience.

Imposed, suppressive silences take their most elaborate and wide-reaching form in cultures of secrecy, but they cannot be reduced to these deliberate acts of concealment. As Gilmour Hamilton's work suggests, they can also be uncovered in cultures that strive to eliminate secrets. Being silenced can be an unintended, and often unacknowledged, consequence of protocols for openness as well as the intentional, but often contested, product of mechanisms of secrecy.

Internal, suppressive silences

Security regimes impose silences on researchers. However, in other situations researchers themselves – either individually or collectively – may choose to stay silent in order to suppress communication about their work. As noted above, even before the Manhattan Project took shape, many US physicists had already chosen a policy of silence, withholding research on nuclear fission from publication in order to prevent the Germans gaining access to the new results. The physicists continued to submit papers to journals in order to evidence their priority claims, but they would request that publication be deferred until after the war. Self-censorship had its difficulties, but from the start of 1940 some significant results were withheld from publication and the system was formalised by the National Research Council later that year (Weart, 1976).

However, a collective voluntary silence is difficult to maintain and, when proposed for reasons of national security or public safety, is unlikely to be a choice that can remain with the research community (Kaiser and Moreno, 2012). In recent years, self-censorship of unclassified research has been considered by microbiologists whose findings could be used in biowarfare, with some researchers wanting to publish papers with key information redacted (Couzin, 2002). One high-profile example, which highlights the ways in which the choice of self-censorship can segue into an imposed system, is the case of research on avian flu. As this case shows, the tensions over the suppression of communication result in a multiplicity of additional communications; a temporary silence may be generated only in the midst of much speech.

In November 2011, the US government's National Science Advisory Board for Biosecurity (NSABB) requested that the journals *Science* and *Nature* refrain from publishing in full two papers on the transmission of the H5N1 influenza virus. The work had shown that a modified variant of the virus could spread between ferrets, which provide a model for human-to-human transmission. The NSABB feared that, if published in full, the studies could enable bioterrorists to create a flu strain that would be deadly to human populations. The Board therefore asked that no details of the methodology or data be released. The authors and journal editors agreed, provided that some other mechanism could be found for communicating this information to legitimate researchers (Butler, 2012). A silence targeted at a general audience was acceptable, this agreement implied; a silence that also encompassed the community of researchers was not.

As a press release explained (NIH, 2011), the NSABB recommended not only deleting information from the papers but also *adding* information to explain the potential benefits and safety measures of the research. Shortly afterwards, leading flu researchers proposed a 60-day moratorium on such research in order to discuss its benefits in a public forum (Fouchier et al., 2012). As one of the authors of the *Nature* paper put it: 'Scientists need to have their voices heard in this debate' (Butler, 2012). The effect of the silence was thus to call forth a great deal of talk; for the scientists, agreeing to stay silent became a means of having their voices heard.

Many scientists disagreed with the decision not to publish. Some argued that the publication restrictions were pointless since the research had already been shared with other researchers and talked about at conferences. They suggested that enough information was already in the public domain to pose a risk (Butler, 2011; Kawaoka, 2012). However, the non-publication did limit the information about the research available to the many journalists who were busy reporting the story. *Washington Post* reporter David Brown, for instance, complained that journalists were having to write reports on the basis of 'woefully inadequate' information. He argued that if the decision not to publish were to become a habit, such silences would breed conspiracy theories and a distrust of scientists' motives. Yet Brown and his editors also silenced themselves, backing off from searching out copies of the censored papers on the internet so that they would not have to face the decision of what to do with the information if they found it (Brown, 2012).

In March 2012, following the release of new US guidelines on the management of dual use research, the NSABB reversed its decision and recommended full publication of the two papers. The *Science* paper was further delayed when the Dutch government requested that the authors, who were based in the Netherlands, apply for an export licence on the grounds that the research fell under the regulations controlling the export of weapons technology. The head of the research team, Ron Fouchier, objected and threatened to publish anyway, but in the end complied and was granted a licence (Frankel, 2012). Research on the H5N1 virus resumed in 2013. The new US guidelines included measures to determine how, and to whom, research judged to be risky should be communicated. Voluntary redactions could now be replaced with silences enforced by the federal government through use of the classification system (Butler and Ledford, 2013).

Similar considerations are also beginning to arise in relation to ethical concerns about research with no immediate dual use implications. In April 2015, the Chinese journal *Protein and Cell* published a paper reporting the use of the CRISPR gene editing technique to modify the genomes of human embryos. The author claimed his paper had been rejected by both *Nature* and *Science* in part on ethical grounds. The journals' choice not to publish was interpreted by some as censorship on extra-scientific grounds (e.g. Gyngell and Savulescu, 2015). As in the avian flu case, rumours about the work prior to publication prompted debate about the need for a moratorium on research of this type (Kaiser and Normile, 2015). The alleged act of silencing was again constructed through heightened levels of discussion.

In cases such as the H5N1 moratorium, collective choice – a voluntary agreement not to publish or to halt further research – can generate a temporary silence, but only by drawing attention both to the silence itself and to its legitimacy. Joanna Kempner and her colleagues argue that these publicly announced silences are aimed at public reassurance (Kempner et al., 2011). They note that in addition to these acknowledged silences, research into socially and culturally sensitive topics is also imbued with unacknowledged silences arising from an internalised set of values that are embedded in the very practice of science. These norms are fashioned from cautionary tales about what happened to scientists who breached them in the past. The mere anticipation of public reprobation, regulatory sanction, time-sapping controversy or loss of funding can be sufficient to silence some lines of inquiry.

Over a third of the scientists interviewed by Kempner said they would not pursue or publish research results that deviated from the accepted dogmas of their discipline (ibid.). Unlike the publicly announced silences, these silences are enacted silently as private decisions reached through reference to unspoken rules.

The internalised silences of modern scientific practice serve to save scientists from potentially career-blocking confrontations. Similarly, discursive omissions that arise in potentially controversial situations can function as a way of bracketing uncertainty without denying it. Choosing silence in such situations can enable action to be taken even in the absence of evidence-based consensus. In Chapter 12 of this volume, Abi Dymond examines how staying silent enables policy to proceed in the face of uncertain knowledge. By choosing silence, those charged with drawing up policy are able to defer decision making. Dymond considers the case of the 'less-lethal' policing weapon TASER, where silences in the policy discourse surrounding the relative safety of the weapon have the effect of devolving responsibility over its use onto police officers on the ground. Officers must temporarily fill these silences through their own actions, weighing up on the spot whether or not to use the weapons, at what strength and for how long. However, silence is reinstated when, after the event, justification of TASER use is diffused once more thanks to the singular circumstances of each decision.

The above examples deal with internally produced silences that are chosen at the level of the discourse community or that have come to be accepted cultural norms within that community. However, individuals may choose to violate the community norms for sharing information, remaining silent when the expectation would be to publish. Kees-Jan Schilt (Chapter 3) considers the case of Isaac Newton's year-long withdrawal from correspondence about natural philosophy. Despite his lasting fame, Newton's publication record is slim in comparison to his contemporaries and he made repeated attempts to leave off correspondence with other philosophers. Rather than attribute such behaviour entirely to Newton's personality, Schilt argues that it was in part the outcome of a specific communication strategy influenced by Newton's other focus of interest, alchemy. Newton, Schilt suggests, approached the dissemination of natural philosophy according to the obscurantist norms of alchemical discourse. In contrast to the rhetoric of openness and disclosure that shaped natural philosophic discourse, the alchemical reader was expected to work hard to extract meanings hidden in the text. The differing expectations of Newton and his readers inevitably led to conflicts, which Newton resolved with silence.

Today, scientists are unlikely to base their publishing strategy on a communicative ideal that privileges private knowledge. However, other pressures can lead modern-day scientists to delay publishing. Today's scientists operate in a competitive environment, competing for recognition and the grants, jobs and students that go with it. In a study drawing on focus groups with fifty-one early- and mid-career scientists in the USA, Melissa Anderson and her colleagues found that increased secrecy was among the detrimental effects of competition, with scientists hesitating over sharing ideas with others and deliberately withholding details from papers to prevent replication (Anderson et al., 2007). Ironically, the pressure to publish brings with it a concomitant incentive to limit communication.

Publication itself may also be delayed. Mario Biagioli (2012) interprets this as a protective response to the risks of publication that arise out of an inevitable moment of instability in establishing priority. Publication is necessary to secure priority, yet it also risks losing priority through leaks by those privy to the process. Researchers and inventors may therefore temporarily guard their ideas with silence even though that silence is ultimately aimed at making the ideas public. In a reward system that prioritises priority, successfully staking a claim becomes more important than gaining immediate publicity; a temporary silence is one mechanism through which to achieve that end.

Internal, generative silences

Whilst some silences are forged with the intention of suppressing communication in order to achieve other ends – such as securing priority claims or protecting national security – other silences are intended to enhance the communicative act. The textual silences of scientific papers afford one such example. As Dacia Dressen (2002) has discussed in her study of geological discourse, genre conventions dictate that much is left out of a technical report. The research narrative, established or tacit knowledge that is assumed to be shared by readers, and the researchers' agency and emotional reactions are all routinely omitted from the scientific paper in its modern form. Such omissions enable scientists to communicate efficiently with each other those aspects of their research that are deemed most salient to building a body of objective knowledge. Whilst also having the effect of excluding readers from outside the discourse community, these 'silential conventions', as Dressen calls them, increase the communicative efficiency of the genre for those within the community.

As Dressen notes, these 'laws of silencing' do not necessarily erase content; rather, authors can use conventional rhetorical devices, such as understatement, to draw the attention of knowing readers to the missing content. Furthermore, the boundaries of such textual silences are dynamic, developing over time; and they can be purposefully transgressed or manipulated by individual authors in response to specific situations.

Textual silences are enacted as collectively agreed conventions. Other generative silences are chosen by individuals who, temporarily or partially, withdraw from interactions in order to develop their ideas. Such withdrawals provide a generative phase in preparation for full and open communication. Communication is delayed, not out of a secretive, competitive urge as is the case for suppressive silences, but in order for the scientist to achieve some clarity of thought so that the communication, when it comes, is clear and compelling. In the generative silence of withdrawal, then, the two meanings of silence come together – the silence of stillness (often accomplished through solitude and isolation) and the silence that is the necessary complement to speech. Where the former has the potential to minimise distractions and increase concentration, the latter marks the beginning of the communicative acts that are to come.

The appeal to withdrawal as a requisite for intellectual labour has a long history. Steven Shapin (1991) has argued that a distinctive feature of early modern science was its fusing of two contrasting ideals – that of the gentleman citizen engaged in public activities and that of the reclusive scholar engaged in private study. The early Royal Society forged a novel combination of these two ideals, continuing to appeal to the rhetoric of solitude ('the hermit's voice', as Shapin puts it) regarding the context of discovery, even as the rhetoric of public display dominated the context of justification.

Whilst rarely acknowledged in modern research settings, the importance of withdrawal to scientific creativity has, at times, been recognised institutionally. For instance, when Robert Oppenheimer went scouting for a site to house the Manhattan Project scientists, he had in mind a place that could function as a 'monk's colony', its isolation not only affording security but also protecting its scientists from distractions in what he envisioned would become a scientific 'Shangri-la' (Hales, 1997: p. 42).

Writing in 1931, Abraham Flexner, the first director of the Institute for Advanced Study in Princeton, had articulated a similar vision: his institute was to be 'simple, comfortable, quiet without being monastic or remote; … and it should provide the facilities, the tranquillity, and the time requisite to fundamental inquiry into the unknown' (Institute for Advanced Study, n.d.).

Three decades later, the idea that scientific creativity benefited from periods of retreat and tranquillity was still influential in the founding of the Aspen Center for Physics. The Center was planned in 1961 as a summer retreat where physicists could be free of distractions. Although promoted as a place for talking, collaboration and the exchange of ideas, these appeals are bookended with references to solitude, isolation, and peace and quiet. Thus the Center's website explains that the Center 'is conducive to deep thinking with few distractions, rules or demands. In our "circle of serenity," physicists work at their own speeds and in their own ways.' Quotes from scientists who have visited the Center attest to the benefits of this 'circle of serenity': 'I desperately needed the peace and quiet provided by the Aspen Center for Physics to just think and reflect'; 'Aspen remains the best place on earth to have a chance to actually think deeply and without interruption'; 'an idyllic sanctuary where scholars abandon their cares, explore in solitude, or more often, in one another's company, where new ideas brew and old ones take flight' (Aspen Center for Physics, n.d.).

It is worth emphasising that to acknowledge the importance of withdrawals and retreats to scientific creativity is not to reinstate the mythology of the lone genius. Both the Institute for Advanced Study and the Aspen Center for Physics aim to facilitate collaboration and discussion between scientists, as the last quote above makes explicit. However, they do this by silencing communication of other sorts. Rather, to attend to the creative role of withdrawal is to reinstate the balance between speech and silence that is required for effective communication. Mara Beller's examination of the history of quantum theory makes this point well (Beller, 1999). Beller is concerned with the way in which new scientific knowledge emerges out of conversations between scientists, conversations that are filled with doubt and uncertainty. She argues that the radically new ideas of quantum theory formed dialogically, as physicists constantly addressed and responded to each others' ideas. Yet these conversations were necessarily punctuated and interrupted with periods of relative silence. Thus Niels Bohr and Werner Heisenberg engaged in intense discussions with each other, but it was only once they were apart that Heisenberg laid the basis for his approach to quantum theory. As Beller puts it, Heisenberg needed time away from Bohr 'in order to strike a proper, uncoerced balance in his own communicative network of cognitive responses' (p. 7). Gaining the communicative control necessary for articulating new ideas requires balancing periods of speaking out with periods of staying quiet.

Perhaps the most famous silence in the history of science is Darwin's long delay before publishing *The Origin of Species*. As Stephen Webster discusses in Chapter 2, Darwin sought a balance between isolation and collegiality, which enabled him to forge trusting friendships that served him well when he found himself under pressure to publish. Webster argues that Darwin can be used as a guide for navigating the pressures of the modern research life, reminding us of the need for retreat and delay punctuated by 'accelerative moments'. Darwin's case is especially instructive because his silence was not easily won, being frequently disrupted and intruded upon, not least by his recalcitrant, complaining body. Yet this bodily uproar also helped him justify his retreat into quietude. With Darwin, then, we see silence as a state that is performed, a physical struggle located in both body and place.

In Chapter 1, Paul Merchant considers the case of another scientist who chose long periods of silence: Joseph Farman, the atmospheric scientist who discovered the hole in the ozone layer over Antarctica in the mid-1980s. Drawing on oral history interviews, Merchant reconstructs the ways in which Farman constituted a silent presence at the British Antarctic Survey where he worked. Farman expected his data to speak for itself; his own silences gave it the opportunity to do so, both by facilitating the actual process of data collection and by allowing the data to gradually accumulate over many years until the measurements could be heard without doubt. Merchant paints a picture of a scientist whose own silence mirrored that of the ozone layer he studied – the taciturn researcher corresponding with reticent nature until, finally, both spoke out, the unambiguity and seriousness of the message emphasised by the silences that had framed it.

Merchant argues that some of Farman's silences were strategic. Nick Verouden, Maarten van der Sanden and Noelle Aarts (Chapter 4) draw attention to another form of strategic silence, where silence is deployed as a way of managing complex collaborations. Scholars of organisation studies have examined the ways in which silence permeates hierarchical workplaces. Verouden and his co-authors look at the case of silences in the less hierarchical context of cross-disciplinary university collaborations. Collaboration is seemingly premised on the drawing out of verbal interactions, but as this chapter shows through an ethnographic analysis of a teaching collaboration at the Delft University of Technology, silence also plays an important role in enabling collaboration to proceed by facilitating compromise and consensus.

In Chapter 6, Alice White presents a further example of chosen silences that were claimed to have generative effects. White examines the layers of silence that coalesced around the role of psychiatrists in British Army selection boards during the Second World War. The introduction of the psychiatric interview in 1942 gave psychiatrists a formal role in the assessment of soldiers put forward for a commission. At one level, the psychiatrists' approach to selection board interviews encouraged talk, drawing the candidate officer into a conversation with the selection panel. Yet they did so by strategically deploying silences aimed at prompting the candidate to reveal aspects of his personality. By remaining silent on certain topics or at certain points during the interview exchange, the psychiatrists believed the interview would furnish greater insights. The psychiatrists' silences were aimed at drawing out revealing talk; however, they saw silence on the part of the candidate as a problem. The psychiatrists', sometimes silent, probing of intimate matters provoked fears that such encounters would be disturbing to the soldiers. White explores how the Army authorities responded to such fears, eventually withdrawing psychiatrists from the interview process. The psychiatrists' chosen, generative silences were now replaced with an imposed, suppressive silence.

External, generative silences

It is telling, perhaps, that it is harder to identify externally imposed generative silences than it is generative silences that are chosen by the silent party or are communally agreed by the discourse community. Nevertheless, two examples – media embargoes and trade secrets – illustrate imposed silences that are often justified on the grounds of their ability to enhance the communication of science, in the first case by improving the communication itself and in the second case by providing the preconditions for the production of the knowledge to be communicated. In both these cases, however, the claims for enhanced communicative efficacy are regularly disputed either by those on whom the silences are imposed or by those who encounter the silences.

The first example concerns the widespread use of media embargoes in the reporting of research news. In the embargo system, scientific journals pre-release information about forthcoming papers to journalists, usually in the form of press releases accompanied by quotes from the researchers and their contact details. In return, the journalists withhold from publishing the news until the embargo date has passed; this usually coincides with the date of publication of the journal. In this way, the scientific journals and their press offices control the dissemination of science news by imposing temporary silences on journalists.

Unlike regimes of secrecy, media embargoes entail a withholding rather than a concealment. Nevertheless, the system is punitive. Journalists who break the embargo are usually punished by being blacklisted from future press announcements. In some cases, even journalists who have sourced a story independently have been removed from a journal's mailing list for covering the story prior to the embargo date (Kiernan, 2006: pp. 30–33). Since losing access to a major source of science news is a significant handicap for a science journalist, violations are relatively rare.

In any case, journalists also benefit from embargoes since the receipt of media-ready press releases minimises the amount of time the journalist needs to invest in a story. Indeed, when the system developed in the mid-twentieth century, the initial impetus came from science and medical journalists, who argued that having a preview of newsworthy papers would help improve the accuracy of their reporting of complex topics (Kiernan, 2006). By improving accuracy, journalists hoped also to improve relations with their sources. Today, the embargo system is still justified on the grounds that it helps ensure the accuracy of science journalism. Some also claim that it promotes more extensive coverage of science in the news media than would otherwise be the case (ibid.). Thus, to the proponents of the system, the pre-embargo silences imposed on journalists serve to enhance the communicative value of the journalists' reporting.

Many scientific journals also impose a similar system of temporary silences on scientists. Under what is known as the Ingelfinger rule – named after Franz Ingelfinger, the editor of the *New England Journal of Medicine* who first articulated the policy in 1969 – journals refuse to publish research findings that have already been reported in the media. Researchers must therefore avoid contact with journalists until after they have published their findings in a journal. Ingelfinger's original motivation was to ensure that his journal maintained competitiveness by only publishing new results, but his rule came to be seen as a way of ensuring that the media did not report research that had not yet been validated by the peer-review process (Toy, 2002).

As with the embargo system, the Ingelfinger rule was justified on the grounds that it would improve the quality of the public communication of science, in this case by ensuring that premature results or poor research would not be reported in the media. Writing in 1977, Ingelfinger observed that 'although [journalists] pride themselves on reporting accurately, there is no assurance that what they report is accurate in the first place' (Toy, 2002: p. 197). The system still holds today, with many journals warning researchers to be cautious about talking to journalists when presenting their work at conferences. If scientists do court media coverage before their findings are published in a journal, they risk having their paper rejected even if it has already been accepted for publication (Kiernan, 2006).

Despite the threat of punitive sanctions, proponents of both the Ingelfinger rule and the embargo system defend the imposition of these silencing protocols as a means of improving communication with public audiences; that is, they are seen as having a generative, rather than a suppressive, function. However, they have also been charged with the opposite. Vincent Kiernan (2006), for instance, argues that not only is there no evidence that embargoes improve the accuracy of science journalism, but they encourage a pack mentality among journalists and distort the media coverage of science by encouraging a focus on research findings rather than the research process, and on a small number of elite journals rather than the full range of scientific research.

Both the Ingelfinger rule and the embargo system could also be seen as counter to current calls for open science. Indeed, the open data movement could be understood as the mirror image of external, generative silence in that, in the hope of improving communication, it imposes a requirement for non-silence regarding those parts of the scientific process that have formerly been kept private. Where the Ingelfinger rule configures publication as the end-point of a closed system, open science breaks all the boundaries of that system, allowing multiple voices to enter at any stage of the research process. In Chapter 11 of this volume, Ann Grand considers the challenges that this imposed un-silencing might entail. Grand argues that simply breaking a silence is not sufficient to ensure communicative efficacy. For data to be meaningful to public audiences, it needs to be addressed to those audiences – a process that requires time, effort and skill. At the same time, researchers have to find ways to compensate for the loss of the benefits of silence that have been discussed above.

My second example of external, generative silences is trade secrets. One of the concerns that Grand highlights in the transition to open science is the ambiguity it creates over the ownership of intellectual property. Such concerns are deeply embedded in modern scientific practice, both through cultural norms pertaining to priority claims and through legal instruments safeguarding the commercial value of research. As noted above, priority claims must navigate a moment of instability in which the risks of publication may outweigh the benefits. Similarly, the legal processes for establishing intellectual property rights (IPRs) frequently entail a movement from the non-disclosure of trade secrets to the disclosure of a patent application. As a draft proposal for a European Commission directive to harmonise trade secrets law across the EU puts it: 'Every IPR starts with a secret' (European Commission, 2013).

The knowledge that might later build into a patentable idea or product may in earlier stages of research and development not qualify for patent protection yet still be of commercial value. In such cases, companies typically use non-disclosure agreements to prevent the circulation of the knowledge to competitors. Such contracts impose silences on employees which are defended on the grounds that investment in the research will only be forthcoming if the commercial value of the research is protected. In other words, it is claimed that the conditions for the knowledge to be created are dependent on silencing mechanisms. For instance, the European Commission directive sees trade secrets as 'a key complementary instrument for the required appropriation of intellectual assets that are the drivers of the knowledge economy of the

21st century' (European Commission, 2013). Whilst trade secrets are primarily invoked for their ability to enhance economic value, they can also be seen as having a communicative effect by creating the conditions that make possible the later communication of a patent.

In an ethnographic study of the control of information flow in scientific settings, Stephen Hilgartner (2012) suggests that trade secrecy is an example of an institutionalised practice through which a 'regime of closure' is enacted. Hilgartner gives the example of a company's secrecy about a new development being accepted as normal, even though both the act and object of concealment are partially revealed through a scientist's comportment and tone. Hilgartner argues that in such situations those involved 'manage a dialectic of revelation and concealment through which knowledge is selectively made available and unavailable to others, often in the same act' (Hilgartner, 2012: p. 268). In such ways, the imposed silences of industry-sponsored research are selectively navigated by those who are expected to reproduce them.

As Hilgartner's example suggests, the silence of a trade secret can itself convey meaning. For instance, Coca-Cola's brand identity is supported in part by the mystery surrounding its secret recipe and Apple selectively releases parts of its secrets to foster an image of innovative creativity (Bos et al., 2015). The value of secrets can therefore derive as much from their status as secrets as from their actual content; a silence can generate a meaning that would be punctured if expressed through language. Vermeir and Margócsy (2012) give the example of alchemists, who guarded their knowledge closely but were often disappointed when they did hear the details of others' work. Often, then, it is the status of secret as secret that has most value and invites desire in others for as long as they don't know what the secret is. Yet this value also requires that these others know that the secret exists. In such instances, the value of a secret requires that the secret be announced even as its content is kept hidden.

The example of trade secrets also illustrates the ways in which the suppressive effects of externally imposed silences can be hard to eliminate. Restricting the flow of information within an organisation can limit, rather than enhance, innovation; and restricting the flow of information externally can curtail collaborations (Bos et al., 2015). Concerns also arise where non-disclosure agreements are used to prevent employees from speaking out on matters of the public interest, or, especially given widespread industry sponsorship of university research, where similar contracts are used to prevent publication of research findings (Gøtzsche et al., 2006).

Epistemological silences

The four-fold typology presented above locates silence at the level of the actors generating or receiving the silence. However, silence can also be encountered at one remove – for instance, as a gap or omission in the historical record that may not have existed in the contemporary discourse, as a consequence of analysts desiring to articulate experiential or tacit knowledge, or as a non-communicative silence rendered meaningful as a result of recontextualisation. Here, rather than examining how silences are maintained and experienced as in the previous examples, the question is how such encountered silences are to be understood and interpreted. In confronting the epistemology of absence, analyses of this category of silence resonate with work in the field of ignorance studies.

In Chapter 9, Brian Rappert, Catelijne Coopmans and Giovanna Colombetti consider the intersection of experienced silence and epistemological silence by examining the ways in which scientists have attempted to fashion accounts of Buddhist meditation practice. In contemplative practices such as meditation, the impossibility of verbalising the lived experience is at its most acute. Yet repeated attempts have been made to do just that, most recently in the guise of neuroimaging experiments on the effects of meditation on the brains of its practitioners. Rappert et al. argue that such accounts are characterised by various forms of indirection, which enable researchers to speak of the meditative *state* whilst remaining silent about the meditative *experience*. In such accounts, scientists negotiate the silences they encounter by generating silences of their own, including silence about which aspects of meditation should remain silent.

The inevitable incompleteness of the historical record offers another form of epistemological silence that researchers must grapple with. Do the silences of the historical record represent the boundaries of what was then known or do they signify nothing more than the limited survival of documents? In Chapter 8, Elizabeth Hind considers the difficulties modern scholars face in interpreting the silences they encounter in Ancient Egyptian mathematical texts. Only a few mathematical papyri survive, and with those that do, holes in the document may compound the already considerable challenges of translating terms that occur in few other documents. The scholar must weigh up whether the silence derives from this modern absence of evidence or from a contemporary absence; and if the latter, whether this signifies an absence of knowledge on the part of Egyptian mathematicians or the opposite, the presence of tacit knowledge. Hind argues that, by severing mathematical texts from their cultural contexts, interpretations of Egyptian mathematics have often assumed that encountered silences should be interpreted as absent knowledge and that this risks incorporating a cultural bias into our historiography.

The epistemological silences that Hind studies derive from the inherent incompleteness of the historical record. Epistemological silences can also derive from historians themselves, as a result of their patterns of inattention. In Chapter 13, Charlotte Sleigh explores the case of Charles Hoy Fort, the writer whose name was appropriated by the International Fortean Society long after his death but who has been largely ignored by historians and literary scholars despite his sustained critique of modern science. Sleigh examines the reasons behind this scholarly inattention to Fort, tracing the confluence of his humorous and provisional writing style with his insistence on the significance of anomalous data and amateur collecting. In his writings, Fort attempted to combat the silencing tendencies of contemporary science with a barrage of words and data, giving voice en masse to data that, if presented individually, would have been silenced. Yet, Sleigh suggests, this voicing was of such excess that, ultimately, it constituted a silence of its own.

Tuning in to silence

We hope that the case studies presented in this volume make the case for the role of silence in the practice and communication of science. As silence studies scholars have argued, silence can have a communicative value, functioning as a dynamic context-dependent sign that is actively produced to convey specific meanings. Scientists, like others, choose silence, experience silence and are silenced, for a variety of reasons with a variety of effects. If we are to understand the ways in which scientists generate knowledge and how this knowledge is brought to bear on social concerns, then we must attend to what scientists don't say, and why they don't say it, as well as to what they do say.

We hope also that attending to silence will encourage some reflection on the communication policies that currently dominate within the scientific community. Without verbal communication, there would, of course, be no science. Policies that encourage collaboration, publication and public engagement therefore play a vital role in furthering scientific research and embedding science within society. Indeed, staying silent can carry risks. The University of East Anglia climate scientists who attempted to stay silent when barraged with Freedom of Information requests from climate sceptics were ultimately unsilenced in a more damaging way when their private emails were leaked. Yet not staying silent can also incur costs. When, in 2011, a team of scientists announced to the media that they had observed neutrinos travelling faster than the speed of light – a finding that would have been difficult to accommodate within special relativity – it cost two of the team leaders their positions after mistakes in the initial measurements came to light. In this instance, a degree of silence would have served the researchers better.

As many of the case studies in this volume show, silences can bring strategic benefits to those who produce them. Directed silences can smooth the progress of research, fostering collaboration and deferring hard-to-resolve questions; temporary silences, such as those brought about by retreats and withdrawals, can provide the conditions for research to take place. Paying attention to silence highlights the ambiguities inherent in all communicative acts and reminds us of the dialectical relationship between speech and silence. Successful communication requires a balance between the two. The logorrhoea of the modern research environment risks upsetting that balance.

References

Acheson, K., 2008. Silence as gesture: rethinking the nature of communicative silences. *Communication Theory*, 18, pp. 535–555.

Anderson, M.S., Ronning, E.A., DeVries, R. and Martinson, B.C., 2007. The perverse effects of competition on scientists' work and relationships. *Science and Engineering Ethics*, 13, pp. 437–461.
Aspen Center for Physics, n.d. Mission. Available at: http://aspenphys.org/aboutus/index.html
Balmer, B., 2012. *Secrecy and Science: A Historical Sociology of Biological and Chemical Warfare*. Farnham: Ashgate.
Basso, K.H., 1970. 'To give up on words': silence in Western Apache culture. *Southwestern Journal of Anthropology*, 26(3), pp. 213–230.
Beller, M., 1999. *Quantum Dialogue: The Making of a Revolution*. Chicago: Chicago University Press.
Biagioli, M., 2012. From ciphers to confidentiality: secrecy, openness and priority in science. *British Journal for the History of Science*, 45(2), pp. 213–233.
Brown, D., 2012. Mutant flu: the view from the newsroom. *Nature*, 485(7396), p. 7.
Bos, B., Broekhuizen, T.L.J. and de Faria, P., 2015. A dynamic view on secrecy management. *Journal of Business Research*, 68(12), pp. 2619–2627.
Böschen, S., Kastenhofer, K., Rust, I., Soentgen, J. and Wehling, P., 2010. Scientific nonknowledge and its political dynamics: the cases of agri-biotechnology and mobile phoning. *Science, Technology, and Human Values*, 35(6), pp. 783–811.
Brummett, B., 1980. Towards a theory of silence as a political strategy. *Quarterly Journal of Speech*, 66(3), pp. 289–303.
Bruneau, T.J., 1973. Communicative silences: forms and functions. *The Journal of Communication*, 23, pp. 17–46.
Butler, D., 2011. Fears grow over lab-bred flu. *Nature*, 480(22 Dec.), pp. 421–422.
Butler, D., 2012. Scientists call for 60-day suspension of mutant flu research. *Nature*, 20 Jan., doi: 10.1038/nature.2012.9873.
Butler, D. and Ledford, H., 2013. US biosecurity board revises stance on mutant-flu studies. *Nature*, 30 Mar. doi: 10.1038/nature.2012.10369.
Cage, J., 1955. Experimental music: doctrine. In: Cage, J., 1973, *Silence: Lectures and Writing*. Middletown, CT: Wesleyan University Press, pp. 13–17.
Cage, J., 1957. Experimental music. In: Cage, J., 1973, *Silence: Lectures and Writing*. Middletown, CT: Wesleyan University Press, pp. 7–12.
Couzin, J., 2002. A call for restraint on biological data. *Science*, 297(5582), pp. 749–751.
Dennis, M.A., 1999. Secrecy and science revisited: from politics to historical practice and back. In: Peppy, J. (ed.), *Secrecy and Knowledge Production*. Cornell University Peace Studies Program, Occasional Paper #23.
Dressen, D.F., 2002. Accounting for fieldwork in three areas of modern geology: a situated analysis of textual silence and salience. Ph.D. thesis. Ann Arbor: University of Michigan Press.
Ephratt, M., 2008. The functions of silence. *Journal of Pragmatics*, 40(11), pp. 1909–1938.
European Commission, 2013. Proposal for a directive of the European Parliament and of the Council on the protection of undisclosed know-how and business information (trade secrets) against their unlawful acquisition, use and disclosure. 52013PC0813. Available at: http://eur-lex.europa.eu/legal-content/EN/ALL/?uri=CELEX:52013PC0813
Frankel, M., 2012. Regulating the boundaries of dual use research. *Science*, 336(6088), pp. 1523–1525.
Fouchier, R.A.M., García-Sastre, A., Kawaoka, Y. and 36 co-authors, 2012. Pause on avian flu transmission studies. *Nature*, 481(26 Jan.), p. 443.

Glenn, C., 2004. *Unspoken: A Rhetoric of Silence*. Carbondale: Southern Illinois University Press.
Goldberg, S., 1995. Groves and the scientists: compartmentalization and the building of the bomb. *Physics Today*, Aug., pp. 38–43.
Gøtzsche, P. C., Hróbjartsson, A., Johansen, H.K., Haahr, M.T., Altman, D.G. and An-Wen Chan, A., 2006. Constraints on publication rights in industry-initiated clinical trials. *JAMA*, 295(14), pp. 1641–1646.
Gray, S.W.D., 2012. Meanings of silence in democratic theory and practice. Annual Meeting of the Canadian Political Science Association, Edmonton, Alberta.
Gross, M., 2010. *Ignorance and Surprise: Science, Society, and Ecological Design*. Cambridge, MA: MIT Press.
Gross, M. and McGoey, L. (eds), 2015. *Routledge International Handbook of Ignorance Studies*. Abingdon: Routledge.
Gyngell, C. and Savulescu, J., 2015. The moral imperative to research editing embryos: the need to modify Nature and Science. *Practical Ethics in the News*. University of Oxford blog, 23 Apr. Available at: http://blog.practicalethics.ox.ac.uk/2015/04/the-moral-imperative-to-research-editing-embryos-the-need-to-modify-nature-and-science/
Hales, P. B., 1997. *Atomic Spaces: Living on the Manhattan Project*. Urbana, IL: University of Illinois Press.
Hilgartner, S., 2012. Selective flows of knowledge in technoscientific interaction: information control in genome research. *The British Journal for the History of Science*, 45(2), pp. 267–280.
Institute for Advanced Study, n.d. There are no excuses in paradise. Institute of Advanced Study, available at: https://www.ias.edu/ias-letter/2008/paradise
Jaworski, A., 1993. *The Power of Silence: Social and Pragmatic Perspectives*. Newbury Park, CA.: Sage.
Johannesen, R.L., 1974. The functions of silence: a plea for communication research. *Western Journal of Speech Communication*, 38(1), pp. 25–35.
Jones, V.C., 1985. *Manhattan: The Army and the Atomic Bomb*. Washington, DC: US Army Center of Military History.
Kaiser, D. and Moreno, J., 2012. Dual-use research: self-censorship is not enough. *Nature*, 492, pp. 345–347.
Kaiser, J. and Normile, D., 2015. Chinese paper on embryo engineering splits scientific community. *Science*, 24 Apr., doi: 10.1126/science.aab2547.
Kawaoka, Y., 2012. H5N1: flu transmission work is urgent. *Nature*, 482(9 Feb.), p. 155.
Keane, J., 2012. Silence and catastrophe: new reasons why politics matters in the early years of the twenty-first century. *The Political Quarterly*, 3(4), pp. 660–668.
Kellogg, D., 2006. Toward a post-academic science policy: scientific communication and the collapse of the Mertonian norms. *International Journal of Communications Law and Policy*. Available at: http://ssrn.com/abstract=900042
Kempner, J., Merz, J.F. and Bosk, C.L., 2011. Forbidden knowledge: public controversy and the production of nonknowledge. *Sociological Forum*, 26(3), pp. 475–500.
Kiernan, V., 2006. *Embargoed Science*. Urbana: University of Illinois Press.
MacLure, M., Holmes, R., Jones, L. and MacRae, C., 2010. Silence as resistance to analysis. Or, on not opening one's mouth properly. *Qualitative Inquiry*, 16, pp. 492.
Merton, R.K., 1973. The normative structure of science. In: *The Sociology of Science: Theoretical and Empirical Investigations*. Chicago: University of Chicago Press, pp. 267–278.

NIH, 2011. Press statement on the NSABB review of H5N1 research. National Institutes for Health, press release, 20 Dec. Available at: www.nih.gov/news/health/dec2011/od-20.htm

Pinto, M.F., 2015. Tensions in agnotology: normativity in the studies of commercially driven ignorance. *Social Studies of Science*, 45(2), pp. 294–315.

Proctor, R. and Schiebinger, L.L. (eds), 2008. *Agnotology: The Making and Unmaking of Ignorance*. Stanford, CA: Stanford University Press.

Quist, S.A., 2002. *Security Classification of Information. Volume 1. Introduction, History, and Adverse Impacts*. Oak Ridge, TN: Oak Ridge National Laboratory. Available at: http://fas.org/sgp/library/quist/

Schiermeier, Q., 2015. Russian secret service to vet research papers. *Nature*, 526(7574), p. 486.

Scott, R.L., 1972. Rhetoric and silence. *Western Speech*, 36(3), pp. 146–158.

Scott, R.L., 1993. Dialectical tensions of speaking and silence. *Quarterly Journal of Speech*, 79(1), pp. 1–18.

Shapin, S., 1991. 'The mind is its own place': science and solitude in seventeenth-century England. *Science in Context*, 4(1), pp. 191–218.

Sobkowiak, W., 1997. Silence and markedness theory. In: Jaworski, Adam (ed.), *Silence: Interdisciplinary Perspectives*. Berlin: Mouton de Gruyter, pp. 39–61.

Stilgoe, J., Lock, S.J. and Wilsdon, J., 2014. Why should we promote public engagement with science? *Public Understanding of Science*, 23(1), pp. 4–15.

Thiesmeyer, L. (ed.), 2003. *Discourse and Silencing: Representation and the Language of Displacement*. Amsterdam: John Benjamins.

Toy, J., 2002. The Ingelfinger Rule: Franz Ingelfinger at the New England Journal of Medicine 1967–1977. *Science Editor*, 25(6), pp. 195–198.

Vermier, K. and Margócsy, D., 2012. States of secrecy: an introduction. *British Journal for the History of Science*, 45(2), pp. 153–164.

Washburn, P.S., 1990. The Office of Censorship's attempt to control press coverage of the atomic bomb during World War II. *Journalism Monographs*, no. 120.

Weart, S., 1976. Scientists with a secret. *Physics Today*, Feb., pp. 23–30.

Wellerstein, A., 2013. The worst of the Manhattan Project leaks. *Restricted Data: The Nuclear Secrecy Blog*, 20 Sep. Available at: http://blog.nuclearsecrecy.com/2013/09/20/worst-manhattan-project-leaks/

Part I
Choosing silence

1 'He didn't go round the conference circuit talking about it'

Oral histories of Joseph Farman and the ozone hole

Paul Merchant

Introduction

This chapter focuses on the use of silence by the scientist whose British Antarctic Survey (BAS) team discovered the 'ozone hole' over Antarctica in the early 1980s: Joseph Farman. Close attention to life story oral history interviews recorded with Farman and his colleagues reveals forms of silent presence and practice that were central to the acquisition and communication of scientific knowledge about damage to part of the planet's atmosphere.

I am not concerned with silence in the way that others using interviews have tended to be: silence as gaps, pauses and omissions in the process of interviewing (Greenspan, 2014; Jesse, 2013; Layman, 2009). I ignore what Farman and others *didn't* say in interview. Instead, I take what *was* said to reconstruct ways in which silence was a part of the way Farman worked, interacted with colleagues, viewed and presented results and, after the revelation of the 'ozone hole', communicated with journalists and politicians.

In other accounts of the discovery of the 'ozone hole', short quotes from private interviews with scientists, including Farman, are used to offer readers a view behind the scenes of science and an apparent closeness to people and facts (Booth, 1994; Roan, 1989). The use of interviews in this chapter could not be more different. I examine the agency of different kinds of silence by attending to several extended extracts from unusually long and detailed life story interviews recorded for National Life Stories' An Oral History of British Science at the British Library, available in full through the British Library sounds website.[1]

The extracts include accounts (his own and others') of Farman in workplaces (Antarctic bases, meeting rooms, television studios, BAS's coffee room) doing and not doing, saying and not saying. In each place, far from being an absence, silence is present and acts – it achieves certain ends. It focuses and blocks, frames and underlines, acts to hide or to attract attention.

Recording ozone

Following public school in Norwich, national service, a natural sciences degree at the University of Cambridge and missile work for de Havilland

Propellers, Joseph Farman joined BAS as a 'scientific officer' in 1956.[2] He soon took charge of atmospheric and geophysical measurements being made as part of the International Geophysical Year of 1957–1958 at BAS bases in Antarctica: Halley Bay and Argentine Islands. The measuring effort included the use of ground-based instruments called Dobson Spectrophotometers to measure concentrations of ozone in the Antarctic stratosphere (Dobson, 1968: 390–403). From 1957 until his retirement in 1990, Farman oversaw the continued measurement of ozone (and many other variables) at these stations, sometimes working in Antarctica but more often by directing others from BAS headquarters in Britain as the head of a research group with a variable title: 'Physics Unit', 'Stratosphere Section' and from 1983 'Chemistry, Radiation and Dynamics Section'. He retired in 1990 and died in 2013.

When using the Dobson Sepctrophotometer to measure Antarctic stratospheric ozone, Farman tended to work alone. Assisted by a photograph of Farman with the instrument at BAS many years after his retirement (Figure 1.1), we can imagine him in a hut in Antarctica, positioning the vertical periscope-like collector, operating levers and dials, and observing a needle in such a way that ozone concentrations in the path between the sun and the instrument could

Figure 1.1 Joseph Farman with a Dobson Spectrophotometer at British Antarctic Survey headquarters, Cambridge, 2011
Source: The British Library. © British Library Board; www.bl.uk/voices-of-science.

be estimated. Using the Dobson instrument, Farman tells us, should become automatic, not needing to be thought about let alone spoken about:

> You have these levers which have prisms on them which enable you to grab hold of two quite narrow bands of wavelength. And since you want four at a time normally what you do is you set little stops on the levers so you can push the levers up to the stops and then push them down to a second set of stops. You can do all this very quickly and in fact what you're trying to do is to get these measurements all done within about thirty or forty seconds, which just needs a lot of practice. [...]
>
> *What difficulties would you encounter in first using the Dobson Spectrometer in getting the measurement?*
>
> Erm [pause], oh, just getting in your mind what you need to do and then getting that firmly enough established so you don't have to think about it any more – so it's quite automatic: you walk up to the instrument, you switch it on, you wait for it to warm up and you choose – well – you have to then read the thermometer [...] because [...] the way you choose the wavelengths is quite strongly dependent on the temperature of the instrument at the time. [...] You have an alternator inside which allows the two wavelengths in succession to form on a detector and then the output from the detector is fed to a galvanometer and if the wavelengths are equal, well of course the galvanometer needle stays still, but on the other hand if it's not in the right place then the needle will affect to one side or the other and then oscillate and so you've got to learn sort of to pick all these signals up and twist your hands at the same time and – it all sounds a bit complicated but it's [laughs].
>
> *So what did you have to turn in order to get the two wavelengths the same? Was it a dial or a lever?* [...]
>
> No, it's a wheel on top which you rotate with your fingers on it and [...] so you just have to fiddle with your hand here, keeping the needle just slightly dithering.
>
> (Farman, 2010: track 4, 8:32–11:28)

Silence is invested in these measurements. The instrument tends to be used well with silent focus; an assistant might disrupt this: 'I think yes most of us preferred to do it on our own; there are one or two things where four hands are helpful but on the whole by the time you've organised someone to do it and you've lost your temper with each other every now and again because you do the wrong thing at the wrong time' (Farman, 2010: track 4, 15:29–15:50). Mistakes can be made in silence, of course, but silence and care in taking measurements seem to fit together here.

Saying nothing

From 1957 until the early 1980s, measurements of concentrations of ozone in the stratosphere above Argentine Islands and Halley Bay showed little variation year to year, apart from the significant intra-annual shifts. Farman, too, was viewed during this period as an undemonstrative, low-key presence in the BAS headquarters in Cambridge. His colleagues' descriptions of him are very consistent: he tends not to talk; when he does he is difficult to understand; he smokes a pipe which produces a lot of smoke (literally he is hidden); he daydreams (though physically present, there is a sense that he is not there). Geologist Janet Thomson, interviewed about this period, remembers the distant, hidden aspect of Farman's personality:

> There was a, usually a sort of [laughs] haze of pipe smoke about this person and he, he was in another discipline on the far side of the building; I really had no contact with him whatsoever. He didn't come to prominence until about '85.
> (Thomson, 2010: track 6, 35:16–35:32)

In the same interview, she elaborates: 'If you hadn't been on base with him or on a ship going down to the Argentine Islands you probably didn't know he existed – just saw this guy cycling by, puffing a pipe and, oh yes that's Joe [laughs]. He was a fairly amiable character but I didn't really interact with him very much at all' (Thomson, 2010: track 8, 5:37–5:56). Similarly, ice core scientist Eric Wolff recalls:

> He's a pipe-smoking, quite taciturn academic type who though, you know, who wouldn't necessarily have gone out of his way to chat to a young whippersnapper, nor would the young whippersnapper have felt that he was entitled to go and talk to Joe.
> (Wolff, 2012: track 4, 35:51–36:06)

At this point in his career, Farman cycles to BAS, manages and analyses Antarctic measurements and smokes. His tendency not to talk about ozone and other measurements is not challenged, even by the employment of assistants and the routine of departmental talks. One of Farman's assistants, Barbara Bowen, employed from 1976 to digitise measurements of all kinds completed at Argentine Islands and Halley Bay, remembers that he tended not to speak about the work to her. Nor would he give talks to the Physics Unit. In fact it appears that other colleagues would have to step in to counteract Farman's silence:

> *And what did Joe tell you about what you were doing, what the importance of it was, what it involved?*
> Not a lot really [laughs] […] I think once I'd been doing it a bit, Brian Gardiner [scientist colleague of Joseph Farman] was, was a bit more sort of normal chatty, you know, Joe was in his – lost in the clouds of the smoke [laughs] pipe smoke and so – thinking half the time – so I think it was Brian would be a bit more sort of [laughs] with the human side of it,

you know, and say, 'well this is all helping to – this is' [...] I think it was probably he if anybody told me what it was all about really and I didn't fully appreciate then, having had several talks by all the people in the different divisions but amongst them, you know, people in our division, like [pause] Joe wouldn't give the talk, Brian would give the talk.

(Bowen, 2010: track 5, 47:36–49:16)

Even in the informal arena of the coffee room Farman was striking for his withdrawn, almost mute demeanour:

We all took a mug [...] the state of some of them! [...] Most of them had been 'down south' which we thought meant Brighton, you know, but of course [laughs] meant Antarctica. So we were [...] ribbed for that, but anyway, it was a very happy, informal coffee time and perhaps we took twenty minutes when we should have only taken quarter of an hour [...] but there was jokes all around all the time pretty well, except if a lot of them had gone south and Joe was left and Munroe [Sievwright] and they were both terribly quiet and rather difficult to talk to; they had – once you got them going they were all right but [laughs] they were terribly hard work to talk to. You know, small talk, they didn't, didn't have small talk really and, well, they, they were quite – they were quite sweet – I was quite fond of them in a way.

(Bowen, 2010: track 5, 17:17–18:26)

Tending not to talk about ozone (or other) measurements, or to take part in the 'jokes all round' with colleagues on a break, Farman was also disinclined to *write* about his team's work. Bowen remembers that Farman was 'notoriously slow' in writing scientific papers (Bowen, 2010: track 5, 23:16–23:18). Indeed, it took twenty years for the first peer-reviewed paper on ozone measurements at Argentine Islands and Halley Bay to appear (Farman, 1977). Its abstract advertises the lack of dramatic change over the period 1957 to 1973: 'Long-term trends in total ozone are shown to be small at both stations.' Ozone did not seem to be saying anything interesting about itself. As a quantity, a constituent, a feature of the upper atmosphere over Antarctica, ozone was silent because the measurements of it were steady.

Farman recalled that this silent steadiness and the long-term scope of the measurements could provoke discouraging comments, as when the chairman of the research council responsible for managing and funding BAS science – the Natural Environment Research Council (NERC) – visited:

He [Hermann Bondi] came to inspect the British Antarctic Survey [...] and we explained what we were doing to him and unfortunately he then produced what one can only say is a fairly typical high table remark; he looked at us very sadly and he said, oh, he said, 'You're making these measurements for posterity are you? Tell me, what has posterity done for you?' And I could have kicked him [laughs] because I had all the — you know – young men who were working on it and we all sort of thought we were doing quite a good job with a rather tedious and time-consuming exercise.

(Farman, 2010: track 2, 32:11–32:55)

There was also pressure from within BAS to discontinue ozone and associated meteorological measurements:

> There was more money in total but in a sense one of the things looked at with a slight sort of suspicion were long term measurements. You know, an awful lot of people say, 'oh, you've measured this for so many years, do you really need to keep on doing it?' Well, what can you say? I mean you can't say, 'oh, but keep doing it for another twenty years, you'll find an ozone hole' [laughs].
> *Who was saying that [...]?*
> Oh, just a general sort of feeling, you know. These were internal meetings usually.
>
> (Farman, 2010: track 10, 41:10–41:42)

Farman's response, we learn from colleagues encouraged to talk about him for a funeral address, relied on a kind of strategic silence:

> Luckily the measurements continued. [...] He fought hard to maintain support for the programme. He had many valuable weapons when dealing with senior management. As Alan Rogers, a BAS colleague, recalls: [...] 'During management meetings Joe would sit at the back of the group steadily puffing away. The density of smoke was directly related to his irritation and exasperation levels, and in particularly trying meetings we would lose Joe behind a cloud of smoke. We suspected Joe's ploy was just to obscure the senior manager from his vision. Joe did not need to say much; we all knew how much he was thinking by how much of him we could see.'
>
> (Harris, 2013: p. 116)

With the more general tendency for negative and simply confirmatory results in science not to be discussed and published, it is tempting to view Farman and ozone at this point as sharing a certain kind of silent presence. The silence of ozone and the silence of Farman are folded up together in an apparent absence of results that concerns managers. Farman and ozone hover undemonstratively in their respective environments.

The ozone hole

The Antarctic stratosphere finally spoke up in the early 1980s. Measurements revealed much lower concentrations of ozone than would be expected in the

Antarctic spring. Farman doubted these early messages from his previously silent subject; his 'first reaction was to check and recheck the Dobson instruments' (Pyle and Harris, 2013: p. 435). He remembered this need to check in terms of a technical distinction between signal and background noise:

> Okay, we [pause] were a bit pedantic and one of the things we always used to do with ozone results was to look at the upper air temperatures; you can be fairly sure that if you get a big change in ozone you're going to see something changing in the temperature graph. That's simply to say that before the paper any change of ozone was indeed dynamic, usually [laughs]. So you had – you had to see something in the upper atmosphere records as well and so if you got one without the other you should scratch your head and perhaps apply a little quality control and flag it and so on and so forth. [...] When people are doing measurements every day, every hour or something of the sort, something goes wrong sooner or later. You know, someone misreads something, someone sort of instead of writing down fifty-nine manages somehow to do it to ninety-five or all the other silly things the human mind [laughs] does, so you must have some quality control. And I don't think anyone could seriously have said that in '82 you would have convinced yourself that there was anything other than some noise about the place.
> (Farman, 2010: track 15, 18:22–19:46)

As Farman worked behind the scenes to determine exactly what his instruments in Antarctica were saying about ozone, his silence continued to cause concern at NERC. Indeed, at some point in the early 1980s, John Woods (then member of NERC Council and chair of Meteorology, Oceanography and Hydrology) was dispatched to sack Farman in view of his very low publication rate. In his own life story interview, John Woods remembers the visit to BAS as follows:

> The work was done by a researcher at the British Antarctic Survey, a man called Joe Farman [...] And he [pause] was a loner by temperament. [...] Joe Farman felt that his measurements were showing ozone depletion as had been predicted. But then he was up against the fact that the big boys [NASA] had looked and it wasn't there, and who was he to – but he was a good enough scientist and meticulous: he checked and checked and checked. But he unfortunately kept very quiet about it; he didn't talk – he didn't go round the conference circuit talking about it, because he'd thought he'd just be laughed out of court. [...] And he wasn't doing anything else, this became a sort of – you know sort of passionate about it. [...] So I went up and he said he wasn't ready to publish. He showed me. He was very open about what he was doing, he showed me it all and it was pretty convincing to me. [...] I told him, 'I've been told to sack you but I'm not going to do it.' [...] I said, 'I'll back you up if you decide to

publish.' [...] And he then published and I didn't have to defend him because his own data and analysis spoke for itself loud and clear.

(Woods, 2011: track 7, 1:49:23–1:55:40)

In a popular account of the discovery of the 'ozone hole', science writer Sharon Roan portrays Farman's silence as an active strategy of secrecy:

> Those curiously low ozone values his research team been detecting each spring over Antarctica had held up again this year 1984. [...] For almost three years Farman had kept mum on this mysterious secret, even quashing the attempts of a young Cambridge Ph.D. student who had made the Antarctic ozone readings part of his Ph.D. thesis.
>
> (Roan, 1989: p. 125)

Farman himself demurs from this interpretation, preferring, in his life story interview, an account of his silence as preferable to the amplification of an unclear signal, or the sounding of a false alarm:

> *The impression you get from reading books like Roan's book,* The Ozone Crisis, *is that the first sort of low results came in 1982 and then there was this long period until the publishing and –*
> No this is all – I mean when you look back you can kick yourself for not thinking something funny was going on but really unless you were watching it every week or so you – I think it would be unkind to say we should have – we couldn't have published any earlier because one would have been in grave doubt about whether – it would have meant an awful lot of statistics. We left it so long that all you needed at that time was grandmother [laughs]. No, I mean it's what I call grandmother's test, that if I show this picture to grandmother will she see what I'm talking about? And the answer is well, yes she should do [laughs]. But, you know, you don't need standard deviations or whatever. In fact I think I invented this new method of presentation. I don't know if you remember but the *Nature* graphs actually show the envelope: the highest and lowest values ever recorded.
>
> (Farman, 2010: track 15, 17:11–18:22)

In other words, he waited until the results would speak for themselves, until it wasn't necessary to do a lot of talking on their behalf, until you could show 'grandmother' a graph without comment. The paper itself was published as a 'letter' to *Nature* (Farman et al., 1985) because he wanted the observation – the trend, the fall – to be communicated quickly, with little fuss:

> 'Letters' are short. The paper is [...] only about four pages [...] and if you want to get something through – I mean, you know, it's a long process by the time you've gone through review and answered all the obvious things

which people are going to say to you: 'don't be silly' and all the rest of it [laughs]. So it had to be short and it really had to go in as letters and not as a bigger article. So I guess the real point about the paper was to be as concise as possible, to put enough graphs in which were moderately sensible.

(Farman, 2010: track 15, 20:18–20:59)

A reference pointed readers to something more wordy: a BAS Report (Farman and Hamilton, 1975), a 'fairly complete discussion of twenty-five years of ozone data with all the supporting facts which show the instruments behaved themselves and so on and so forth' (Farman, 2010: track 15, 22:38–22:49). The graphs in the 'letter' were designed to be read without elaboration – to be seen as statements in themselves. In his interview, referring to one of the graphs, Farman says:

When they sort of fall out of the envelope really what anyone says about statistics, you have to sort of say to yourself, well, these at least are unprecedented values [laughs]. […] These are much lower than any value seen in the rest of the dataset. […] I decided that it was much more dramatic if you actually put in the highest and the lowest because, you know, if these had fallen in here somewhere one might have been a little bit uncertain but, since they dropped outside it, the obvious thing to do seemed to be say, well, here's the highest we've seen before, here's the lowest we've seen before and, you know, *gosh, these are low aren't they*? [laughs] That's what I sort of say with grandmother, you know, you don't really need to start talking about distributions and frequency distributions and such like things; it's perfectly obvious what's going on.

(Farman, 2010: track 15, 23:38–25:12)

The publication's clarity was designed to prompt further action, not discussion (to silence critical review):

I mean it was fairly obvious that there would be a certain amount of hesitation in accepting it shall we say [laughs] […] the main thought was to make it as simple and clear as one possibly can […] I don't know, we were sort of – you know the main thing was to get the, the facts published in such a way that if possible people wouldn't be arguing too much about whether it could possibly be true or not but were prepared to accept it and [pause] start investigating it in great detail as soon as was possible.

(Farman, 2010: track 15, 26:26–36:59)

The aim – as expressed here – was to close off talk of the wrong kind, especially discussion of: (a) whether or not the data was showing a recent dramatic reduction in ozone concentrations in the stratosphere over Halley Bay and Argentine Islands; and (b) whether or not this reduction could be caused by changes in the movement of air in this bit of the stratosphere ('dynamic changes'). Having closed off this talk of the wrong kind, only certain kinds of further investigation and action remained: inspection of satellite data, instrumented research flights through the Antarctic stratosphere to investigate chemical depletion of ozone and representations to policy-makers and the public.

Talking about the ozone hole to the media and politicians

With the publication of the 1985 *Nature* 'letter', Joseph Farman's prestige at BAS rose. At a time when long term monitoring projects were out of favour, an unlikely kind of science and an unlikely individual came to represent the institution. Janet Thomson remembers:

> Well it was a bit ironic in a way that the value of those datasets was that they were long term monitoring projects and by that time [...] monitoring and basic surveying work was regarded a bit of a no-no. So suddenly this project that had been the result of long term monitoring sort of became high profile so everybody thought, gosh what a joke, you know [laughs], here we are trying to stop that sort of thing. [...] And I don't think Joe was a particularly high profile scientist within the organisation because he was on this sort of rather basic routine project that had been going on forever and people rather forgot that *he* existed as well as *it* and his team, so I think it was rather in the background before suddenly the, the great results started to come out and, and he was being pursued by the media. Which again was a bit of a joke because, you know, he always had this pipe in his hand and he muttered behind this cloud of smoke and [laughs] you sometimes wondered what he was talking about.
>
> (Thomson, 2010: track 8, 3:29–5:28)

Barbara Bowen remembers that Farman's appearance on BBC Radio 4's *Today* programme focussed strictly on the communication of the science, rather than on institutional advertising:

> When the discovery came out [...] when he was interviewed on the radio one morning on *Today* [...] whoever the interviewer was – said he was from the [...] University of Cambridge [...] and Joe didn't bother correcting this, he just went on to say – because he'd only two or three minutes – to say his important announcement and how important this ozone depletion was and blah, blah, blah, and then that was the end of the interview and Dick Laws got rung up immediately by Sir Hermann Bondi, then Chairman of NERC, saying, 'What does Joe Farman mean by not correcting the interviewer about his establishment – place of work', saying that he was British Antarctic [Survey] and not the University of Cambridge, so he got a rocket for that and Dick Laws told him off and he was persona non grata for a bit [laughs].

Joe didn't care; what he wanted to say was the important thing about the ozone [laughs] and bless his heart, you know, it *was* the important thing to say but, oh dear, that wasn't the thing to do at all. [...] So he, he got it in the neck about that, however, I think Joe just treated with disdain and ignored it really and just went his own sweet way [laughs], bless him. So anyway [...] he stuck to the main point and issue and – and that was that.

(Bowen, 2010: track 5, 23:20–25:46)

Bowen's account interprets Farman's performance as having a deliberate focus: the message and argument are prioritised over other utterances. Silence operates to frame and emphasise what Farman *did* say. In his own accounts of talking to the media about the 'ozone hole', Farman himself could not be clearer about what he would talk about, and what he would not:

I spent a whole television interview, which went on for three quarters of an hour I think, and they wanted me to [pause] what shall we say, be forthright about Mrs Thatcher's views on the environment and things like that. And I said politely that it wasn't my job to criticise the prime minister, and if they persisted in asking this question I'd just cover my face up and so they agreed and we started off and then just as you were sort of talking sense they suddenly switched back to try and get it. And I spent, you know, most of that flipping hour's interview putting my hands in front of my face and saying, 'You've forgotten what I told you. [laughs] I'm not here to openly criticise the prime minister, you know, it's not my job. I'll tell you about ozone, I'll tell you about what policies should be, I'm not going to say whether she's going fast enough or not, it's not a personal matter, it's the advice she gets from elsewhere which causes it all.'

(Farman, 2010: track 15, 1:22:00–1:23:12)

In these media interviews, silences have significant agency. They are not simply markers of stress, haste or forgetfulness. They act to shield scientific explanation against the attempts by journalists to draw out talk of other kinds. In the television studio, by not responding, and hiding behind hands (echoing a tendency at other times to hide behind clouds of smoke) Farman controls the content of what can be broadcast. Elsewhere, in a presentation to a House of Commons Select Committee in the late 1980s, Farman uses talk to silence others and to draw attention to messages that he felt were self-explanatory:

Do you remember what you said or showed on that occasion?
[...] It was one sheet of foolscap of I hoped quite sensible English [...] and some graphs showing, you know, the thing. And I can remember to this day that – I mean first of all I'd been told I shouldn't, I shouldn't say anything until I was spoken to as it were and the head of NERC and my

director would take the questions and if they thought there was something technical they might hand it on to me. But luckily the chairman sort of addressed me straight away, so I, I started talking as rapidly as possible to keep these other people at bay [laughs]. And I can remember sort of looking at the Committee and they were sitting there with these pieces of paper in front of them but with a glazed expression over their faces. And so I sort of banged on the table [bangs on table] and I said, 'Oh, for God's sake wake up and look at that piece of paper I've given you' [laughs].

(Farman, 2010: track 15, 1:19:25–1:20:32)

We have seen that in the composition of the *Nature* 'letter' and in certain public appearances, Farman attempts to control what is said about ozone with the hope that silent attention to a trend in data is all that is needed to be persuasive. As he moves further into political spaces, he finds that his ability to control what is and isn't said about stratospheric ozone fades. He becomes a more passive spectator, observing complicated interactions which he remembers for their negative uses of talk: talk without intent, interrupted explanations and absurd chatter:

So we had this Save the Ozone Layer conference in which we had six initial talks: [...] Sherry Rowland started off with talking about what he and Molina had decided about it all, then I spoke and then Bob Watson spoke [...] whereupon Mrs Thatcher suddenly climbs to her feet and takes her entourage out, leaving Bob sort of in mid- mid-sentence [laughs] gasping somewhat. She had to go to a hospital where there'd been a terrible accident or something and have pictures taken talking to them. [...] And then she appeared later in the day and gave some press conference and told everyone to go out and buy these wonderful new refrigerators from ICI which were going to be full of hydrofluorocarbons 134! [...] And – and while she was away the heads of delegations from all around the world [...] solemnly got up and made statements about what needed to be done and so on and so forth. And I thought, very interesting, they're all saying quite often stronger statements than *I* say. And then suddenly it sort of dawned on me, well, actually they weren't going to be asked to sign anything at this particular conference; this was a build-up for next year for the protocol. Then when they came back next year, well, they didn't quite speak in the same way they had at this particular conference [laughs]. In fact they didn't promise to do very much at all. [...] That wasn't quite the end [...] it so happened that [pause] Virginia Bottomley was environment minister, wasn't she, at the time [...] She came and grabbed me at the end of the meeting to meet the prime minister you see, oh dear. Well, the prime minister says a few kinds words and then [...] she says to me somewhat sadly, 'Did I make a mistake in telling everyone to buy this new refrigerator from ICI?', you see.

So, what can a poor man do but say, 'Well, actually I don't think they'll be on the markets for a year; what you saw this morning was a prototype and they haven't yet gone into mass production' [laughs].

'Oh, get rid of this man; I don't want to' – so you get passed off and I find myself face to face with Denis [husband of the prime minister], you see, that's what happens to people when she – [laughs]. So Denis is busy studying the marble floor and says to me, 'Tell me Mr Farman, and what are you going to turn your brilliant mind to now that the ozone question is settled?', you see [laughs]. Bloody hell, I'd had enough of this! And so I thought – well – so I said, 'Oh, I don't know, there are one or two problems we're up against aren't there? Let me see, there's the balance of payments isn't there? There's the council tax, there's – ' and he has the grace to sort of smile gently and we discover we can talk about rugby [laughs], and that's what life at some levels is all about. It's somewhat sad I'm afraid.

(Farman, 2010: track 15, 1:23:40–1:28:31)

He comments further on the politicians who spoke differently at this and a subsequent meeting:

No, it's really quite staggering that, you know, with a piece of paper to sign politicians don't commit themselves but if they're speaking at a preparatory conference or something of that sort they can all get up and preach and make it sound very exciting.

(Farman, 2010: track 15, 1:25:44–1:25:59)

The use of the word 'preach' here ought to be set against Farman's memories of churchgoing as a child:

People were using all these wretched things but when you sort of kept saying you use these three letters together G-O-D. I haven't yet fathomed out what on Earth you mean by it, it doesn't sort of – you know, it's not a concept which somehow I've got any real control over. [...] And so all the words which are being flung at you over the pulpit and so on, it might just as well be a meaningless noise. You know, you might as well stand up and say, blah, blah, blah [laughs].

(Farman, 2010: track 1, 28:32–29:36)

Talk detached from substance, understanding, meaning becomes just noises made by the mouth – not worth listening to. It follows that Farman's view of proper science communication is disturbed by any slips in the correspondence between what is claimed and what is measured or in another way known. He is therefore critical of over-talking, overstating, talking-up:

I suppose I'd better go on record somewhere as pointing out that, please, if you ever read the World Health Organisation's view of the Montreal

Protocol just don't believe it: they will tell you that the Montreal Protocol has prevented n cases of eye trouble in the tropics and various other things, and the answer is what on Earth do you mean? There hasn't been any ozone depletion in the tropics. There hasn't been any […] documented increase of UV in the tropics – what the hell has the Montreal Protocol got to do with saving all this vast number of malignant melanomas and all the rest of it, you know. […] As far as I know no-one has yet demonstrated anywhere in the world that any harm has been done to anyone by ozone depletion, or any organism. All the krill round Antarctica are very sensible creatures; they live under the ice, got the most wonderful sun screen there is, you know, and if they're in the open water, when the sun comes up they go down. […] We don't want to lose the ozone layer – I'm as keen as anyone to get rid of CFCs and put the ozone layer back to where it was – but you don't fight battles by misleading the public. Once they, you know, get known that this is the sort of thing which goes on, you can't even start a story any more.

(Farman, 2010: Track 15, 1:33:57–1:35:54)

Here Farman suggests that talking-up closes down discussion. Saying too much cuts off the opportunity to speak with any impact in future. Excessive talk destroys its own purpose. Other environmental problems are affected too: 'I mean it's the trouble with climate change. It's been so overdone' (Farman, 2010: track 15, 1:35:55–1:35:59).

Conclusion

The case of Joseph Farman illustrates how long term measurements may result, for extended periods, in observations that do not seem to be saying anything – that are 'silent'. Scientists engaged in such measurements may lack prominence in their own organisations. This may not concern them personally, but lack of published findings might draw the attention of scientific managers and funders, rendering them vulnerable to permanent scientific silencing. There may be similar long term environmental monitoring projects in other organisations that did not find their own 'ozone hole', and in fact there is an instructive contrast to be made between Farman's experience and that of the UK Marine Biological Organisation's long term project: the Continuous Plankton Recorder Survey (CPRS), which was suspended in the late 1980s but has since been saved.

The CPRS had collected data on plankton communities in the Atlantic and North Sea since the 1930s. In his life story interview, oceanographer Bob Dickson speculates on why the CPRS was stopped:

1988/89 was when it was actually closed and at the time I suspect […] that the CPR team itself, which continued to carry out the analyses but on a – you can see from the time series of analyses made, that they'd been kind of slowly tapering down since the '70s and I think they had begun to feel unappreciated. Also to begin to feel that they had developed some sort of 'laager' mentality, where they were not going into the canteen for coffee; they were having it at their bench in a little cluster in the corner. And I'm almost hinting that [pause] people might not even have known that they were still there. They were very quiet – very clever people – but they were very quiet about it and their head was down, and when it came to the point of saving money, I can imagine [a manager/administrator/assessor] saying, 'What about the survey? How much do they spend? Can we save on them? I don't seem to have seen any of them around recently.'

(Dickson, 2011: track 8, 44:05–45:27)

Here silence seems non-strategic and dangerous, leading to a kind of blending in or fading from view – a becoming invisible and stopping. Bob Dickson saved the CPRS by talking it up, praising its virtues, lobbying. In the case of Joseph Farman's survey of stratospheric ozone concentrations over Antarctica there is an obvious difference: it produced a result that could be made to speak for itself. But there is also a difference in the operation of silence. Sebastian Grevsmuhl has argued, as part of a wider account of visualisation technologies in the development of the 'ozone hole' metaphor, that the BAS measurements were trusted because Farman's team had kept themselves apart from a wider global effort of data collection and standardisation:

> the *refusal* to participate in the (to some eyes compromised) international network was precisely the reason for BAS' credibility. By building their own institutional and material culture, their own standards and techniques, BAS could successfully establish the credibility of the only long-term dataset of atmospheric ozone measurements on the Antarctic continent. In other words, the British research group could, unlike any other, speak with legitimate authority on the Antarctic ozone values.
>
> (Grevsmuhl, 2014: p. 33)

This chapter suggests that the 'maverick institutional culture' (Grevsmuhl, 2014: pp. 32–33) of BAS might depend more than Grevsmuhl is able to discern on the kinds of *individual* silence revealed (however partially) by close attention to extended oral history interviews. However, this refocusing on the individual does not necessarily lead away from an appreciation of the historical, practical agency of silence. It is possible to detect in obituaries of Farman the argument that he was listened to because he had not spoken up before:

> Joe did not naturally seek the limelight, so it was remarkable that after the publication of the ozone hole paper, he entered the policy arena so keenly and effectively. [...] Whether Joe talked to industry representatives or to journalists, it was always important to him to educate them about

the science. In fact, it was his scientific integrity that gave him such credibility as a public advocate in the development of the Montreal Protocol.

(Pyle and Harris, 2013: p. 435)

The direct appeal is to 'scientific integrity', but there is some sense here ('Joe did not naturally seek the limelight') and elsewhere ('he didn't go round the conference circuit talking about it') that a lack of spurious talk allows in time a more effective intervention, because the break from silence is so obvious.

Notes

1 See: http://sounds.bl.uk/Oral-history/Science. I provide catalogue numbers and time-codes for each interview extract. The interviewer's questions are shown in italics.
2 Note that until 1962 BAS was called the Falkland Islands Dependency Survey.

References

Booth, N., 1994. *How Soon Is Now: The Truth About The Ozone Hole*. London: Simon and Schuster.
Bowen, B., 2010. Life story interview for 'An Oral History of British Science', British Library catalogue reference C1379/18.
Dobson, G.M.B., 1968. Forty years' research on atmospheric ozone at Oxford: a history. *Applied Optics*, 7(3), pp. 390–403.
Dickson, B., 2011. Life story interview for 'An Oral History of British Science', British Library catalogue reference C1379/56.
Farman, J.C. and Hamilton, R.A., 1975. *British Antarctic Survey Scientific Report*, number 90.
Farman, J.C., 1977. Ozone measurements at British Antarctic Survey stations. *Philosophical Transactions of the Royal Society of London B*, 279, pp. 261–271.
Farman, J., 2010. Life story interview for 'An Oral History of British Science', British Library catalogue reference C1379/07.
Farman, J., Gardiner, B. and Shanklin, J., 1995. Large losses of total ozone reveal seasonal ClOx/NOx interaction. *Nature*, 315, pp. 207–210.
Freund, A., 2013. Towards an ethics of silence? Negotiating off-the-record events and identity in oral history. In: Sheftel, A. and Zembrzycki, S. (eds), *Oral History off the Record: Towards and Ethnography of Practice*. New York: Palgrave Macmillan, pp. 223–238.
Greenspan, H., 2014. The unsaid, the incommunicable, the unbearable, and the irretrievable. *The Oral History Review*, 41(2), pp. 229–243.
Grevsmuhl, S.V., 2014. The creation of global imaginaries: the Antarctic ozone hole and the isoline tradition in the atmospheric sciences. In: Schneider, B. and Nocke, T. (eds), *Image Politics of Climate Change: Visualisations, Imaginations, Documentations*. Bielefeld: Transcript Verlag, pp. 29–53.
Harris, N., 2013. Edited version of the funeral address for Joseph Farman given by Neil Harris in Chapel. *The Letter*, Corpus Christi College, p. 116.

Jesse, E., 2013. Considering silence. In: Sheftel, A. and Zembrzycki, S. (eds), *Oral History Off the Record: Towards and Ethnography of Practice*. New York: Palgrave Macmillan, pp. 219–222.

Layman, L., 2009. Reticence in oral history interviews. *The Oral History Review*, 36(2), pp. 207–230.

Pyle, J. and Harris, N., 2013. Joe Farman (1930–2013): discoverer of the ozone hole. *Nature*, 498, p. 435.

Roan, S., 1989. *The Ozone Crisis: The 15-Year Evolution of a Sudden Global Emergency*. Chichester: Wiley.

Thomson, J., 2010. Life story interview for 'An Oral History of British Science', British Library catalogue reference C1379/20.

Wolff, E., 2012. Life story interview for 'An Oral History of British Science', British Library catalogue reference C1379/70.

Woods, J., 2011. Life story interview for 'An Oral History of British Science', British Library catalogue reference C1379/64.

2 Darwin's silence
An anatomy of quietude

Stephen Webster

> The weather is quite delicious. Yesterday after writing to you I strolled a little beyond the glade for an hour and a half and enjoyed myself ... At last I fell asleep on the grass and woke with a chorus of birds singing around me, and squirrels running up the trees and some woodpeckers laughing...
> Charles Darwin, letter to Emma Darwin, 28 April 1858
> (Burkhardt and Smith, 1991: p. 84)

Introduction

Charles Darwin (1809–1881) has a deserved reputation as someone who knew how to hold his tongue. Discussing in his autobiography his tendency to delay publication, he explains how sixteen years passed between his first observations on sundews and the eventual book *Insectivorous Plants*. Such procrastination, he wrote, 'has been a great advantage to me; for a man after a long interval can criticise his own work, almost as well as if it were that of another person' (Neve and Messenger, 2002: p. 81). Significantly for the topic of this chapter, Darwin remembered also that those years of fruitful work began one summer when he was 'idling and resting near Hartfield' and noticed insects being trapped by the sticky plants (ibid.: p. 80). The suggestion is clear: by idling, he got an idea.

Idling, delay and procrastination – and their virtues – form the organizing theme of this chapter. As an umbrella term I shall use the term 'quietude', defined by the *Oxford English Dictionary* as 'the state or condition of being quiet or calm'. My broad aim is to offer a perspective of Darwin that emphasizes his own interest in the quieter ways of science. As we shall see, Darwin's ambitions to work steadily and without event were frequently thwarted. I will use his frustrations in this regard as a way of casting light on the contemporary and growing debate about the form, and especially the pressures, of the research life. We can sketch the matter thus: to judge by commentary on the research life, the academic resource of quietude has become an endangered species, if not yet completely extinct. Conversely, scratch any scientist – perhaps any researcher – and all will agree that the charming Darwinian moment quoted above, so clearly a well-judged moment of pause and refreshment, points to an important truth. In the development of science, no-one should neglect the importance of lolling about.

In the spirit of the theme, my argument is roundabout. The chapter is largely a venture into some well-known byways of Darwinian mythology, foraging for insights into his undeniably contemplative sensibility. Yet the wider aim is to make these Darwinian selections gesture also to contemporary anxieties about speed and competitiveness in the research life. In short, if Darwin's troubles drive this chapter, it is the anxieties of today's researchers that provide the background hum.

First, though, a word about the richly suggestive term 'quietude'. In relation to the academy, I invoke it here as a benign and generative style of work that stands in contrast to the rush to publish, to institutional exaggeration and to professional self-aggrandisement. Fortunately those particular vices, though so common as to seem the norm, do draw criticism. Blaise Cronin, editor of the *Journal of the American Society for Information Science and Technology*, signals his unease this way: 'At the risk of sounding like a fogey, there is something to be said for deliberative writing and deferred gratification ... Perhaps what academia needs now is a Slow Writing movement akin to the Slow Food, Slow Cities movement' (Cronin, 2013).

For Cronin, it is haste that is the problem: 'I am even more persuaded of the need for both up-and-coming as well as established scholars to take their foot off the accelerator.' He sees an alarming degradation of quality. Manuscripts are written too quickly with little attention to bibliographic correctness and 'junior scholars are now running like rats on a treadmill' (ibid.). He quotes approvingly Anthony Grafton's recommendation that 'slow scholarship – like slow food – is deeper and richer and more nourishing than the fast stuff' (Grafton, 2010), and he suggests that the daily fight to secure funding and priority is leading not only to poor work but also, occasionally, to chicanery. Nor is Cronin a lone voice here: his complaint that the benefits of steady scholarship are at risk from simpler, measurable forms of productivity joins a drumbeat of concern that is now quite assertive.

In pushing Darwin forward as someone who can illuminate the current debate about the culture of science, I am not suggesting that he was a scientific monk who can inspire us by his dogged isolation. On the contrary, the relevance to us of Darwin's quietude lies in his ambivalent attitude: he might have enjoyed the silence of his study, but he was also ambitious, knew the value of communication and anxiously sought the approval of others. This mix very likely resonates too with the contemporary scholar and exploring such resonance is a major ambition of this chapter. As a retiring nineteenth-century country squire, Darwin could take steps only distantly possible or relevant for a scientist today (such as building an eight-foot wall in front of his house). Nevertheless, I shall argue that his almost constant struggle between the norms of scientific productivity and his personal style of doing science is highly instructive today. For quietude is a relative term, taking a position that contrasts with noise and excess. Darwin's determination to preserve a kernel of silence, while still maintaining a busy domestic and professional identity, provides, I will suggest, a thought-provoking model for today's scientists.

'Struggle' is a highly appropriate word to apply to Darwin's working life. His was a somewhat turbulent character and his life was full of incident. He combined a desire simply to be left alone to pursue his work, with an acute watchfulness over how others viewed his ideas. And he was harried by any number of domestic demons, including worries about money, the upbringing of his children, and – particularly – ill health. Perhaps the best way to understand the significance of Darwin's interest in quietude is this: that such peace and quiet that he did obtain did not come easily and was always the companion of stress and disturbance. Darwin did indeed construct an impressive fastness in the Kent countryside and he carefully moderated incursions from the outside world. But internal disruption, in the form of his noisy and numerous children, his rebellious gastro-intestinal system and his itchy skin, and his generally lively sensibility, would never allow him to settle down as a true contemplative. Even in death, with a village funeral and burial next to his dear brother Erasmus planned, large forces intervened: scientists and politicians, intellectuals and deans together achieved something quite different, a ceremony in Westminster Abbey. His wife and absolute confidante, Emma Darwin, missed the funeral, preferring to stay at home.

The race to the top

Before turning to Darwin more fully, I will sketch out aspects of the debate about research life that have relevance for our exploration of quietude, noting that some slightly different questions are interwoven in this contentious area. For example, the hunch that scientific productivity is now over-revved is expressed fairly frequently in academic commentary. This is not a debate we can characterize as the province of science's dispossessed. At a meeting of the Royal Society held to discuss the future of scholarly communication, Sir Mark Walport, chief scientific adviser to the UK government, asked: 'Should we reduce the volume of research and strengthen the infrastructure?' (Royal Society, 2015: p. 23). The same report also notes a comment that: 'The primary motivation of young scientists is to publish in high status journals ... and this constitutes a very profound cultural problem' (ibid.: p. 6). And in the introductory session, Sir Paul Nurse, president of the Royal Society, asked delegates to consider: 'How can we reform the culture of science to tackle the causes of misconduct?' (ibid.: p. 4).

Throughout the report, research culture is presented as somewhat fraught, with power dynamics painted as a determining factor. The issue of justice, for example in career progression, seems one of the drivers and has attracted particular scrutiny from the geneticist Peter Lawrence. His concern is with the distribution of academic credit and his diagnosis is bleak. Graduate students, he says, 'are like boosters on space rockets, they accelerate their supervisors into a higher career orbit, and, when their fuel is spent, fall to the ground as burnt-out shells' (Lawrence, 2002: p. 836). Lawrence's concerns about justice are a reminder that people's lives and not just the quality of science are at stake here. Plainly moral issues hover about these debates.

Is it simply the rush to publication and to priority that is the problem? Is haste, as Cronin implies, the villain? For Filip Vostal, the advocates of scholarly quietude need more nuance in their approach, for a simple hostility to speed might miss the point. He notes the importance of what he calls 'accelerative moments of inspiration' and he suggests that scholarship – even the best variety – is not necessarily slow. More broadly, he suggests, slowness cannot plausibly be seen as the organizing centre of academic life. For Vostal, what is important, and what is so under threat from the academy's galloping managerial charge, is 'scholarly time autonomy' – the ability of researchers to deploy their temporal resources as they see fit (Vostal, 2013). The sense that managerialism is implicated here is indeed a very widespread perception. This particular colouring of university life provoked 120 British academics to write to the *Guardian* newspaper complaining of a profound erosion of basic academic norms. For the signatories, who by their seniority clearly speak from experience, the hazards are mounting: 'Innovation, creativity, originality and critical thought, as well as notions of social justice, [are] being threatened by forces of marketization demanding "competitiveness" and "efficiency" in teaching and research' (Lesnick-Oberstein et al., 2015).

Similar alarms were expressed by the Nuffield Council on Bioethics when it investigated and reported on the research culture of UK science. Summarizing the views of the many scientists they consulted, the Council notes some general complaints. A culture of short-termism, they say, results in fewer new ideas, a decrease in the time available to plan good research, greater adherence to 'safer' research topics and people cutting corners in research (Nuffield Council on Bioethics, 2014: p. 26). If we look for some unifying features underlying these themes, some worries seem dominant. There are too many pressures; people publish too quickly; and they pile into the best journals as though life itself depends on it. Adopting for the moment an attitude of opposition, it is hard to escape the thought that quietude and the reflective instinct are the qualities most able to counteract these baleful aspects of research culture. If so, the conclusion is clear: when quietude vanishes, the body politic of science begins to suffer.

Let us return now to the life and work of Charles Darwin. Can we press him into service as an opponent of haste in the scientific method? Could this establishment figure, famous in his own day and foundational in ours, revalorize delay and contemplation – become, as it were, a guru of scientific slowness? I suggest he is indeed an exemplar, a mythical hero, for those who value the measured pace.

Some important problems must first be cleared away before we interpret Darwin in this manner. The most obvious I have mentioned already: Darwin cannot simply be categorized as a reclusive contemplative, or as someone blithely unflustered by a glitch in his productivity. Take, for example, the final

52 *Choosing silence*

leg of his *Beagle* voyage, when Darwin complained to his sister that the fastidious Captain FitzRoy had postponed the return to England in order to make one final set of measurements in South American waters. As he told her, 'this zig-zag manner of proceeding is very grievous', for the delay put off the hour when Darwin could unload his crates and set to work. That work, he suggests, he planned to expedite with vigour – for 'a man who dares to waste one hour of time, has not discovered the value of life' (Burkhardt and Smith, 1985: p. 503). Here, certainly, Darwin is reminding his family that his easy-going days are over. If, once, his father had been 'very properly vehement against my turning into an idle sporting man, which then seemed my probable destination' (Neve and Messenger, 2002: p. 29), now, with the *Beagle* voyage over and busy scholarship ahead, he had gained 'happiness and interest for the rest of his life' (Burkhardt and Smith, 1985: p. 505).

Another objection to casting Darwin as a master of scientific silence lies in the accepted norms of historiography. Isn't my strategy a Whiggish sleight-of-hand, where I project the lifestyle of this wealthy Victorian squire onto the stressed landscape of our own research community? More generally, is it ever feasible to invoke a dead scientist as a moral guide for the living? There are two sets of defences here. The first relies simply on Alasdair MacIntyre's famous diagnosis of the interminable nature of contemporary ethical debate (MacIntyre, 1981). For MacIntyre, today's moral framework is no framework at all. Instead it is formed piecemeal and in disorderly fashion from fragments of moral programmes laid out by, and inherited from, the prophets of the past. Today we live out our ethical lives by sifting, as though through rubble, remnants from history. It is in that spirit that we can be justified in looking to Darwin for some guidance on the life scientific. My second justification makes use of the fact that Darwin and Darwinism have always been pushed and pulled by those with political and moral agendas. Even when Darwin was alive his theories were being drawn into arguments about the 'naturalness' of slavery, the futility of charity and the divisions of the human species (Desmond and Moore, 2009). These arguments, developed now into scientific declarations on gender relations, human culture and governance, ring on still (Gould, 1996; Rosenberg, 2000; Rose and Rose, 2001; Ridley, 2015). This chapter continues the habit.

A quiet place to work

Although quietude to an extent is portable, Darwin demonstrates the important principle that constructive peace and quiet – the mood that pays a creative dividend – often has its favoured spot. We all have such places, and Darwin was particular too. It is significant that among the Darwin biographies, the second volume of Janet Browne's has the title *The Power of Place*, signalling the vital role of long term stability in Darwin's creativity (Browne, 1995: 2002). Darwin himself, in his autobiography, mentions that as a boy he sat for hours 'reading the historical plays of Shakespeare, generally in an old window in the thick walls of the school' (Neve and Messenger, 2002: p. 20). And he wrote that when very young and keen on fishing, he would 'sit for any number of hours on the bank of a river or pond watching the float' (ibid.: p. 9).

Darwin's attitude to the state of contemplation, and to busy activity, is best described as oscillatory. His attitude to conversation – to scientific chat – provides a good example. We can start with Darwin's work space on the *Beagle*, an undeniably cramped cabin beneath the poop deck that doubled also as ship's library and chart room. Letters Darwin wrote before the voyage, justifiably anxious about which guns, shirts and dissecting pins he should pack, centred very heavily on the nature of this space, which FitzRoy had decreed should – mostly – be his. The particular worry was over books, the things that together with his collections would ground his life as a professed scientist for the next five years. Here there arose a very notable example of the way the contemplative scholar, buried in the institution, relies just as much on honest and supportive leadership as on austere self-discipline. FitzRoy, though fussy and insecure and abrupt, clearly saw the seriousness of Darwin's purpose and waved him on: bring as many books as you want and we'll talk about them over dinner. Indeed, it is sometimes forgotten that it was Darwin's conversational abilities, as much as his science, that got him the berth on HMS *Beagle*. To the fastidious Captain FitzRoy, Darwin was precisely the person he needed on board. It made sense to have a congenial companion on board to counterpoise the severe demands of the *Beagle*'s central project, the surveying of the South American coastline. In this FitzRoy was merely being prudent. The captain understood himself to be a man of quick temper, but he also feared a more radical upsetting of his mental balance as the voyage wore on. The previous captain of the *Beagle* had committed suicide – it was this which secured FitzRoy his captaincy. More particularly, FitzRoy was disturbed by what he saw as sinister signs in his own family: his uncle, Lord Castlereagh, had committed suicide while foreign secretary. Darwin was on board in part for his human quality as a companion in conversation, particularly scientific conversation. Darwin was 22, FitzRoy 26.

Darwin's youthful spell on the *Beagle* is a reminder that scholarly quietude should not be seen as the territory of the ageing academic, the end state of a distinguished career. To that idea Darwin's book *The Journal of the Beagle* gives immediate correction. It is one of the finest explorations of a young person's excitement at a travel adventure, yet it contains also every sign of measured thought, of reading and writing. Darwin on the *Beagle* had time to notice things:

> There was a glorious sunset this evening and is now followed by an equally fine moonlit night. I do not think I ever saw the sun set in a clear horizon. I certainly never remarked the marvellous rapidity with which the disk after having touched the ocean dips behind it.
>
> (Keynes, 1988: p. 21)

Along with this he was physically daring and decisive. The crew called him 'philosopher' but knew him to be strong and alert. There is a good story of

Darwin's ability to suddenly spring into action. Darwin and the crew were resting on a beach in Chile, the boats pulled up away from the sea. They were admiring a glacier when 'a large mass fell roaring into the water ... we saw a great wave rushing onwards and instantly it was evident how great was the chance of [the boats] being dashed into pieces' (Keynes, 1988: p. 140). According to Captain FitzRoy, who was present, Darwin was instrumental in saving the boats: 'Had not Mr Darwin, and two or three of the men, run to them instantly, they would have been swept away from us irrecoverably' (ibid.: p. 140, fn. 1). The next day, sailing into a great expanse of water, and impressed by his companion's ability to deal with discomfort and risk, FitzRoy named the area Darwin Sound.

The *Beagle* years, then, were a striking mix of action and of thought, and of course were extraordinarily creative. For Darwin the voyage was 'by far the most important event in my life, and has determined my whole career' (Neve and Messenger, 2002: p. 42). Shortly after arriving back, in 1837, he opened his first – secret – notebook on transmutation. If Darwin was silently thinking about evolution, it was nevertheless the benefits of talk that drove him to settle in London. He needed to make contact with illustrators and publishers, and he had to twist the arms of zoologists and botanists who could help him with his *Beagle* collections. By the time he returned to England, Darwin's reputation was growing. He would foster his good start in science and get a secure hand-hold on the metropolitan professional scene, even though that meant an end to fresh air and wild life. As he told Robert Henslow, his Cambridge teacher and mentor: 'I assure you I grieve to find how many things make me see the necessity of living for some time in this dirty odious London' (Burkhardt and Smith, 1985: p. 512).

Yet this abrupt and reluctant transition from ship life to a London existence of work, noise and haste was intellectually productive. Darwin's brother Ras lived close by and liked to talk about ideas. He held lively dinner parties where the conversation moved fluently between politics, philosophy and science, and where arguments were developed and challenged. From Darwin's point of view, one of the most important of the guests at Ras's table was Hensleigh Wedgwood, a relative. Wedgwood was a philologist and interested in the evolution of language. Conversations about the descent of languages, one from another, and their similarities and divergences, were important to Darwin as he gently pondered the transmutation of species.

Once Darwin had married Emma Wedgwood, the couple instituted a more retiring lifestyle. They retreated from the social circuit, began to have babies (they'd eventually have ten) and developed a relationship so strong that it must be considered a primary cause of Darwin's achievement. Within a couple of years, London had become intolerable for them and they bought a large house on the grass chalklands of Kent, funded by Darwin's father. Darwin was happy about the apparent remoteness of his house: 'I never saw so many walks ... the country is extraordinarily rural and quiet' (Burkhardt and Smith, 1986: p. 324). Yet Darwin chose this place also for its ready access to

London and his colleagues. He knew that his science required communality as well as isolation. It was just a matter of getting the mix right.

Darwin's admiration for the silent aspect of this mix made him prone to exaggeration. He told his friend and cousin William Fox (who lived on the Isle of Wight) about the remoteness of Down: 'we are absolutely at [the] extreme end of the world'. It was typical of Darwin to further justify his interest in quietude by giving a gloomy bulletin on his health, which had much reduced his ability to 'stand mental fatigue or rather excitement, so that I cannot dine out or receive visitors, except relations with whom I can pass some time after dinner in silence' (Burkhardt and Smith, 1986: p. 353). Yet, at the same time he was also describing its convenience to an old neighbour from Upper Gower Street, London: 'It is 16 miles from London Bridge and I hope to be in town pretty often' (ibid.: p. 328). Down House occupied exactly the right orbit in relation to London – it was distant enough to discourage casual trips in either direction and so reduced the froth of social life. On the other hand his relatives, and colleagues he wanted to see, could get there if they were prepared to take the train from London Bridge.

As soon as the Darwins moved in, the builders started making the necessary adjustments for an uninterrupted working life by making the home more defended. For while the country around Down House would provide Darwin with all he would ever need in terms of fresh air and rural inspiration, the actual house was problematic. He was happy with his work room ('Capital study 18 x 18', he told his sister) but the way the village road ran so close to the house was unnerving. 'I have determined to have a 6 feet 6 inch wall ... the whole length', he wrote. Otherwise people would be peeping in. Nor would that completely solve the problem, he thought. Where the road ran alongside the house, it would have to be dug out and lowered by a full two feet. It was a large job, but an important one: 'The publicity of the place is at present intolerable' (Burkhardt and Smith, 1986: p. 360).

To the contemporary visitor, Darwin's home, now under the wing of English Heritage, is highly evocative of his tussle between the benefits of quiet and those of communication. The centre of it all seems to be Darwin's study, certainly a shrine of research: the place where one of the great works of science was produced. The neatness of the books and of the filing systems is suggestive of Darwin's naval experience. The atmosphere is scholarly and ordered: you sense a long-drawn out project, a life work. There is an atmosphere of efficiency and productivity too: his chair is on wheels, to make it easy for him to move between his two work tables. The study faces the road rather than the garden, with nothing for Darwin to look out for other than the post, and possibly those disruptive passers-by.

Yet a sympathetic viewing of the house suggests that the study, though plainly important, was always at risk of disturbance and interruption. The staircase is just across the corridor and is equipped with a wooden slide for the children to slither down. There is a billiard room and in the drawing room a piano. The piano was for Emma, the billiard table for everyone. Leisure and enjoyment seem an important element in the Darwin amalgam. He once wrote to his son William: 'We have bought a stunning book on billiards, costing 21s, and it has nearly 200 diagrams of various strokes and accounts of famous games. Altogether the table has been a splendid purchase; only I hope it will not make you lads a set of black-legs' (Burkhardt and Smith, 1991: p. 263).

The dining room, too, suggests an interest in companionship. There was a large family to feed (the Darwins had eight surviving children), but valued guests too. His brother Ras and scientific colleagues – notably the botanist Joseph Hooker – were, especially in the early years, persistently invited down for talkative weekends. The house was big enough, and the welcome warm enough, for families to come along too. 'I am in truth delighted at the thought of your coming here. Cannot Mrs Hooker come along too?', he wrote to Hooker. And he was solicitous in removing obstacles for his visitors. 'I will send you and Mrs Hooker back on Monday by carriage, so as to escape [the] horrid omnibus journey' (Burkhardt and Smith, 1991: p. 96). Down House, then, is not best seen as simply a well-defended retreat. It was constructed by Darwin to afford him the quiet and routine he needed but it wasn't isolationist or anti-social. His expressive aspects, his interest in his children and his friends, are evident everywhere.

As Janet Browne puts it, with Darwin: 'His life and his science were of a piece' (Browne, 2002: p. 10). Further, as Browne suggests, the reader who aims to understand Darwin will find themselves gaining traction at a bewildering variety of points and levels: his 'life can be explored from the inside looking out ...' (ibid.: p. 13). Thus, while Darwin's *The Origin of Species* (Darwin, 1859) seems the central text, importance too must be given to those vital accessories of his work: the snuff box in the corridor, the raised chair on casters for rowing about his study and the potty in the corner that made his retching less disruptive. It is the 'inside-out' aspect of the Darwin project that makes him so instructive a figure to the partisans of slowness. Throughout Down House we can see the elements that link to set the pace of Darwin's work: potty, snuff box, microscope, dinner table and out-of-doors cloak.

Darwin's knowledge, his study makes clear, is closely dependent on the stillness of microscopy and of writing. But his knowledge, too, is ambulatory and marked by the pedestrian instinct. The point is made by his son Francis, who on writing an account of his father's manner of working, gave great emphasis to Darwin's midday walk and its unvarying quality. First came a call on the greenhouses, for 'a casual examination' of his germination experiments and the insectivorous and climbing plants. Then he went on to the Sandwalk: 'In earlier times he took a certain number of turns everyday, and used to count them by means of a heap of flints, one of which he kicked out on the path each time he passed' (Darwin, 1887: p. 115).

The Sandwalk, extant today, is an elliptical path, perhaps 400 metres long, that skirts a small copse at the far end of Darwin's garden. Darwin had it laid out for precisely this purpose, complete with a small bench and pagoda for taking shelter or simply pausing. It is a place of contemplation. Even to reach it, one passes along a narrow path which runs between two grassland fields. One of these fields draws the gaze back towards the house, to the village of Downe and thus to habitation. The other field leads, it seems, only to an uninhabited expanse of Kent's rolling, chalk formations, richly interspersed with woodland, and not a person in sight. Once in the Sandwalk, the path is hard and flinty and the light subdued.

The actual copse, to which the path forms a perimeter, is mixed woodland. It is dominated by deciduous trees, but toned with small conifers, holly, hawthorn and scrub. It is a place where, however densely thoughtful Darwin might be, he might be surprised by a similarly quiet, but much smaller, animal. Darwin is reported to have been long-legged and able to move softly: according to his children, his physical quietude could be remarkable. On one occasion he stood so still for so long that some young squirrels ran up his leg. Yet it does not seem that on these walks he was simply lost in his thoughts, dead, as it were, to the outer world. In seeking out birds' nests 'he had a special genius'; and as the Sandwalk was also a favourite playground for the children it is significant that their incursion drew no irritation from Darwin, who 'liked to see what we were doing and was ever ready to sympathise in any fun that was going on' (Darwin, 1887: p. 115).

A good time to publish

In 1858 Charles Darwin wrote to Syms Covington, his assistant during the *Beagle* years. 'I have been preparing a work for publication which I commenced 20 years ago ... I have to discuss every branch of natural history ... the work is beyond my strength and tries me sorely' (Burkhardt and Smith, 1991: p. 95). This was not his first attempt to turn his thinking on natural selection into something that could be published. Fourteen years earlier, in 1844, he had prepared a draft essay explaining his ideas, and had put it aside, along with instructions about it being published in the event of his death. Even in 1858, as Darwin suggested to Covington, the project was not going well and publication seemed an ever-receding horizon. The sluggish pace was linked to Darwin's concept of good work – to his sense of scientific virtue. The animating impulse to get writing had come from his long-time friend and guide Charles Lyell, who told him the time was right for publication, that interested others were working on similar ideas and that his priority might be at risk: 'I wish you would publish some small fragment of your data ... and so out with the theory and let it take date – and be cited – and understood' (Burkhardt and Smith, 1990: p. 89). This was in 1856. Characteristically Darwin had straightaway been troubled by this idea of publishing quickly just to assert priority. He admitted he 'certainly should be vexed if any one were to publish my doctrines before me' (ibid.: p. 100); nevertheless, as he was to put it, 'it yet strikes me as quite unphilosophical to publish results without the full details which have led to such results' (ibid.: p. 109). Darwin, in other words, didn't want to rush to publish; he wanted to do a good job that properly represented his years of work.

There can be no precision about when Darwin started those decades of reflection on evolution. Even during his time as a student in Edinburgh, when he came under the influence of the zoologist Robert Grant, he was discussing the possibility of evolution (or transmutation, as it was then known). The idea that species were not fixed was a family affair anyway. Darwin's grandfather Erasmus (1731–1802), a poet much smitten with the possibilities afforded by technology and political progress, had given his own version of the evolutionary march in his 1794 book *Zoonomia*. The grandson's own more systematic thinking got under way during his *Beagle* years. He had, in his poop cabin, Charles Lyell's recently-published *Principles of Geology*, which argued that even the most dramatic structures we see around us – high mountains or deep valleys – are produced by ordinary action, such as rivers and earthquakes (Lyell, 1830). Liberal-minded about the possibility that species were not fixed, Darwin took these ideas and began to move them into biology.

During this long process, there were, however, some critical, accelerative moments. If there was a sudden moment when a mechanism of natural selection became imaginable, it was quite soon after the *Beagle* voyage:

> In October 1838, that is, fifteen months after I had begun my systematic inquiry, I happened to read for amusement Malthus on *Population*, and it at once struck me that favourable variations would tend to be preserved, and unfavourable ones to be destroyed ... Here, then I had at last got a theory by which to work.
>
> (Neve and Messenger, 2002: p. 72)

But if for Darwin there were moments of acceleration, interruptions and pauses loom very large too. Illness was a very particular problem for Darwin and it frequently brought work to a halt. Yet, in our quest to segregate the various aspects and benefits of quietude, there are subtleties about Darwin's frailty worth noticing. As Janet Browne has argued, the constancy of his complaints, so much a theme of his letters, invites the thought that in an important way his illness is part of his work. All Victorian intellectuals noted the impact of their particular maladies, but with Darwin the habit is very far advanced (Browne, 1998). Even on the first page of *The Origin* Darwin rather sombrely invites us to be aware of his fragility: 'My work is now nearly finished; but as it will take me two or three more years to complete it, and as my health is far from strong, I have been urged to publish this abstract' (Darwin, 1859, p. 65).

Whatever the cause of his troubles, the symptoms were often severe. Largely they were gastro-intestinal in nature but he also suffered at times from boils, eczema and migraine. While there has been speculation that an infection gained during the *Beagle* years may be to blame, the literature on the exact nature of Darwin's problems emphasizes their highly psychological aspect (Bowlby, 1991). Darwin was clear that the rather dramatic physical collapses he suffered through his life were related to stress and an over-busy life. This became a theme early on in London, when he began to refuse invitations so as to secure his quiet evenings at home. When in March 1837 he was asked to become one of the two secretaries of the Geological Society, he declined, giving an impressive number of objections. These included an ignorance of English geology and an ignorance of all languages. He explained to Henslow – his most important teacher and advocate – that he didn't know 'how to pronounce even a single word of French, a language so perpetually quoted.' He worried that the responsibility, 'so early in my scientific life, ... though no doubt a great honour for me, would be the more burdensome.' But primary among the issues, Darwin told Henslow, was the question of his health. He had been advised by his doctor to give up all writing for a while: 'Of late, anything which flurries me knocks me up afterwards and brings on a bad palpitation of the heart' (Burkhardt and Smith, 1986: p. 51). He was sure his symptoms were a sign of some fatal and hereditary flaw in his make-up and he was equally sure they made him poor company. In later life he decided against going to Henslow's funeral because he was worried that, as he would have to stay the night, the sound of his retching would disturb the other guests. For Darwin, his ill health and his style of work over many years joined to make quietude a central desire. He liked to remind his friends of the fact. As he wrote to his fellow pigeon fancier W. B. Tegetmeier: 'If you are inclined to come, I shall much enjoy seeing you, but my health has lately been so bad that I am physically incapable of talking for long' (Burkhardt and Smith, 1991: p. 154).

Illness, then, is one of the sinews in Darwin's anatomy of quietude. So too is his preparedness to prolong his studies by diverting his attention to new topics. This happened, most notably, with the barnacles, when Darwin set out on an eight year analysis of the taxonomy and anatomy of the entire class *Cirripedia*. This drastic act, a paradigm example of the slow scholar's willingness to explore further, was prompted by his confidante Joseph Hooker and came at a sensitive time. By 1844, his 200-page draft manuscript complete yet hidden away, Darwin identified Hooker as a likely acolyte and dropped some heavy hints. Hooker seemed open-minded: he was young and recently returned from a long expedition, and he duly became an important vector in Darwin's long journey to publication. He could talk with pleasure to Darwin about transmutation and they corresponded with vigour. With Hooker, Darwin could explore all his doubts, and certainties, about his ideas and probe for information.

Hooker is responsible for one of the most important delays in the Darwinian saga. In their correspondence, Hooker was contemptuous of scientists who merely speculated and were weak in their practical knowledge of animals and plants. Darwin, always open to the idea that there was more to know, wondered whether he himself was vulnerable to such criticisms. He had collected, identified and theorized – but was he thorough in his knowledge? Had he ever studied any group of organisms in depth? In particular, how much did he know about variation within a species? Quite by chance Darwin was at this moment looking at an unusual barnacle from his *Beagle* years. He decided that this was the animal he would spend some time on. And so he set off down a detour that was previewed to last two years and in fact ran to eight. It was a programme of work that would establish him as the world's pre-eminent expert on the *Cirripedia*, while his most important insight as yet remained largely hidden.

Darwin's long sojourn with the barnacles is an important episode in the history of the life sciences (Stott, 2003). It is richly interesting too in terms of scientific silence. In 1844, we have seen, Darwin did not feel able to publish. Leaving aside Hooker, scientists he admired were deeply sceptical about transmutation and inclined to doubt material explanations of speciation. These critics dismissed the French transmutationist zoologist Jean Baptiste Lamarck (1744–1829) and they were scathing of the popular science sensation of the decade, the anonymously published evolutionary text, *Vestiges of the Natural History of Creation* (Chambers, 1844). Whatever might be the religious sensibilities of these critics, they mounted their attacks on the grounds of scientific evidence, and this is what made Darwin nervous. At the same time Darwin's reticence was being deepened by his observations of political turmoil in England. The village of Downe (as it is now spelled) was a world away from civic unrest, but Darwin was a committed follower of political tides and anxious about his wealth, which relied on stocks and shares, and therefore on stability. He saw that radicals were using evolutionary ideas to naturalize the claim that 'lower forms' need not always stay that way. Darwin feared for his science should his ideas get mixed up with these political currents.

The barnacles gave him cover: here was time to further stabilize his understanding. His steady life, although challenged by collapses into ill health, alarms over children or his money, and by the deaths of his father and his siblings, remained intact. Emma made sure it had protected status. Through the barnacles, he could continue his project and ignore the work that would one day be necessary, a very large book synthesizing all his ideas and data into one long argument. Yet the diversion, if such it was (van Wyhe, 2007), turned out to be vital for Darwin's final accounting of evolution. The barnacles became important because his complete knowledge of the class, and of the evolutionary trends within it, duly gave him a profound understanding of the plenitude of variation within individual species. If we corral Darwin as a believer in the scientific potency of steady and unsung work – view him as a champion of quietude – then his barnacle project seems especially significant. Darwin didn't feel his work was completely convincing and looked for a way to pause. Aided by friends, and with a strong sense of his own autonomy, he turned aside and looked at something he suspected, given time, might be important.

Finally, of course, the need to publish became unavoidable. The motivating impulse arrived (inevitably for so committed a letter writer) in the post bag and was far more frightening than a hint from Lyell or Hooker. It is one of the best-known stories in the history of science: the young collector Alfred Russel Wallace, travelling in the Malaysian Archipelago and prostrated by malaria, comes to his own understanding of an evolutionary theory essentially the same as that of Darwin. Thus conceived in stress rather than in quiet, the paper is sent to Darwin, with a request to forward it on to Lyell and thus to publication.

The letter was a severe jolt. In spite of warnings from Lyell that others were thinking about evolution, Darwin had settled into his normal way of working, where thoroughness and regularity is everything, and where publication is simply something that happens at the end. Wallace's letter, then, was a challenge to Darwin's style. 'I always thought it possible that I might be forestalled, but I fancied that I had grand enough soul not to care; but I found myself mistaken and punished' (Burkhardt and Smith, 1991: p. 129). When his priority was indeed challenged, Darwin found that he minded.

During this crisis, Darwin's anguish was much exacerbated by his youngest child Charles Waring contracting, and dying from, scarlet fever; another child, Etty, was ill with diphtheria. Darwin found he simply did not know what to do. He could not act and so sought to put the matter in the hands of his friends and colleagues, Charles Lyell and Joseph Hooker. 'I would far rather burn my whole book than that he or any man should think that I had behaved in a paltry spirit ... I am worn out with musing', he told Lyell (Burkhardt and Smith, 1991: p. 117).

If Darwin's working method had been akin to a slow and steady river, with many pools and eddies, now it became a cataract. After a quickly-arranged meeting at the Linnean Society, where Wallace's letter of June 1858 was read alongside Darwin's unpublished essay of 1844, Darwin's priority was established. Having been knocked so hard on the head, he dropped his plans for 'a big book' and decided to write something quickly – an 'abstract', as he called it. In terms of quietude Darwin had been snapped from one state to another, from an apparently endless process of compilation to one where he had to write vigorously, briefly and with publication firmly in mind.

Someone to turn to

The saga of the Darwin–Wallace paper at the Linnean prompts questions about the probity of the arrangement developed by Hooker and Lyell. It is clear that Darwin was fortunate not only in the advice of friends, but also in the way the advice was swiftly followed by action. Wallace had no say in the matter. Eight thousand miles away, he knew nothing of the plan. In effect any claim he had to priority was silenced by the swift work of Darwin's people. For the purposes of my argument, however, the significance of the Hooker–Lyell campaign lies elsewhere. The point to consider is the level of trust that existed between all these men. Wallace trusted Darwin to send on his article to Lyell and he trusted Lyell to facilitate its publication if he judged it valuable. In turn, Darwin, utterly incapacitated by the shock of Wallace's letter, by the moral dilemma it presented and by the illness of two of his children, could manage one action only – to place the matter in the hands of his friends.

The story carries contemporary meaning, pointing out how professional communication depends on trust. The question arises: do the pressures of the research life, described earlier in this chapter, impact adversely on this precious commodity? My descriptions of quietude may summon images of isolation and solitude, but we are reminded through the life of Darwin that collegiality is part of the constellation that governs scientific production. Darwin depended on routine and isolation in order to work, but throughout his career the dynamics of friendship and trust were vital too. When in 1844 he felt he wanted to tell Hooker about his ideas, and compared his confession to that of owning up to a murder, his sense of relief was huge. To put it another way, though Darwin was rich enough, and interested enough, to shape his work in a way that suited him, he still relied on others. His relations with Hooker and with Lyell were long-standing and steady: they could be depended upon. Thus, when Darwin was felled by the letter from Wallace there was no hiatus, no disturbing silence from Hooker and Lyell. They knew what to do.

We have seen how in Darwin quietude is accompanied by the social impulse. The Wallace affair points to a third element, namely trust, as an important aspect of the craft of science. If trust be considered the form of understanding that arises when social relations combine with a reflective sensibility, then we can see how Darwin's delays link to his productivity. To put the point bluntly, slowness rather than haste may prove to be the more productive, because it more naturally allies with trust. The danger faced by contemporary science – by scientists – now seems obvious. For exaggerated competitiveness and hyper-productivity quickly reduces human communication and annihilates trust in the process. It must do, for this is a game of winners and losers. In those circumstances, the contemplative sensibility simply cannot thrive.

Darwin was fortunate. He had someone to turn to. Though inclined to hide away, over many years he had drawn his colleagues into his way of thinking. At a time of crisis they were prepared to reassure and guide him. Understanding each other perfectly, a vertiginous change of pace was possible and a solution was found that quickly established Darwin's priority. The speed they suddenly attained is worth emphasizing. Darwin told Lyell of the Wallace letter on 18 June 1858 and the Darwin–Wallace papers were read to a meeting of the Linnean Society on 1 July 1858. Two weeks later Darwin took his family to the Isle of Wight to recuperate, renting a villa in Shanklin just next to the sea. The family could relax and it was here that the new publication took shape in his mind. 'It will be longer than expected … I am extremely glad I have begun in earnest on it', he told Hooker (Burkhardt and Smith, 1991: p. 141). Once again at peace, and with his children below on the beach, Darwin began to write.

References

Bowlby, J., 1991. *Charles Darwin: A New Biography.* London: Pimlico.
Browne, J., 1995. *Charles Darwin: Voyaging.* London: Jonathan Cape.

Browne, J., 1998. I could have wretched all night: Charles Darwin and his body. In: Lawrence, C. and Shapin, S. (eds), *Science Incarnate: Historical Embodiments of Natural Knowledge*. Chicago: University of Chicago Press, pp. 240–286.

Browne, J., 2002. *Charles Darwin: The Power of Place*. London: Jonathan Cape.

Burkhardt, F. and Smith, S., 1985. *The Correspondence of Charles Darwin, Vol. 1: 1821–1836*. Cambridge: Cambridge University Press.

Burkhardt, F. and Smith, S., 1986. *The Correspondence of Charles Darwin, Vol. 2: 1837–1843*. Cambridge: Cambridge University Press.

Burkhardt, F. and Smith, S., 1990. *The Correspondence of Charles Darwin, Vol. 6: 1856–1857*. Cambridge: Cambridge University Press.

Burkhardt, F. and Smith, S., 1991. *The Correspondence of Charles Darwin, Vol. 7: 1858–1859*. Cambridge: Cambridge University Press.

Chambers, R., 1844. *Vestiges of the Natural History of Creation and Other Evolutionary Writings*, ed. Secord, J. Chicago: University of Chicago Press, 1993.

Cronin, B., 2013. Slow food for thought. *Journal of the American Society for Information Science and Technology*, 64(1), p. 1.

Darwin, C., 1859. *The Origin of Species by Means of Natural Selection or the Preservation of Favoured Races in the Struggle for Life*. Harmondsworth: Penguin Books, 1968.

Darwin, F. (ed.), 1887. *The Life and Letters of Charles Darwin, Vol. 1*. London: John Murray.

Desmond, A. and Moore, J., 1991. *Darwin*. London: Penguin Books.

Desmond, A. and Moore, J., 2009. *Darwin's Sacred Cause: Slavery and the Quest for Human Origins*. London: Penguin Books.

Gould, S.J., 1996. *The Mismeasure of Man*. New York: Norton.

Grafton, A., 2010. Britain: the disgrace of the universities. *New York Review of Books*, 9 March. Available at www.nybooks.com/blogs/nyrblog/2010/mar/09/britain-the-disgrace-of-the-universities/. Accessed 7 September 2015.

Keynes, R.D. (ed.), 1988. *Charles Darwin's Beagle Diary*. Cambridge: Cambridge University Press.

Lawrence, P., 2002. Rank injustice: the misallocation of credit is endemic in science. *Nature*, 415, pp. 835–836.

Lesnick-Oberstein, K., Burman, E. and Parker, I., 2015. Let UK universities do what they do best – teaching and research. *Guardian*, 6 July. Available at: www.theguardian.com/education/2015/jul/06/let-uk-universities-do-what-they-do-best-teaching-and-research. Accessed 7 September 2015.

Lyell, C., 1830. *Principles of Geology*. London: Penguin Books, 1997.

MacIntyre, A., 1981. *After Virtue: A Study in Moral Theory*. London: Duckworth.

Neve, M. and Messenger, S. (eds), 2002. *Charles Darwin: Autobiographies*. London: Penguin Books.

Nuffield Council on Bioethics, 2014. *The Culture of Scientific Research in the UK*. Available at: http://nuffieldbioethics.org/wpcontent/uploads/Nuffield_research_culture_full_report_web.pdf. Accessed 7 September 2015.

Ridley, M., 2015. *The Evolution of Everything: How Ideas Emerge*. London: Fourth Estate.

Rose, H. and Rose, S. (eds), 2001. *Alas Poor Darwin*. London: Vintage.

Rosenberg, A., 2000. *Darwinism in Philosophy, Social Science and Policy*. Cambridge: Cambridge University Press.

Royal Society, 2015. *The Future of Scholarly Scientific Communication*. Available at: https://royalsociety.org/~/media/events/2015/04/FSSC1/FSSC-Report.pdf. Accessed 7 September 2015.

Stott, R., 2003. *Darwin and the Barnacle*. London: Faber and Faber.

van Wyhe, J., 2007. Mind the gap: did Darwin avoid publishing his theory for many years? *Notes and Records of the Royal Society of London*, 61(2), pp. 177–205.

Vostal, F., 2013. Should academics adopt an ethic of slowness or ninja-like productivity? In search of scholarly time. *The Impact Blog*, London School of Economics. Available at: http://blogs.lse.ac.uk/impactofsocialsciences/2013/11/20/in-search-of-scholarly-time/. Accessed 7 September 2015.

3 'Tired with this subject ...'
Isaac Newton on publishing and the ideal natural philosopher

Cornelis J. Schilt

Introduction[1]

In early 1673, Isaac Newton announced his intention to withdraw from all correspondence about natural philosophy. Just a year before, the young Cambridge don had submitted his first publication to the Royal Society and had agreed to have it printed in the *Philosophical Transactions*. The essay, titled 'New Theory of Light and Colours', introduced the world of natural philosophy to a new methodology of scientific discourse and experimentation. For Newton, it was quality that mattered, not quantity. Written in a rather concise style, the essay put forward a theory that seems plausible to us now: that light consists of differently coloured rays, each with their own angle of refraction. It was not so clear-cut to Newton's contemporaries though. Within a few weeks Newton found himself the recipient of letters from various national and international natural philosophers. Evidently, he was not pleased with the tone of many of these letters, which questioned his methods, his accuracy and even his honesty. One year and one month after he had submitted his 'New Theory' to the Royal Society, Newton wrote:

> I desire that you will procure that I may be put out from being any longer fellow of the R Society. For though I honour that body, yet since I see I shall neither profit them, nor (by reason of this distance) can partake of the advantage of their Assemblies, I desire to withdraw.
> (Turnbull, 1959: p. 262)

Henry Oldenburg, the Royal Society's secretary and the recipient of these words, appeared rather baffled by Newton's resignation. Writing on the back of Newton's letter, he remarked that he was 'surprised at his resigning for no other cause, than his distance, which he knew as well at the time of his election' (Turnbull, 1959: p. 263).[2] But Newton did not immediately lay down his quill. He kept corresponding about mathematics and in a letter to Oldenburg, written in June 1673 and directed at the Dutch philosopher Christiaan Huygens, he once more set about explaining himself 'a little further in these things ... since there seems to have happened some misunderstandings between us'

(Turnbull, 1959: p. 294). However, in case Oldenburg might have misunderstood his intentions, he ended the letter saying:

> I must, as formerly, signify to you, that I intend to be no further sollicitous about matters of Philosophy. And therefore I hope you will not take it ill if you find me ever refusing doing any thing more in that kind, or rather that you will favour me in my determination by preventing so far as you can conveniently any objections or other philosophicall letters that may concern me.
>
> (Turnbull, 1959: pp. 294–295)

'And then silence', as biographer Richard Westfall (1966: p. 303) characterized the period that followed. 'The effort – which had begun the summer before – to cut his ties with the world succeeded at last.' Indeed, Newton's correspondence record shows a remarkable gap. A single, short letter on mathematics in September 1673, followed by almost a year of silence. When Oldenburg forwarded another response to the 'New Theory' from the Liège-based English Jesuit, Francis Linus, in early October 1674, he did not receive Newton's reply until December: 'I am sorry that you put yourself to the trouble of transcribing Fr. Linus's conjecture', Newton wrote, 'since (besides that it needs no answer) I have long since determined to concern my self no further about the promotion of Philosophy' (Turnbull 1959: p. 328).

Traditionally, historians of science have understood Newton's temporary withdrawal from the scene of natural philosophy as the result of the rather harsh nature of the debate following his early optical publication and of Newton's own short-temperedness. After all, from the start Newton framed the replies he received with words like 'improper', 'rudeness' and 'ungrateful', and he often responded in a manner that could be described with similar words (Turnbull, 1959: pp. 264, 282, 292). When, at the end of 1675, Newton decided to write another essay to satisfy his critics once and for all, it took only another year before he regretted breaking his silence:

> I see I have made my self a slave to Philosophy, but if I get free of Mr Linus's buisiness I will resolutely bid adew to it eternally, excepting what I do for my privat satisfaction or leave to come out after me. For I see a man must either resolve to put out nothing new or to become a slave to defend it.
>
> (Turnbull, 1960: pp. 182–183)

However, I am not convinced that Newton's withdrawal can be explained solely from character traits and the nature of the debate. As Rob Iliffe (1995: p. 160) writes, 'there tend to be as many psychological portraits as historians, and explanations of his specific decisions to suppress or publish elements of his natural philosophy vary as much as accounts of his personality'. Rather, it seems that Newton held a particular attitude to the exchange of scientific knowledge, and his entire approach is reminiscent of that other major interest of his in his Cambridge period: alchemy.

In this chapter I will explore Newton's communication strategy as evidenced in his correspondence. By examining what was at stake, I will draw out both Newton's attitude to experimental philosophy and his ideas about scientific communication. I will finish by making connections between this approach to communication and his alchemical programme, drawing on Newton's private notes to demonstrate how Newton attempted to fashion himself into a true alchemical adept.

All was light ...

In the early 1670s, when Newton's star had yet to rise, Robert Boyle was by far England's most famous natural philosopher – one whose approach to publishing contrasted strongly with Newton's, as we will see. In an essay published in the middle of Newton's year of silence, he reflected on two decades of publication and public activities as a natural philosopher:

> And first, if [one] should think to secure a great Reputation, by forbearing to couch any of his Thoughts or Experiments in Writing, he may thereby find himself not a little mistaken. For ... he will not avoid the Visits and Questions of the Curious. Or, if he should affect a Solitude, and be content to hide himself, that he may hide the things he knows; yet he will not escape the solicitations that will be made him by Letters. And if these ways of tempting him to disclose himself, prevail not at all with him to do so, he will provoke the Persons that have employ'd them; who finding themselves disoblieg'd by being defeated of their Desires, if not also their Expectations, will for the most part endeavour to revenge themselves on him, by giving him the Character of an uncourteous and ill-natur'd person ...
> (Boyle, 1674: pp. 184–185; see also Shapin, 1991: p. 203)

One of the key figures of the early Royal Society, Boyle had been caught up in vigorous debates about the epistemological status of philosophical experiments, many of which involved his newly invented air pump.[3] He had likewise published extensively on all things natural philosophical, clearly indicating the nature of his research with titles like *New Experiments Physico-Mechanical: Touching the Spring of the Air and Their Effects* (1660), *Experiments and Considerations Touching Colours* (1664) and *New Experiments Touching Cold* (1665). Following Francis Bacon's example, most of Boyle's works contain large-scale experimental reports, often accompanied by a long introduction discussing both theory and experiment.

In *Touching Colours*, for example, we find Boyle addressing a fictional pupil, Pyrophilus, who is 'passionately addicted ... to the delightful art of limning and painting' as his name, 'friend of fire', suggests (Boyle, 1664: p. 1).

Through a series of theoretical considerations, followed by sixty-five numbered experiments, Boyle invites Pyrophilus 'to enquire seriously into the Nature of Colours, and assist [him] in the Investigation of it'. He explicitly states that he intends not to provide his pupil with an accurate and particular theory of colours, for 'that were to present you with what I desire to receive from you; and, as farr as in mee lay, to make that study needless, to which I would engage you' (Boyle, 1664: p. 2).

This does not mean that Boyle does not voice his own preferences. After all, the book is meant to communicate his experimental results and inferences to the larger audience of *virtuosi* involved in the study of nature. But he offers up his position tentatively. For instance, dismissing a number of topical explanations for the nature of light and colours, Boyle states that he 'encline[s] to take Colour to be a Modification of Light' but immediately adds that:

> though this be at present the Hypothesis I preferr, yet I propose it but in a General Sense, teaching only that the Beams of Light, Modify'd by the Bodies whence they are sent (Reflected or Refracted) to the Eye, produce there that Kind of Sensation, Men commonly call Colour; But whether I think this Modification of the Light to be perform'd by Mixing it with Shades, or by Varying the Proportion of the Progress and Rotation of the Cartesian Globuli Cælestes, or by some other way which I am not now to mention, I pretend not here to Declare.
>
> (Boyle, 1664: p. 90)

Throughout *Touching Colours*, we hear Boyle employing a similar, rather cautious tone. His unwillingness to prescribe to Pyrophilus the exact nature of light seems a reflection of his views about experimental certainty. There is only so much (or little) one can know through experiment, he says, that we should be careful not to make too bold claims about nature when interpreting experimental results:

> For whensoever I would Descend to the Minute and Accurate Explication of Particulars, I find my Self very Sensible of the great Obscurity of things, without excepting those which we never see but when they are Enlightned, and confess with Scaliger, Latet natura hæc, (says he, Speaking of that of Colour) & sicut aliarum rerum species in profundissima caligine inscitiæ humanæ.
>
> (Boyle, 1664: p. 92)[4]

As is apparent from Boyle's reflections ten years later, this cautious attitude had not always served him well in terms of avoiding dispute. Nor would Newton's contrasting approach as he entered the scene in the early 1670s. Newton's works were to have a lasting impact upon science and society. However, had it not been for the enduring stimulus of others – sometimes much to his chagrin – Newton might never have published more than a few mathematical tracts and would have remained in virtual obscurity. Yet, paradoxically enough, it was precisely because of this stimulus and his subsequent first publication that Newton almost resigned his membership from the Royal Society in preference for a secluded silence.

Unlike Boyle, the young Isaac Newton was not at first very interested in publication at all. Born on Christmas Day 1642, by the time he finally submitted his 'New Theory of Light and Colours' to the Royal Society Newton was already 29 years old. In comparison, by that age Robert Boyle had written half a dozen of books, with many more to follow. When Newton died aged 84, he had published no more than two books, albeit it in various editions, and a few papers. But even at the time of Boyle's lament about the burdens of publication, having published just the one paper, Newton could already fully sympathize with such concerns. He too had experienced 'the visits and questions of the curious' and because he did not respond to their expectations, he had also experienced their 'revenge'. As he put it in his letter to Oldenburg, the experience had persuaded him 'to be no further sollicitous about matters of Philosophy'.

Yet the reasons Newton provided for his resignation from the Royal Society do seem somewhat mysterious, as Oldenburg himself clearly felt. After all, when Newton submitted his 'New Theory' to Oldenburg in early February 1671/72, his name was already becoming known within the walls and letters of the Royal Society. He had been appointed Lucasian professor of mathematics at Trinity College, Cambridge, in 1669 and had since been involved in an active correspondence on various mathematical topics. The reflecting telescope he presented to the society in 1671 was seen as a masterpiece of ingenuity and craftsmanship; despite being only seven inches long, it had the same magnifying capacity as a normal telescope of several feet (see Hall and Simpson, 1996). The resulting exchange over techniques for polishing concave metal mirrors involved in the design of reflecting telescopes also demonstrated Newton's intimate knowledge of material cultures. As we shall see, the debate that followed the 'New Theory' was lively and involved scholars from all over Europe. The Royal Society had definitely profited from Newton.

So what *did* cause Newton to write his letter of withdrawal? What happened during those thirteen months that made Newton decide to give up on natural philosophy? Newton biographer Gale Christianson saw as prime reason for Newton's resignation from the Royal Society the nature and tone of the correspondence that followed the publication of his 'New Theory' (see Christianson, 1984: pp. 149–202). As evidence, Christianson cites a letter to the mathematician and information broker John Collins, where Newton says:

> I could wish I had met with no rudeness in some other things. And therefore I hope you will not think it strange if to prevent accidents of that nature for the future I decline that conversation which hath occasioned what is past.
>
> (Turnbull, 1959: p. 282)

70 *Choosing silence*

According to Christianson, it was this 'rudeness' that annoyed Newton most. It seems as if Newton had anticipated a different sort of debate, or at least one with a different tone. As we will see, Christiansen is right that Newton was definitely not amused by the criticism he received. Things did not seem to go as planned. However, I think there is much more to this than Newton's alleged short temper and that of his correspondents. In fact, only a few actually criticized Newton, albeit it rather vocally. As Henry Oldenburg reported, when Newton's 'New Theory' was read to the Royal Society:

> it there mett both with a singular attention and an uncommon applause, insomuch that after they had order'd me to return you very solemne and ample thankes in their name (which herewith I doe most cheerfully) they voted unanimously, that if you contradicted it not, this discourse should without delay be printed.
>
> (Turnbull, 1959: p. 107)

Newton contradicted it not, and thus the eightieth issue of the *Philosophical Transactions* contained Newton's maiden publication. The debate that subsequently arose contains important keys to the understanding of Newton's disillusionment, even if this does not reduce to a matter of his correspondents' alleged rudeness, and it is therefore worth exploring in more detail. First, however, it is necessary to take a closer look at the publication that caused so much upset.

A new theory of light and colour

Always on the lookout for copy for the *Philosophical Transactions*, Henry Oldenburg must have loved the mathematical correspondence that arose between Isaac Newton, John Collins and David Gregory in the late 1660s and early 1670s, in which the young Lucasian professor demonstrated his intimate understanding of geometry. Although the letter containing his reply to Newton has not survived, Oldenburg was clearly also delighted when Newton indicated that he had a significant discovery to reveal. In a letter of 18 January 1671/72, Newton explained to Oldenburg that it was:

> a Philosophical discovery which induced mee to the making of the said Telescope, & which I doubt not but will prove much more gratefull then the communication of that instrument, being in my Judgment the oddest if not the most considerable detection which hath hitherto beene made in the operations of Nature.
>
> (Turnbull, 1959: pp. 82–83)

The discovery, presented in the 'New Theory of Light and Colours', involved two important questions: the composition of light in terms of colours and the nature of light itself. Newton connected these, arguing that white light was a composite of all other colours as well as being corpuscular, consisting of particles. Most importantly, and contrary to Boyle's preferred hypothesis, Newton insisted that colour was a property of light and not of the reflected object:

> Colours are not qualifications of light derived from refractions or reflections of naturall bodies as 'tis generally beleived, but originall & connate properties, which in diverse rayes are divers, some rayes are disposed to exhibit a red colour & noe other, some a yellow & noe other, some a green & noe other & so of the rest. Nor are there only rayes proper & peculiar to the more eminent colours, but even to all their intermediate gradations.
> (Newton, 1672: p. 3081)

Likewise, differently coloured rays refracted under different angles when led through a prism. Boyle's research was meant primarily to demonstrate the reality of prismatic colours and hence he had designed his experiments with that aim in mind. His light beams travelled only short distances before reaching a surface, which was not enough to make the divergence noticeable (Westfall, 1962: p. 346).[5] Newton's experiment, on the other hand, involved placing a prism over 22 feet from the wall on which the light's image was projected (Newton, 1672: p. 3077).

Using the standard narrative form also adopted by Boyle and other contemporaries, Newton began his essay recalling how:

> in the beginning of the year 1666 ... I procured me a Triangular glass-Prisme, to try therewith the celebrated *Phaenomena of Colours*. And in order hereto having darkened my chamber, of the Suns light, I placed my Prisme at its entrance, that it might be refracted to the opposite wall. It was at first a very pleasing divertisement, to view the vivid and intense colours produced thereby; but after a while applying my self to consider them more circumspectly, I became surprised to see them in an oblong form; which, according to the received laws of Refraction, I expected should have been circular.
> (Newton, 1672: pp. 3075–3076)

As Patricia Fara (2015: p. 5) remarks: 'Right from the beginning, the unknown young Newton casts himself in a superior position as the lynx-eyed detective, the expert on lenses who decided to amuse himself by playing with colours but who rapidly realized that major scientific issues were at stake.' Almost casually, Newton introduces the prism as his preferred instrument. Nowadays this might seem obvious, but at the time the prism was seen as a plaything, used for mere amusement, with a likewise unclear epistemological status (see Schaffer, 1989). One did not buy prisms in a specialist craftsman's shop, but, for instance, at annual fairs. From the experiments recorded in his notebooks, we know that Newton had already used prisms before 1666, probably as early as 1664. In fact, it is highly unlikely that Newton bought a prism at the beginning of 1666 since he had left Cambridge for his native Woolsthorpe because of the plague that threatened Cambridge, only to return in March 1666 for about three months.[6] Indeed, despite Newton's description of the introduction to his 1672 essay as 'historical' (Turnbull, 1959: p. 97), it was not meant as a factual representation of the pathway to discovery. Instead, it is a carefully crafted narrative, 'artfully naïve', downplaying years of trial-and-error experiments in an early-career attempt at self-fashioning and identity construction (Fara, 2015: p. 5; see also Dear 1985: pp. 154–156).

However, halfway through, Newton abruptly quits the narrative:

> I shall now proceed to acquaint you with another more notable difformity in its Rays, wherein the Origin of Colours is unfolded: Concerning which I shall lay down the Doctrine first, and then, for its examination, give you an instance or two of the Experiments, as a specimen of the rest.
>
> (Newton, 1672: p. 3081)

Newton's audience might have found it rather hard to digest the sudden change of tack. Only those present at the meeting in early February, where Newton's letter was being read, knew that at this point the printed version differed significantly from Newton's original letter, which ran thus:

> I shall now proceed to acquaint you with another more notable difformity in its Rays, wherein the Origin of Colours is unfolded: A naturalist would scearce expect to see the science of those become mathematicall, & yet I dare affirm that there is as much certainty in it as in any other part of Opticks; For what I shall tell concerning them is not an Hypothesis but most rigid consequence, not conjectured by barely inferring 'tis thus because not other wise or because it satisfies all phaenomena (the Philosophers universall Topick) but evinced by the mediation of experiments concluding directly & without any suspicion of doubt. To continue the historicall narration of these experiments would make a discourse too tedious & confused, & therefore I shall rather lay down the Doctrine first, and then, for its examination, give you an instance or two of the Experiments, as a specimen of the rest.
>
> (Turnbull, 1959: pp. 96–97)

These were bold statements: in his first ever publication, Newton not only claimed to have found a new theory of light and colours, he also insisted it was the one and only explanation for the observed phenomena because it followed directly from the experiments 'without any suspicion of doubt'. Likewise, he stated that research into the nature of colours was as mathematical as any other form of optics and that his conclusions should not be seen as mere hypotheses, a word to which I will return shortly. This was far removed from the cautious and modest tone that Boyle had employed in his *Touching Colours* and it might well be that Oldenburg removed these lines from the printed edition in an effort to make Newton's theory more digestible.

The fact that Newton's 'New Theory' suddenly adapted the form and language of mathematics was as ambitious as it was unorthodox, and possibly as momentous as the discovery itself (Iliffe, 2004: pp. 437–438). However, Newton's audience struggled most with the fact that he seemingly derived all his conclusions from a single experiment, which he labelled *experimentum crucis*:

> I took two boards, & placed one of them close behind the Prism at the window, so that the light might passe through a small hole made in it for that purpose & fall on the other board which I placed at about twelve foot distance, having first made a small hole in it also for some of that incident light to passe through. Then I placed another Prism behind this second board so that the light trajected through both the boards might passe through that also & be again refracted before it arrived at the wall.
> (Newton, 1672: p. 3078)

In his influential *Micrographia* (1665), Robert Hooke had coined the term *experimentum crucis* to describe an experiment of vital importance. Newton, however, gave the concept an entirely new meaning: one crucial experiment was enough to prove a theory. As he would state in one of the debates he found himself engaged in, 'it is not number of Experiments, but weight to be regarded; & where one will do, what need many?' (Turnbull, 1960: p. 76; see also Schilt, 2016: pp. 59–62). It was exactly this idiosyncratic *experimentum crucis* that would cause Newton so much trouble in the following months.

Storm and silence

By the time the 'New Theory' was printed in the *Philosophical Transactions*, Newton had already received a number of responses. The first response, from Oldenburg, reported on the very positive reception by those present at the Royal Society meeting where the 'New Theory' was read. The second, barely a week later, was a detailed exposition given by Robert Hooke and also delivered to the Royal Society, explaining why Newton's 'Excellent Discourse' did not contain 'any undeniable argument' that convinced him of the certainty of Newton's 'Hypothesis of salving the phaenomena of Colours thereby' (Turnbull, 1959: p. 110). Referring to the many experiments he had performed as valid support for his own wave-theory, he did not so much question Newton's observations as his experimental methodology:

> But how certainesoever I think myself of my hypothesis, which I did not take up without first trying some hundreds of experiments; yet I should be very glad to meet with one Experimentum crucis from Mr Newton,

that should Divorce me from it. ... I cannot only salve all the Phaenomena of Light and colours by the Hypothesis, I have formerly printed and now explicate that by, but by two or three other, very differing from it, and from this, which he hath described in his Ingenious Discourse. Nor would I be understood to have said all this against his theory as it is an hypothesis, for I doe most Readily agree with him in every part thereof, and esteem it very subtill and ingenious, and capable of salving all the phaenomena of colours; but I cannot think it to be the only hypothesis; not soe certain as mathematicall Demonstrations.

(Turnbull, 1959: pp. 110–111, 113)

Newton's immediate response was very short: '[H]aving considered Mr Hooks observations on my discours, [I] am glad that so acute an objecter hath said nothing that can enervate any part of it'. But he promised Oldenburg that 'You shall very suddenly have my answer' (Turnbull, 1959: p. 116). It took almost four months for that answer to appear. Newton was obviously not amused by Hooke's 'unintelligible' observations (Turnbull, 1959: p. 176). The response, when it finally came, was long and detailed, and the one word that Newton kept repeating and invariably underlining was 'hypothesis'. That word carried a different meaning in Newton's days, as is clear from his own words: 'conjectured by barely inferring 'tis thus because not otherwise or because it satisfies all phaenomena'. Newton would later famously state 'hypotheses non fingo' – 'I feign no hypotheses' – in the 'General Scholium' attached to the second edition of the *Principia Mathematica* (Newton, 1713: p. 484). Hypotheses without any experimental support whatsoever had no place in Newton's natural philosophy (Shapiro, 2002: p. 228).

Newton clearly stated that his new theory should not be treated as such a hypothesis; he had proven, directly and without doubt, through the application of a particular experiment, that white light was composed of all other colours and that differently coloured rays refracted under different angles. However, as is clear from the outset, Hooke treated Newton's theory as any other hypothesis. Likewise, he did not accept the status of Newton's *experimentum crucis*, or any single conclusive experiment for that matter, but instead insisted upon the accepted practice of a series of experiments, one building upon the other to gradually strengthen the evidence for a particular hypothesis. This also explains why Hooke did not bother to reproduce Newton's experiment, for what would he have gained from it?

Between Newton's response to Oldenburg and his detailed answer to Hooke, Newton had a short exchange with the Jesuit Ignace-Gaston Pardies, a respected French natural philosopher who taught at the College of Louis-le-Grand in Paris (Philips, 1913). In his short life, Pardies published on mathematics, mechanics and sundials, and he left a beautifully illustrated celestial atlas in manuscript which was published shortly after his death. Pardies objected that the oblong figure that Newton described in his first experiment could be explained by laws already known, a claim which he tried to substantiate mathematically. It is not clear whether Pardies attempted to reproduce Newton's *experimentum crucis*, but, like Hooke, he claimed that existing hypotheses could sufficiently explain the results of that experiment and that therefore 'non videtur necesse occurrere ad aliam Hypothes in aut admittere diversam illam radiorum frangibilitatem' – 'no other Hypothesis as that of diverse refrangibility of the rays is necessary' (Turnbull, 1959: p. 133).[7] Newton's reply followed soon, demonstrating how he had already anticipated these objections in the 'New Theory'. After providing short but detailed answers to Pardies's main suggestions and even instructions on how to reproduce particular experimental details, he ended his letter with:

> I am content that the Reverend Father calls my theory an hypothesis if it has not yet been proved to his satisfaction. But my design was quite different, and it seems to contain nothing else than certain properties of light which, now discovered, I think are not difficult to prove, and which if I did not know to be true, I should prefer to reject as vain and empty speculation, than acknowledge them as my hypothesis.
>
> (Turnbull, 1959: p. 144)

Although the letter conveys a rather firm tone, I fail to find in it a 'deeply irritated Newton', as Christianson does, and likewise do not quite see why Christianson portrays Pardies as 'shaken' by the reply (1984: p. 46). Newton's answer to Pardies is firm but polite, and Pardies graciously admits that Newton met all his objections. He even apologizes for using the term hypothesis, 'the first word that came to mind' (Turnbull, 1959: p. 168). Westfall (1963: p. 86) likewise describes Pardies's comments as 'not in the least unintelligent ... moreover, dignified and respectful'; indeed, Newton wrote that 'R Patris veròcandor in omnibus conspicitur' – 'the Reverend Father's sincerity is manifest throughout' (Turnbull, 1959: p. 96).[8]

When Oldenburg sent Pardies's initial letter to Newton, he also included a note from the Dutchman Christiaan Huygens, Europe's leading natural philosopher. In that note, Huygens remarked that Newton's 'New Theory' was 'fort ingenieuse ... mais il faudra voir, si elle est compatible avec toutes les Experiences' – 'quite ingenious ... but it remains to be seen whether it is compatible with all the Experiments' (Turnbull, 1959: p. 135). Alas, with a second and third note that followed in the summer and autumn of 1672, Huygens became more and more careful with his praise, until he finally submitted his full response to the Royal Society in January 1672/73. That full response amounted to no more than a single paragraph and must have been extremely unsatisfactory to Newton. Seemingly without having tried to perform the *experimentum crucis*, Huygens simply stated that any hypothesis explaining in mechanical fashion the colours yellow and blue would be sufficient to account for the phenomena Newton describes. And, as Oldenburg's accompanying introduction read, '[Huygens], like Mr Hook, calls [your doctrine of Light] an Hypothesis or an opinion, notwithstanding your printed Defence against that Appellation' (Turnbull, 1959: p. 255).

Both Westfall and Christiansen agree that this was the last straw that led Newton to resign his membership from the Royal Society. Pardies's letters were tedious and the correspondence with Robert Hooke annoying, but the fact that Europe's leading scholar took a year before dismissing Newton's theories without even trying to reproduce his findings must have been very painful. Yet Newton did not immediately cease all correspondence and even found the dignity to respond to Huygens twice. In between a narrative description of where Huygens erred in his ideas about the nature of white light and another detailed and systematic explanation of his own theories, 'since there seems to have happened some misunderstanding between us', Newton confronted Huygens. He referred him to the correspondence with Hooke, which had been printed in full in the *Philosophical Transactions*, writing that 'it was a little ungratefull to me to meet with objections which have been answered before, without having the least reason given me why those answers were insufficient' (Turnbull, 1959: p. 292).

It is after Newton's second letter to Huygens in June 1673 and a letter on mathematics to Collins at the end of the summer, that we find the one-year silence in Newton's correspondence. There was no communication whatsoever, even on mathematical matters. It was mainly because of the tenacity of others, like Henry Oldenburg and John Collins, that Newton resumed his semi-public correspondence through the Royal Society. This eventually resulted in a second optical publication called 'An Hypothesis Explaining the Properties of Light Discoursed of in my Severall Papers', which he submitted in December 1675. Within weeks Newton was again entangled in another series of disputes with Linus, Hooke and others. When Hooke suggested that they might continue their discourse in private correspondence, Newton responded saying:

> I was formerly tired with this subject, & have not yet nor I beleive ever shall recover so much love for it as to delight in spending time about it; yet to have at once in short the strongest or most pertinent Objections that may be made, I could really desire, & know no man better able to furnish me with them then your self.
>
> (Turnbull, 1959: p. 416)

However, we have no evidence of a private correspondence between Hooke and Newton and the letter's famous phrase, 'If I have seen further it is by standing on the sholders of Giants', has been read by many historians as mere mockery of Hooke's stooped posture.[9] Reading the ensuing correspondence with his critics, it is highly likely that by early 1676 Newton was not just 'formerly' tired with this subject, but 'again' tired with this subject, for the same reasons as before. And just as in 1673 and 1674, by the end of the decade Newton's output in terms of correspondence had dwindled to a bare minimum.

Deliberations

Let us take a moment of reflection. Newton is 29 years of age when he decides to finally publish his theories, after almost a decade of optical studies. He seems eager to demonstrate his new-found knowledge about the nature of light and colours, and he has great expectations of the Fellows of the Royal Society. But the responses of Hooke, Pardies and Huygens are disappointing. None of them seem willing to accept the new rules of the game. Neither does any one of them try to reproduce the *experimentum crucis*, with the possible exception of Pardies. But even if they had attempted the experiment, as later commentators would, Newton's description of the experiment left much to be desired. In contrast to his notebooks and the lecture notes he composed in the early 1670s, Newton did not provide his audience with a detailed experimental set up or illustration, nor did he explain to his audience how he arrived at the *experimentum crucis*, nor why it would prove his theory beyond all doubt.[10] As Fara (2015: p. 5) says: 'Rewriting his repeated trials, mistakes and cul-de-sacs as a logical and systematic investigation, he makes his conclusions sound inevitable and hence unchallengeable – and by concealing vital details, such as the type of glass or shape of the prisms, Newton renders replication extremely difficult.' It also did not really help that Newton was ambiguous about the experiment himself and whether it could really replace the usual large series of experiments (Schaffer, 1989: p. 85).

With a new theory, a new instrument, a new form of evidence and proof, a new experimental epistemology, all topped off with a new interpretation of the term hypothesis, it is no wonder that Newton got into debates. What was he thinking? Jan Golinski (1998: p. 117) remarks that:

> Newton was engaged in articulating a new model of the relations between experimental fact and theoretical interpretation at the same time that he was seeking to have his claims accepted as facts. In this situation, rhetorical decisions evidently embraced more than just questions of style. Designations of what were factual claims and construals of their content were themselves constructed with available rhetorical resources and were subjects of negotiation in the ongoing debate.

As will soon come clear, Newton had deliberately cast his 'New Theory' in a form that proved to be unintelligible to most, if not all, of his audience. And, as it turned out, the construals of the content of his claims were indeed negotiable. Soon after his initial replies to Hooke, Pardies and other 'curious sollicitors', and well before Huygens's final reply, Newton decided to recast his 'New Theory' as eight Queries, to which he added:

> Touching the Theory of Colours I am apt to believe that some of the experiments may seem obscure by reason of the brevity wherewith I writ

them which should have been described more largely & explained with schemes *if they had been intended for the publick.*

(Turnbull, 1959: p. 212; italics added)

This statement seems at first incredibly puzzling. Surely Newton was writing for the public, was he not? After all, he did not object to Oldenburg's request to have his 'New Theory' printed in the *Philosophical Transactions*. But we should not forget that the Royal Society was still in its infancy; in 1672 its membership counted only 216 fellows, a figure that had dropped below 150 near the end of the century (Hunter, 1982: p. 124). And out of these 200-odd fellows, only a handful, like Boyle, Hooke and Huygens, were interested in optical theories and capable of conversing in intelligent debate about these matters. In reality, Newton was writing to a small, almost private audience. His much later published *Opticks* (1704) would be far more elaborate, with many drawings and diagrams, clearly written for a wider audience.

That Newton considered the fellows of the Royal Society a special category is also evident from the letter he wrote in answer to Oldenburg's request for permission to publish the 'New Theory'. Having expressed his gratitude for his election as a member of the Royal Society, Newton continued:

> I doe not onely esteem it a duty to concurre with them in the promotion of reall knowledg, but a great privelege that instead of exposing discourses to a prejudic't & censorious multitude (by which means many truths have been bafled & lost) I may with freedom apply my self to so judicious & impartiall an Assembly. As to the printing of that letter I am satisfyed in their judgment, or else I should have thought it too straight & narrow for publick view. I designed it onely to those that know how to improve upon hints of things.
>
> (Turnbull, 1959: pp. 108–109)

He more or less confesses that he has deliberately crafted a 'straight and narrow' discourse designed for a select and expert audience. Unfortunately, it seems Newton made a rare miscalculation here, because the Fellows found it impossible to 'improve upon the hints' that he had provided them with, pushing him into tedious debates about the validity of his observations and conclusions. As Iliffe (1995: p. 162) points out, by the early 1680s, when the optical debate finally petered out, Newton had still not managed to convince any of his adversaries about the heterogeneous character of white light.

I think we can safely say that Newton overestimated the philosophical qualities of his fellow members of the Royal Society. Or, more precisely, he did not realize how much his own qualities differed from theirs. As historian Rupert Hall put it: 'Newton's experimental study of refraction and its associated colour-phenomena, already codified in the optical lectures, not to say the precision of his geometrical treatment, was so far beyond anything known to his contemporaries that their criticisms were largely beside the point' (Hall,

1993: p. 65; see also Stuewer (1970) for an analysis of the quality of Newton's experimental observations).

On the other hand, this was a deliberate construction on Newton's part, which somewhat counterbalances the emotional narrative that many Newton biographers seem so fond of. This was not a naïve attempt by a young hothead that failed miserably because he did not know how to deal with scientific criticism. Indeed, as Iliffe (1995: p. 162) writes: 'Newton's reputation was a deliberately cultivated one, a public pose carefully fashioned to suit his own interests.' In similar fashion, I would argue that Newton was deliberate about every word of his 'New Theory'; with every phrase he had a particular purpose in mind. For instance, he makes it clear from the outset that he adheres to a corpuscular theory, something which Hooke immediately denied. Westfall (1962: p. 353) labels Newton's reference to the materiality of light 'an inattentive moment' and points to later statements by Newton where he, according to Westfall, 'refused correctly to mix his experimental demonstrations of the property with discussions of its source'. But to me another reading of this episode seems more convincing. Newton did not mix up experiment and theory; in his frame of reference, the properties of light and its nature were intrinsically entwined and he deliberately pointed out that connection. However, during the ensuing correspondence he understood this was not so clear to others and decided to strip from his theories all such ontological implications. This was a hint that others did not know how to 'improve upon'.

When we study the debates that Newton had with Hooke, Pardies, Huygens, and later on with others like the Jesuits Francis Linus, John Gascoigne and Anthony Lucas, we can detect two sorts of responses by Newton. When his methodology, accuracy or honesty is at stake, we find him replying with vigour. These are sometimes fierce reactions involving strong language. For instance, in his reply to Francis Linus, via Oldenburg, Newton doubted whether Linus had actually tried the first experiment that Newton described in the 'New Theory', involving the oblong figure on the wall:

> I thought an answer in writing would be insignificant because the dispute was not about any ratiocination, but my veracity in relating an experiment, which he denies will succeed as it is described in my printed letters: For this is to be decided not by discourse, but new tryall of the experiment. What it is that imposes upon Mr Line I cannot imagin, but I suspect he has not tryed the experiment since he acquainted himself with my Theory, but depends upon his old notions taken up before he had any hint given to observe the figure of the coloured image.
> (Turnbull, 1959: pp. 356–357)

John Gascoigne, who took over from Linus after the latter's death, insisted that Linus had repeated the experiment but had failed to see any image that matched Newton's description, except, as Linus had written in one of his own letters, when 'the sunne eyther shined through a white cloud, or enlightened some such clouds neere unto it' (Turnbull, 1959: p. 318). Gascoigne claimed that:

> Mr Line ... hath try'd it again and again, and called divers on purpose to see it, nor ever made difficulty to shew it to any one ... [w]e think it probable he hath tried his experiment thrice for Mr Newton's once, and that in a clearer and more uncloudy sky than ordinarily England doth allow. ... So ... unless the diversity of placeing the Prisme, or the bigness of the hole, or some other such circumstance, be cause of difference betwixt them, Mr Newton's experiment will hardly stand.
> (Turnbull, 1959: p. 394)

Mr Newton was obviously not amused by Gascoigne's accusations, directly questioning his philosophical integrity. His indirect reply, to Oldenburg, contains a repetition of his first experiment and his initial reply to Linus, and his willingness to repeat the experiment at a meeting of the Royal Society, which ironically had to be postponed because of a cloudy sky. He adds:

> All I think that they can object to you is that you were at a stand becaus you could not ingage me in the controversy, & to me that I had no mind to be ingag'd: a liberty every body has a right to & may gladly make use of, sometimes at least, & especially if he want leisure or meet with prejudice or groundles insinuations. But I hope to find none of this in Mr Gascoin. The hansome genius of his present Letter makes me hope it for the future.
> (Turnbull, 1959: p. 411; see also Westfall, 1966: pp. 305–306)

I cannot help but smile at the thinly veiled sarcasm in that last line. But then there are also replies of a much more friendly nature, where Newton is tutoring his reader. Anthony Lucas for instance, for all his critique, had one redeeming feature: a measurement. He had tried to recreate Newton's observation of the oblong figure cast at the wall, which according to Newton was five times its width; whereas Lucas found an image that was at most three and a half times as long. Newton wrote:

> Concerning Mr Lucas's other experiments, I am much obliged to him that he would take these things so far into consideration, & be at so much pains for examining them; & I thank him so much the more because he is the first that has sent me an experimental examination of them. By this I may presume he really desires to know what truth there is in these matters.
> (Turnbull, 1960: p. 79)

Although Newton found many faults in Lucas's reasoning and experimental practice, he seemed genuinely grateful that Lucas at least demonstrated he had tried out Newton's experiments and returned quantitative results.

These exchanges between Newton and his readers provide us with valuable insights into Newton's ideas about what natural philosophy should be, how the natural philosopher should conduct his research and how he should communicate his research. It is important to understand that Newton first communicated his 'New Theory' after six or seven years of experimenting and revising his ideas. As his notes and drafts show, he performed a vast number of experiments investigating the nature of light and colours, and was lecturing on these experiments in Cambridge by the end of the 1660s. By 1672 he must have felt certain that his ideas were correct and beyond the status of a mere hypothesis, and that they should be communicated to a wider audience. As regards the fullness of that communication, it seems that Newton had wholly different ideas from Boyle. Instead of providing an exhaustive account of all his experiments, as Boyle did in *Touching Colours*, dealing with all possible and topical theories to provide guidance for his students, Newton is much more restrictive, making it clear from the outset that there is only one possible explanation that satisfies the experimental results.

That word – results – matters here. Boyle gave an entire history of experiments, of trial and error, in such a way that Pyrophilus would not have to repeat Boyle's path but merely observe and learn from it, to take up where Boyle left off. But Newton wants his readers – his students – to actively engage in experiments, following his lead to arrive independently at the same conclusions as he did. They need to repeat the entire process of reaching the optimal experimental set up. He does not provide them with full experimental details, because his readers should figure out these details for themselves. This is Newton's pedagogical strategy. He wants his readers to 'improve upon hints of things', because it is not just results that matter, but the process that leads up to those results. And natural philosophy is not for everyone, only for those who 'really desire to know what truth there is in these matters'. This also means that the natural philosopher should do his own experimental research and not just copy others. Likewise, communication about these experiments should be based upon practice; before criticizing, first try hard to reproduce and understand. But where did this pedagogical attitude come from?

From light to darkness

As we have seen, it was not at all common practice to communicate philosophical discoveries in such a way that the reader had to do all the hard work himself. There was no such precedent within the field of natural philosophy. There was, however, a precedent in another field in which Newton had become very interested by the mid-1660s: alchemy. As his lists of purchases show, he bought all the necessary equipment to perform alchemical experiments, which would keep him occupied for most of his Cambridge life until he moved to London in 1696 (Newton, 1662–1669: ff. 8r-v; see Schilt, 2016: p. 75). Likewise, Newton's library contained scores of alchemy books and tracts, some of them very hard to come by, showing how deeply he was involved in alchemical networks and exchanges (Harrison, 1978; Kassell, 2011). The manuscript evidence shows that he had access to many more of these obscure sources, which he must have borrowed from other alchemists in order to copy them.[11] In fact, Newton's withdrawal from the scene of natural philosophy by the end of the 1670s might well be explained as a deliberate strategy in order to devote all his time to alchemy and to his religious studies, which, as Iliffe (1995: p. 163) notes, were 'neither suited to nor acceptable in a public sphere'. It is this period that Westfall refers to as the 'years of silence' (1980: pp. 335–401).

In past decades a number of scholars, most notably Betty Jo Teeter Dobbs, have tried to understand Newton's alchemical interests as part of a well-structured and coherent programme (Dobbs, 1975; 1991; see also McGuire and Rattansi, 1966), with inseparable connections between his natural philosophy and esoteric studies (cf. Iliffe, 2004). Although today historians no longer think these to be as clear as they seemed to Dobbs, we can draw relations between Newton's alchemy and, for instance, his optics, as William Newman has shown (Newman, 2010; see also Newman, 2002 and Figala, 2002).

One of the key aspects of alchemy that Newton was interested in, was the exact nature of the philosopher's stone and how to create it. The name 'stone' is somewhat misleading here; what alchemists were looking for was a certain pure form of mercury, which was said to be able to transmute metals into other metals, even into gold. Alchemical knowledge was not free; a student of alchemy had to study hard and deep to decipher the obscure language in which alchemical adepts would communicate their findings. In one of Newton's manuscripts, for instance, we find him copying from a tract by a Basilius Valentinus (probably a pseudonym) called *De microcosmodeque magno mundi mysterio, et medicina hominis* (Newton, n.d.a: f.1r). The first lines are:

> The Philosopher speaketh thus.
> Bright glorious king of all this world, o Sun,
> Whose progeny's upholder is the Moon,
> Both whom Priest Mercury does firmly bind,
> Unles Dame Venus favour you do find,
> Who for her spous Heroic Mars hath ta'ne.
> Without her aid what ere you do's in vain.

These were the sort of texts that Newton sought to understand, and he did so methodically and diligently. He studied deeply the alchemical books in his possession, as witnessed by the many dog-ears and annotations they contain. Those dog-ears tell us what passages Newton was particularly interested in. Not surprisingly, many of these passages refer to the pure mercury that alchemists were looking for and how to create it. Others refer to the process of acquiring true alchemical knowledge. They urge the reader to do exactly what Newton did – study hard, study deep, try every experiment, judge for yourself whether the author's conclusion makes sense, and only then comment upon it.[12] Likewise, in his copying from books and manuscripts we find similar examples.

For instance, taking notes from an introduction written by George Starkey to a tract by George Ripley, called *The Twelve Gates*, Newton writes:

> I made not 5 wrong experiments in it before I found the truth & though in some particular turnings of the encheiresis I erred oft yet by those errors I knew my self a Master & 2 1/2 years of a rude vulgar ignorant I became a true Adept.
>
> (Newton, n.d.b: f.3r)[13]

These are not Newton's own words; they come more or less directly from Starkey's introduction. But the fact that Newton wrote them down shows his interest in acquiring and disseminating alchemical knowledge and his preferred method: through practice. In the line preceding the text that Newton copied, the master says to his pupil: 'trust me, for I speak knowingly, the art is both true and easie; yea so easie, that if you did see the experiment, you could not believe it' (Starkey, 1678).

Unfortunately for Newton, his Royal Society 'students' found his experiments far from easy and started providing critical commentaries straightaway. But even if his experiments had been easy, Newton's pedagogical strategy seemed bound for failure. Despite the innovative use of a mathematical discourse, which gradually became accepted methodology, those elements of his 'New Theory' that really mattered to his contemporaries were hidden from view. By applying an obscurantist approach, Newton had written his natural philosophy in the form of an alchemical discourse, proposing a hypothesis that others could not verify. Had not Robert Boyle in his *Sceptical Chymist* strongly argued against such practice? 'For ... 'tis most commonly far easier to frame objections against any propos'd hypothesis, than to propose an Hypothesis not lyable to objections' (Boyle, 1661: p. A6-v). It is no wonder that Newton found himself from the outset besieged by sceptical remarks regarding the validity of his hypothesis and the need for replicating his *experimentum crucis*, or his initial experiment for that matter. Indeed, instead of celebrating the success of his crucial experiment, Newton must have lamented his crucial error in assuming too much from his audience. His subsequent attempts to repair matters with his replies could not outdo the damage done and Newton would never again use the term *experimentum crucis*.

Some concluding remarks

We have witnessed the rollercoaster ride that Newton found himself on within moments of publishing his first scientific paper, the 'New Theory of Light and Colours'. It is an episode traditionally framed in terms of hot-headedness, a clash of characters, and epistemological expectations. All these were definitely present; yet there is more than meets the eye. Newton was not the naïve lad from Lincolnshire who did not know how to write a proper paper. Nor did he carelessly insert a line on the nature of light without overseeing the implications. He clung to a deliberate pedagogical strategy that he derived from his other major passion during those years: alchemy. It was a strategy that put emphasis on the replication of experiments based upon a sufficient number of hints, without providing full disclosure. The private process of unravelling nature's secrets mattered as much as those secrets themselves and Newton likewise modelled his paper along alchemical lines. Unfortunately, he made the tactical error to assume too much at once from his audience; had he led them gently by the hand, this episode might have taken an entirely different turn. As things stood now, Newton decided to withdraw from public communication about most things natural philosophical.

When in 1684 a young Edmund Halley visited Newton with an intriguing question about the motion of falling bodies, nobody could have suspected that within the span of three years Newton would rise like a phoenix from the ashes of natural philosophy, the subject matter that he had sworn to abandon. His *Principia Mathematica* at once made him Europe's leading natural philosopher, eclipsing even Boyle and Huygens. Hooke was nowhere to be seen, nor the French or English Jesuits. As to why Newton returned to the scientific arena is a question that remains unanswered, and is possibly unanswerable. We simply don't know. Newton published his second major work, *Opticks*, only after Hooke's death, at the instigation of some of his friends. Just like his second paper and, not entirely incidentally, his non-published optical lectures, *Opticks* follows a much more classical route of experiments, demonstrations, and conjectures. His audience might not have known how to 'improve upon hints of things', but he did. It seems he learned his lesson well.

Postscript

In a rare letter on alchemical matters, in the middle of the optical controversies, Newton commented upon the behaviour of a particular alchemist in a letter to Oldenburg. He wrote:

> But yet because the way by which Mercury may be so impregnated, has been thought fit to be concealed by others that have known it, & therefore may possibly be an inlet to something more noble, not to be communicated without immense damage to the world, therefore I question not but that the great wisdom of the noble Authour will sway him to high silence till he shall be resolved of what consequence the thing may be either by his own experience, or the judgment of some other that througly understands what he speaks about, that is of a true Hermetic Philosopher, ... there being other things beside the transmutation of metals ... which none but they understand.
>
> (Turnbull, 1960: p. 2)

That particular alchemist is, perhaps surprisingly, Robert Boyle, who was no stranger to alchemy himself and corresponded and collaborated with George

Starkey (Newman and Principe, 2002; Newman, 2004: pp. 271–283; Principe, 2011). The attitude that Newton preaches here, of a 'true Hermetic philosopher', is reminiscent of Newton's own attitude as a true *natural* philosopher, the attitude he had hoped to convey to his audience. Study deep, study diligently, try to thoroughly understand what you speak about. Or else, keep to 'high silence'.

Notes

1. A note on spelling and dates. Original spelling has been kept, with abbreviations like ye, yt, and wch, and alchemical shorthand expanded. Dates in this paper follow the Julian calendar that was in use in England throughout Newton's life. For example, Newton was born on Christmas Day 1642, which according to the Gregorian calendar by then adopted by most of Europe refers to 4 January 1643. With the Julian New Year commencing 25 March, all dates between and including 1 January and 24 March are given with double years, as was common practice in Newton's days. Thus 15 February 1667/68 refers to our year 1668, whereas it was still 1667 in England.
2. The letter that Oldenburg sent in reply is lost, but from these notes and other evidence we know that Oldenburg responded to Newton with kindness, even offering to waive Newton's membership fees. See Turnbull (1959: p. 263) and Hall and Tilling (1977: p. 387).
3. See Shapin and Schaffer (1985) for a full account of Boyle's air pump and the replication and interpretation of experimental results.
4. The quotation is taken from J. C. Scaliger's *Exotericarum Exercitationum* (*Exoteric Exercises*, Frankfurt, 1557), book XV, exercise 325, paragraph 4 (p. 1031) which in the original reads 'Latet enim natura haec, sicut & aliarum rerum species in profundissima caligine inscitiae humanae', meaning: 'For the nature of these things, just like of other sorts of things, is hidden in the deepest darkness of human ignorance.'
5. In an off-hand remark, Westfall labels the first part of Boyle's *Touching Colors* 'a fascinating study in Baconian futility. Only when its loving accumulation of facts was replaced by a systematic plan of experimentation inspired by a new idea was the riddle of color solved' (Westfall, 1962: p. 346).
6. See Westfall (1980: pp. 156–158) and Hall (1993: p. 40) for a detailed analysis of the chronology of Newton's early optical research. See also Guerlac (1983).
7. Christianson (1984: pp. 45–46) claims that 'Pardies had failed in several attempts to duplicate the *experimentum crucis* and concluded that Newton's results lacked scientific merit', but I cannot find any reference to Pardies's attempts, nor do I agree with Christianson's interpretation.
8. Christianson describes this letter as 'an even more chilling blast ... scolding the Jesuit for his sloppy work and for "philosophizing" instead of "establishing properties by experiments."' (1984, p. 46). Again, I think this is exaggerating.
9. The interpretation most likely stems from Manuel (1968: pp. 145–146). See also Christianson (1984: pp. 195–196). Richard Westfall (1980: p. 274 n.106) disagrees: 'As Newton said once before in regard to Hooke, he avoided oblique thrusts. When he attacked, he lowered his head and charged.'
10. For Newton's optical lectures see Shapiro (1984) and Hall (1993: pp. 45–59). See Hall (1993: pp. 33–45) and Schilt (2016: pp. 55–62) for an account of Newton's early optical research. See also Westfall (1962; 1966), Guerlac (1983) and Shapiro (2002).
11. For Newton's alchemical manuscripts, see *The Chymistry of Isaac Newton*, www.chymistry.org, a project headed by William R. Newman devoted solely to Newton

alchemical studies. The project site contains images and transcriptions of the majority of Newton's alchemical manuscripts, and a wealth of other resources.
12. See Schilt (2016: pp. 57–58, 71–76) for some examples from Newton's own books.
13. Newton took these notes from a book that he owned, Eirenaeus Philalethes [i.e. G. Starkey]'s *Ripley Reviv'd: Or, an Exposition upon Sir George Ripley's Hermetico-Poetical Works. Containing the Plainest and Most Excellent Discoveries of the Most Hidden Secrets of the Ancient Philosophers, That Were Ever Yet Published* (London, 1678); its present location is unknown (Harrison, 1978: p. 243). See Newman and Principe (2002) for the identification with George Starkey; see also Newman (1994).

References

Boyle, R., 1661. *The Sceptical Chymist or Chymico-Physical Doubts & Paradoxes*. London.

Boyle, R., 1664. *Experiments and Considerations Touching Colours*. London. Available at www.gutenberg.org/files/14504/14504-h/14504-h.htm. Accessed 24 March 2015.

Boyle, R., 1674. *The Excellence of Theology*. London. Available at: tp://quod.lib.umich.edu/e/eebo/A28966.0001.001?rgn=main;view=fulltext. Accessed 2 December 2013.

Christianson, G.E., 1984. *In the Presence of the Creator: Isaac Newton and His Times*. New York: The Free Press.

Dear, P., 1985. Totius in verba: rhetoric and authority in the early Royal Society. *Isis*, 76(2), pp. 141–161.

Dobbs, B.J., 1975. *The Foundations of Newton's Alchemy or 'The Hunting of the Greene Lyon'*. Cambridge: Cambridge University Press.

Dobbs, B.J., 1991. *The Janus Faces of Genius: The Role of Alchemy in Newton's Thought*. Cambridge: Cambridge University Press.

Fara, P., 2015. Newton shows the light: a commentary on Newton, 'A letter … containing his new theory about light and colours …'. *Philosophical Transactions of the Royal Society A*, 373, doi: 10.1098/rsta.2014.0213.

Figala, K., 2002. Newton's alchemy. In: Cohen, I.B. and Smith, G.E. (eds), *The Cambridge Companion to Newton*. Cambridge: Cambridge University Press, pp. 370–386.

Golinski, J., 1998. *Making Natural Knowledge: Constructivism and the History of Science*. Cambridge: Cambridge University Press.

Guerlac, H., 1983. Can we date Newton's early optical experiments? *Isis*, 74(1), pp. 74–80.

Hall, A.R., 1993. *All Was Light: An Introduction to Newton's Opticks*. Oxford: Oxford University Press.

Hall, A.R. and Simpson, A.D.C., 1996. An account of the Royal Society's Newton telescope. *Notes and Records of the Royal Society of London*, 50(1), pp. 1–11.

Hall, A.R. and Tilling, L. (eds), 1977. *The Correspondence of Isaac Newton, Volume VII: 1718–1727*. Cambridge: Cambridge University Press.

Harrison, J., 1978. *The Library of Isaac Newton*. Cambridge: Cambridge University Press.

Hunter, M., 1982. *The Royal Society and Its Fellows 1660–1700: The Morphology of an Early Scientific Institution*. 2nd edn. Chalfont St Giles: British Society for the History of Science.

Iliffe, R., 1995. 'Is he like other men?' The meaning of the *Principia Mathematica*, and the author as idol. In: Maclean, G. (ed.), *Culture and Society in the Stuart Restoration*. Cambridge: Cambridge University Press, pp. 159–176.

Iliffe, R., 2004. Abstract considerations: disciplines and the incoherence of Newton's natural philosophy. *Studies in the History and Philosophy of Science*, 35, pp. 427–454.

Kassell, L., 2011. Secrets revealed: alchemical books in early-modern England. *History of Science*, 49, pp. 61–87.

Manuel, F.E., 1968. *A Portrait of Isaac Newton*. Cambridge, MA: Harvard University Press.

McGuire, J.E. and Rattansi, P.M., 1966. Newton and the 'Pipes of Pan'. *Notes and Records of the Royal Society of London*, 21(2), pp. 108–143.

Newman, W.R., 1994. *Gehennical Fire. The Lives of George Starkey, and American Alchemist in the Scientific Revolution*. Cambridge, MA: Harvard University Press.

Newman, W.R., 2002. The background to Newton's Chymistry. In: Cohen, I.B. and Smith, G.E. (eds), *The Cambridge Companion to Newton*. Cambridge: Cambridge University Press, pp. 358–369.

Newman, W.R., 2004. *Promethean Ambitions: Alchemy and the Quest to Perfect Nature*. Chicago and London: University of Chicago Press.

Newman, W.R., 2010. Newton's early optical theory and its debt to chymistry. In: Jacquart, D. and Hochmann, M. (eds), *Lumière et vision dans les sciences et dans les arts, de l'antiquité du XVIIe siècle*. Geneva: Librairie Droz, pp. 283–307.

Newman, W.R. and Principe, L.M., 2002. *Alchemy Tried in the Fire: Starkey, Boyle, and the Fate of Helmontian Chymistry*. Chicago: University of Chicago Press.

Newton, I., 1662–1669. *Fitzwilliam Notebook*. Cambridge: Fitzwilliam Museum.

Newton, I., 1672. New theory of light and colours. *Philosophical Transactions of the Royal Society*, 80, pp. 3075–3087.

Newton, I., 1704. *Opticks*. London.

Newton, I., 1713. *Philosophiae Naturalis Principia Mathematica*. 2nd edn. London.

Newton, I., n.d.a. *Verses at the End of Basil Valentine*. Keynes Ms. 63. King's College Library, Cambridge.

Newton, I., n.d.b. *On Ripleys Preface to His Gates*. Keynes Ms. 52. King's College Library, Cambridge.

Philips, E.C., 1913. Gaston-Ignace Pardies. In: Herbermann, C.G., Shahan, T.J., Pallen, C.B., Pace, E.A., MacErlean, A.A. and Wynne, J.J. (eds), *The Catholic Encyclopedia*. Vol. 11. New York: Robert Appleton Company/The Encyclopedia Press.

Principe, L.M., 2011. In retrospect: The Sceptical Chymist. *Nature*, 369, pp. 30–31.

Royal Society, 1662. *The Charters of the Royal Society, First Charter* (translation). Available at: https://royalsociety.org/~/media/Royal_Society_Content/about-us/history/Charter1_English.pdf. Accessed 4 April 2015.

Schaffer, S., 1989. Glass works: Newton's prisms and the uses of experiment. In: Gooding, D., Pinch, T. and Schaffer, S. (eds), *The Uses of Experiment*. Cambridge: Cambridge University Press, pp. 67–104.

Schilt, C.J., 2016. 'To improve upon hints of things'. Illustrating Isaac Newton. *Nuncius*, 31(1), pp. 50–77.

Shapin, S., 1991. 'The mind is its own place': science and solitude in seventeenth-century England. *Science in Context*, 4, pp. 191–218.

Shapin, S., and Schaffer, S., 1985. *Leviathan and the Air-Pump: Hobbes, Boyle, and the Experimental Life*. Princeton, NJ and Oxford: Princeton University Press.

Shapiro, A.E., 1984. *The Optical Papers of Isaac Newton: Volume 1. The Optical Lectures 1670–1672*. Cambridge: Cambridge University Press.

Shapiro, A.E., 2002. Newton's optics and atomism. In: Cohen, I.B. and Smith, G.E. (eds), *The Cambridge Companion to Newton*. Cambridge: Cambridge University Press, pp. 227–255.

Starkey, G. (as Eirenæus Philalethes), 1678. *Ripley Reviv'd: Or, an Exposition upon Sir George Ripley's Hermetico-Poetical Works. Containing the Plainest and Most Excellent Discoveries of the Most Hidden Secrets of the Ancient Philosophers, That Were Ever Yet Published*. London. Available at: www.rexresearch.com/riplrevv/rip lyrevv.htm#preface. Accessed 11 April 2015.

Stuewer, R.H., 1970. A critical analysis of Newton's work on diffraction. *Isis*, 61(2), pp. 188–205.

Turnbull, H.W. (ed.), 1959. *The Correspondence of Isaac Newton, Volume I: 1661–1675*. Cambridge: Cambridge University Press.

Turnbull, H.W. (ed.), 1960. *The Correspondence of Isaac Newton, Volume II: 1676–1687*. Cambridge: Cambridge University Press.

Westfall, R.S., 1962. The development of Newton's theory of color. *Isis*, 53(3), pp. 339–358.

Westfall, R.S., 1963. Newton's reply to Hooke and the theory of colours. *Isis*, 54(1), pp. 82–96.

Westfall, R.S., 1966. Newton defends his first publication: the Newton–Lucas correspondence. *Isis*, 57(3), pp. 299–314.

Westfall, R.S., 1980. *Never at Rest: A Biography of Isaac Newton*. Cambridge: Cambridge University Press.

4 Engineers at the patient's bedside

The case of silence in inter-institutional educational innovation

Nick W. Verouden, Maarten C. A. van der Sanden and Noelle M. N. C. Aarts

Introduction

Innovation in science and technology is increasingly linked with interdisciplinarity. Encouraging this trend depends in part on cutting-edge educational programmes that revise, reinvent and redesign curricula as interdisciplinary vehicles, establishing and re-establishing relations between traditional fields and areas of expertise (Stone et al., 1999; Casey, 1994). Such programmes are valuable because they can overcome 'silo' mentalities and equip prospective students with the skills and knowledge necessary for understanding and solving complex societal problems (Stone et al.,1999; McFadden et al., 2010).

Although these programmes are very promising, their development and implementation also brings challenges. The literature on curriculum development shows that many programmes have struggled to achieve true integration (McFadden et al., 2010; Stone et al., 1999). Dam-Mieras et al. (2008), in their study of an international master's programme in sustainable development and management developed collaboratively by nine universities, observed that universities have their own experts and own programmes and that the 'not invented here' argument influences how details about new programme are discussed. Focussing on innovative online instruction courses, Xu and Morris (2007) found that the absence of group cohesiveness between faculty and project coordinators can hinder the collaborative course development process and affect the quality of the end product. Stone et al. (1999) emphasize that faculty members and administrators work at cross-purposes and view each other's initiatives with suspicion. Given the importance that scientists, academic institutions and policy makers ascribe to innovation, along with their assumption that such innovation is a sure result of interdisciplinarity, it is essential to gain a better understanding of how curriculum development in academic education actually works.

For this chapter, we consider how processes of connecting and interrelating could add to our understanding of the problems and dilemmas that arise in developing and implementing such programmes. Scholars of innovation, in science and technology and beyond, have explained that innovation is not some abstract algorithm: it relies on interaction and collaboration between multiple actors with different expertises, visions, priorities and investment (Van Bommel et al., 2011; Leeuwis and Aarts, 2011; Akrich et al., 2002; Fonseca, 2002). This process of interacting is very difficult, however, and creates many tensions. This is revealed by studies that show the lurking problems of connecting previously unconnected people around new ideas and technologies. These studies show how innovation processes become defined by competition for scarce resources, protracted negotiations over priorities and interests, and dynamics of inclusion and exclusion (Leeuwis andAarts, 2011; Pretty, 1995; Van Bommel et al., 2011). Fonseca (2002) hence explains that innovation always creates a paradoxical situation, in which organizations, in their search to accelerate change and adapt to and find solutions for external challenges and demands, unavoidably create new and unpredictable interactional patterns.

Given that interacting is a complicated matter in innovation processes, a key question within the management of innovation literature is how we can account for the way relevant actors connect, or fail to connect (Akrich et al., 2002). In this respect, verbal communication is often cited as an essential mechanism for effectively connecting important actors and social groups around innovative ideas, products, or technologies (Van Bommel et al., 2011). In turn, the markers of effective verbal communication as a frame for innovation are seen to be openness, dialogue, and the ability to cooperate and be reflective on one's thoughts and actions (Stilgoe et al., 2013). Thorp and Goldstein (2010), writing about university innovation, describe conversations as the fertile ground from which innovation grows and urge us to make time and space for those conversations. Dialogue and openness are seen as indicators of the quality of interaction, and process transparency as a decisive component of academic innovation. By being open or transparent in discussing issues and problems, actors build confidence that negotiation is 'real' and not a cover-up for private backroom deals (de Bruijn and ten Heuvelhof, 2008).

Although there is a wealth of research on communication for innovation, most scholarly work focuses on what is exchanged verbally, on how actors collate all the relevant evidence, put it on the table and discuss it openly. As of yet, silence is absent from these studies of communication for innovation. Building on recent organizational and strategy scholarship, in which silence is approached as an intricate concept with powerful functions and meanings in social interaction (Van Assche and Costaglioli, 2012; Carter et al., 2008; Henriksen and Dayton, 2006; Panteli and Fineman, 2005; Tucker and Edmondson, 2003; Jaworski, 2005; Morrison and Milliken, 2000), we suggest that silence merits much more attention in analyses of academic innovation. This chapter therefore explores the role of moments of silence during interactions within networks developing and implementing educational innovation.

The structure of this chapter is as follows. We start by looking at the literature on dynamic innovation networks and communication and complement these insights with scholarship on silence within organization studies. After briefly introducing our approach, we present the findings of a study of an inter-institutional and interdisciplinary joint bachelor's programme that was implemented at the interface of health and technology. The purpose of the study was to better understand the significance of moments of silence in developing and implementing this programme. We end with the implications of our findings for steering in the context of interdisciplinary innovation.

A dynamic and relational perspective on innovation

Innovation processes have been much studied over the years. The traditional approach, dating from the 1950s, conceptualizes the application of new scientific and technological knowledge as an instrumental and linear process. In this view, specific strategies are followed, introducing planned innovations through a sequential process of stages or phases, all of them designed as part of a structured approach (Rogers, 1962). Each of these stages must be completed, as part of a ground plan laid down beforehand. Decision-making is a crucial aspect of this process and requires analysis, clear goals and good information (de Bruijn and ten Heuvelhof, 2008).

While this linear view of innovation remains dominant (Rogers, 1962), thinking about the subject has changed considerably over the past decades (Leeuwis and Aarts, 2011). Early models are now criticized for failing to recognize the full complexity of the background underpinning innovation and for overstating its manageable character (Fonseca, 2002). In contrast to the prevalent idea that innovation stems from the behaviour of individual geniuses and heroic entrepreneurs, who supposedly beat their own path as they develop and design their final products and deliver them to the public (Nicolini, 2010), we now see innovation as behaviour that is rooted in its social context and that must be explained as such. The direction and pace of innovation, according to this contemporary model, is determined by the continuous interaction and negotiation of multiple actors positioned within networks and inter-organizational projects and influenced by continuous interference from specific policies and external circumstances (Fonseca, 2002; Gladwell, 2000; Leeuwis and Aarts, 2011).

Within innovation research there is a strong interest in how such relationships come into being in multi-actor networks. From an actor-network perspective, Akrich et al. (2002) picture this process as including the constant recruitment and involvement of different kinds of actors who are potentially resistant to the introduction of innovation in society and who often fight tooth and nail to impose their own views on each other (Czarniawska and Sevón 1996; Akrich et al., 2002; Nicolini, 2010). As they write: 'An innovation in the making reveals the multiplicity of heterogeneous and often confused decisions made by a large number of different and often conflicting groups, decisions which one is unable to decide a priori as to whether they will be crucial or not' (Akrich et al., 2002: p. 191). In this view, innovations are introduced by attracting the interest of an increasing number of allies, gaining their support and convincing other important actors and parties of their relevance.

For example, Nicolini's (2010: p. 1) study of cardiac telemedicine, an innovative technology that provides health care at a distance shows how the acceptance of this technology involved establishing 'a space within the existing texture of medical practices by enrolling in successive waves a range of allies and support'. In order for the technology to become 'an appropriate and legitimate way of delivering care' (ibid., 2010: p. 13), multiple actors (doctors, nurses, managers and health officials) first had to see its value and support it. For this to happen, it constantly had to be 'translated and metamorphosed into something else' in order to respond to conflicting demands and expectations. As this shows, making the broader public accept an innovation does not resemble a simple trajectory – 'the picture of a curve which ends' – but is better described as something far more complex, a 'rhizome that extends itself by growing new branches and rooting itself in new ground' (Nicolini, 2010: p. 12).

As this illustrates, the process of innovating is not the work of single actors and does not follow blueprints that are laid out beforehand (Van Bommel et al., 2011). While individual contributions may be an important driver in the process, innovations become meaningful through linking and connecting the competing interests, wishes and dreams of multiple actors and interests groups within a network, a process filled with uncertainty and unpredictability.

Communication and conversations for innovation

Thus far, we have shown that understanding innovation requires us to focus on the complex interweaving of separate contexts of visions, perspectives and interests into a network. This does not necessarily mean that actors depend on a common practice or vision, but there does need to be some kind of 'overlapping consensus' in which parties with very different views and stakes agree that there is enough unity for them to accept definitions and ideas (Nussbaum, 2006). While we argue that silence is of crucial importance in creating room for manoeuvre, studies of communication in innovation usually highlight the significance of verbal communication in bringing about intended changes.

Leeuwis and Aarts (2011) explain that traditional models that conceptualize communication as an intermediary function, and prioritize the 'transfer' of information between individual senders and receivers (e.g. Rogers, 1962), do not capture the complex communicative interactions taking place between multiple actors in multiple sites. From an interactional framing perspective, 'meaningful innovation is dependent on changes in discourses, representations and storylines that are mobilized by interacting social actors' (Leeuwis and Aarts, 2011: p. 27). By framing vocabularies and arguments, participants seek to influence ongoing definitions of issues, identities and processes (Dewulf et. al., 2009). They do this by proposing – or challenging – specific definitions and interpretations in order for others to start seeing the value of innovation from their perspective. As Gladwell (2000) has shown, when the innovation becomes the object of conversation, ideas get confirmed and this leads to their further introduction and distribution throughout the network. As a result, innovative ideas, be it messages, food, a movie, or product, become graspable, interesting and even memorable – or, as Gladwell calls it, 'sticky'.

At the most basic level, this framing process occurs in face-to-face interaction between actors, but these actors are also included in broader conversational networks that span multiple interactional settings. The ideas, plans or strategies that people talk about in one setting are interconnected with what they talk about in another. According to Krippendorff (2008: p. 156), conversations are therefore the outgrowth of preceding conversations and fuel subsequent conversations. As Krippendorff writes, 'conversations leave behind their own histories of what happened, available to all who contributed to them, which serve as the expanding ground for future conversation.' To ensure that the right actors are clustering around an innovation, it becomes necessary to influence what interested parties, distributed across various sites and moments, say about innovation. Managing innovation becomes a matter of deliberately shifting conversations in the network and providing a new conversational context in which to interact (Ford, 1999).

The role of silence in complex relational processes

In broad terms, then, studies of innovation see innovation as the building of coalitions through the framing of messages and the creation of conversational contexts and networks that are likely to be of benefit to it (Leeuwis and Aarts, 2011; Krippendorff, 2008; Fonseca, 2002; Ford, 1999). Yet, despite the importance of verbal exchanges, it is widely recognized that people's behaviour is not only shaped by what they talk about, but also by that about which they stay silent (Morrison and Milliken, 2000; Panteli and Fineman, 2005; Carter et al., 2008). Although, to our knowledge, there are no studies focusing on silence in educational innovation processes, scholars have examined its role in numerous organizational and strategic contexts, showing how silence facilitates as well as complicates connections between actors. In this section, we review some of these studies and indicate their relevance as an additional perspective for understanding educational innovation.

Studies of ritual and performance in social interaction examine silence to reveal people's strategic attempts at realizing their goals and managing collaborative working relationships with co-workers. In their interactions with others, people speak about certain subjects, but not others, based on judgements and decisions related to regulating their appearance. From a symbolic-interactionism perspective, for instance, Goffman (1959) has shown that silence is used by participants to conceal their own feelings and perspectives in order to maintain a veneer of consensus. This makes us attentive to the discrepancy between people's behaviour in public, where many things are expected of them, and 'backstage', where they can speak more freely. Indeed, a roaring silence in the meeting rooms can be juxtaposed with people talking about things at lunch. As Goffman (1959: p. 119) himself writes, 'employees will often grimace at their boss, or gesticulate a silent curse, performing these acts of contempt or insubordination at an angle such that those to whom these acts are directed cannot see them.' This line of inquiry shows that silence has a performative character; that is, it can be deployed as a strategy for not disclosing ideas and feelings with the aim of influencing and sustaining interactional processes.

In practice, there is usually not a sharp distinction between what is concealed and what is disclosed. Silence and talk are inextricably tied to each other. Studies of secrecy insightfully demonstrate how the dynamic interrelationship between keeping, telling and revealing secrets is essential for understanding organizational functioning and performance (Vermeir and Margócsy, 2012; Rappert, 2010; Bok, 1989). Although closed procedures are conventionally viewed with scepticism, Bok (1989, p. 18) notes that secrecy and silence are as 'indispensable to humans as fire, and as greatly feared. Both enhance and protect life, yet both can stifle, lay waste, spread out of all control.' In the light of the many decisions that are necessary at the various operating levels of organizations and institutions, it is essential to balance openness and secrecy, sometimes articulating issues very publicly, sometimes managing them discreetly.

For example, in their study of the Dutch Health Council, Bijker et al. (2009) describe how the preparation and communication of scientific advice and policy is marked by a contradiction between the obligation of transparency and public consultation, on the one hand, and the need to resist the power of lobby groups, on the other. The authors illustrate how the Dutch Health Council deals with this paradox by constantly shifting between procedural forms of transparency (information that is openly communicated, such as summaries of minutes intended for public display) and substantive forms of transparency (scientific information that is only discussed backstage, such as council meetings that are closed to outsiders). By constantly switching between openness and closure, the Health Council is able to control how views are ventilated and discussed, for example making it possible for the Council's members to develop ideas without the interference of dissenting voices and lobby groups. Seen in this way, temporarily excluding certain voices is an important condition for maintaining the democratic function of the Health Council. This research thus illustrates that what is talked about and kept silent constantly interact with and influence each other.

Scholars of strategy have furthermore argued that this interplay between silence and talk is also continuously changing. These scholars have tied silence to the unpredictable and unfolding nature of the strategic process. According to Carter et al. (2008), strategy occurs in an unfixed and highly dynamic interactional context in which numerous competing voices seek to define their strategies. They argue that silence can be used to examine the absent, but nevertheless crucial, aspects of strategy making. Simply put, for every decision, agenda point and actor present at a meeting, there is an issue that remains unspoken and undecided, something that is omitted from the agenda, and someone who is not present or has not been invited. Because competing voices always seek to define their strategy, what is silenced at one moment can still become potentially relevant at other moments throughout the process, undercutting the power and authority of official discourses and strategies. Van Assche and Costaglioli (2012) hence propose seeing the strategic planning processes as a 'series of sites of silence' in which plans are constantly being reinterpreted and redirected. Each interactional site will highlight new combinations between silence and talk, and has the potential to reshape concepts and plans (ibid., 2012: p. 42). In drawing attention to the unfolding nature of silence and talk, strategy scholars thus make us attentive of the fact that silence is part of a layered and unfolding series of interactional patterns that are always changing.

The latter point is also addressed by the sociologist Niklas Luhmann (1994), who ties this emerging dimension of silence to the self-organizing nature of complex systems. Drawing on the theory of autopoiesis, as developed by Humberto Maturana and Francisco Varela (1988), Luhmann argues that social systems, like societies and organizations, consist of in-principle closed subsystems that try to survive and while doing so disturb each other. Systems generally constitute and reinforce their own boundaries through communication, but varieties of speech also actively interact with silence. When it is an act of communication, silence has the capacity to connect: 'Society can also include silence within communication – for example, in the sense of attentive silence, in the sense of an eloquent silence, or in the sense of "*qui tacet consentire videtur*" [one who is silent appears knowing]' (Luhmann, 1994: p. 34). However, Luhmann also refers to silence in a second sense, as an expression of the impossibility of communication. This silence causes threatening or chaotic elements from the wider context to remain unspoken and thus marginal, confirming and reproducing discourses in ongoing chains of interaction (Ford, 1999). In Luhmann's words: 'Every system coproduces that which, as environment, does not enter into the system, and this may then be called (!) "silence" – though silence in a second sense: silence without the ability to connect' (Luhmann, 1994: p. 34). Silence thus has a structuring function; that is, while it helps to protect the system from chaos, it inevitably draws limits to what can be communicated, causing interdependencies between different parts of the network that cannot connect.

Studies such as these show that silence is not the absence of meaning and intention (Ephratt, 2008), but a performative action through which actors strategically seek to define the course of the interaction process. In order to grasp these dimensions, these studies emphasize that we must not only pay close attention to what is said and what is not, but also how these categories are dynamically constitutive of each other and continuously change across different actors and networks. Finally, these studies emphasize that silence does not only forge productive exchanges and interactions between people, but also unintentionally constrains these interactions and thus the way parts of the network connect with each other and their environment (Luhmann, 1994).

Although the literature reviewed here provides important insights into the role of silence in various relational and interactional processes, the role played by silence in the specific context of educational innovation remains unclear. From our theoretical considerations, we assume that silence also applies here. As previously mentioned, innovation processes by their very nature require the building of mutual interdependencies, but at the same time consist of opposing and shifting positions and movements. We suggest that exploring educational innovation through the lens of silence allows us to see more clearly how actors are trying to promote and achieve intended educational change (Ford, 1999), but also how the numerous interactions this requires can unintentionally complicate these efforts. As such, a focus on silence can provide an additional approach for attending to Nicolini's (2010) call to make visible the hard work of connecting required in the innovation process. In order to explore these issues, our study has been guided by the following questions: which moments of silence can be identified in complex multi-actor settings in the educational innovation process, and what are the unintended consequences of these silences for developing and implementing innovation across interactional contexts?

A case study of innovation in education

To generate more insights into the role of moments of silence in the course of the interaction process through which educational innovation is developed and implemented, we used an interpretive case study approach (Flyvbjerg, 2006). Interpretive case studies are particularly useful for capturing the ambiguities and subjective experience of social phenomena and are therefore appropriate to the study of silence, which is difficult to capture 'objectively'. The case study presented here examines an interdisciplinary bachelor's programme that was developed and implemented at the interface of technology and health. The programme was a joint initiative between a university of technology and two academic medical centres. Its primary aim was to educate professionals in both medicine and technology in order that they could apply this knowledge in healthcare settings.

In today's shifting medical landscape, healthcare problems increasingly require the application of medical tools and instruments for prevention, diagnosis and treatment. A medical specialist is increasingly expected to operate and interpret complex technology such as MRI scanning, echography, surgery and robot technologies. A central feature of this arena is the way it combines the time-honoured imperative of diagnosing and treating illness with the ever-greater impact of complex technologies. Responding to the growing need for such professionals, the joint programme set out to educate clinical professionals with knowledge of technology. These professionals could function as part of an interdisciplinary medical treatment team and independently carry out certain medical practices and diagnoses. For example, such a person might assist the preparation for surgery by simulating the operation with biomechanical computer models or by supporting surgeons with operating robots, but also by carrying out simple diagnostic procedures at the patient's bedside.

In addition to its educational aim, the programme was developed in the light of wider strategic ambitions of the three academic institutions to work together and form an alliance in numerous areas of research and education. Hence, the programme also served as a tangible and high profile example of this coalition.

Given the main aim to develop this educationally ambitious and institutionally prestigious programme as a collaborative project, we considered it offered an interesting opportunity to examine the role of silence in the course of developing and implementing educational innovation. For the programme to become a reality, many complex interdependencies and relational processes had to be formed, which created ample opportunities for silence to become significant. First, a strategic internal network had to be established between the three academic intuitions. However, three-way institutional partnerships are not yet very common in higher education. They create many internal, administrative challenges in terms of facilitating institutional and faculty coordination; for instance, requiring additional support from senior administrators and faculty and staff development (Holley, 2009). Institutions may also have quite different missions and governance policies and cultures, which can result in an ever-growing complexity in the decision-making procedures needed to develop shared educational needs and benefits, budgets, joint quality standards, identification of support, and so on (Holley, 2009). While actors were clear in wanting to cooperate, the everyday reality required different kinds of professionals to engage in a daunting and complex collaborative process.

To build a successful interdisciplinary programme, it was also necessary to activate stakeholders and other interest groups and shape a climate favourable to its reception (Akrich et al., 2002). These wider networks differ from internal networks in that actors are not necessarily aware of one another. They do not engage in collective action and do not necessarily have a shared purpose (Knight, 2002). In our case, this network consisted of the wider scientific community, policy actors and medical interest groups, who did not naturally accept the new programme. Activating this network was especially complex because the programme wanted to include its trained professionals in the BIG-register. This register organizes qualifications and entitlement to carry out specific medical treatments. The plan to extend the register made the programme controversial because it anticipated changes to the existing medical landscape and tinkered with responsibilities traditionally reserved for medical professionals such as doctors, physiotherapists and nurses. Previous studies, such as those examining career structures for radiographers (Coleman et al., 2014), show that the acceptance of new professions is often met with resistance from established groups and professions, who fear for their career opportunities and their professional identities. Creating an external environment favourable to the programme thus involved dealing with these established groups and tied interests.

Implementing the new educational programme also required the shaping of relations at the work-floor level. The programme aimed to develop interdisciplinary modules (combining knowledge from diverse medical and technical fields such as anatomy, physiology, pathology, mechanical engineering, electrotechnology and natural sciences) that encompassed multiple teachers across faculty and institutional boundaries, all with their own educational philosophies and perspectives on teaching. To build a curriculum, broader internal support from departmental heads was needed within and beyond the traditional biomedical disciplines and professions, as well as teachers able to develop and teach the actual courses.

From the above, it is clear that developing and implementing the joint programme constituted a significant challenge. It inherently implied 'a re-configuration of relationships within and between networks, and possibly the formation of new networks and/or the demise of existing ones' (Leeuwis and Aarts, 2011: p. 30). To illustrate the role silence played in these processes, we draw on thirty-four in-depth interviews conducted in 2013 with a diverse range of participants, including senior professors, deans, medical specialists, researchers and support staff members. The interviews examined the institutional and wider political context in which the programme was developed, the complex processes of internal decision-making between the three institutions and the realization of the programme at the work-floor level.

In the next section, we discuss silences in the three contexts addressed above. We first describe how silence was used to create a favourable external climate for innovation and then discuss its role in developing the programme within a small network of actors from the three participating academic institutions. We conclude by discussing the implication of silence for implementing the programme at the work-floor level. In our specific focus on silence, we recognize that we have left out other issues and information that were also important to the process (Flyvbjerg, 2006). Such choices notwithstanding, we consider the instances of silence presented here as indicative of the dilemmas and tensions that arose in this educational innovation process.

Navigating a political minefield

As is often the case with very innovative academic projects, the collaboration did not emerge from the need for funding, but rather because some researchers saw it as an excellent opportunity to launch a new project that could lead to the development of a bold and imaginative idea (Shrum et al., 2007). Highly innovative projects are also typically too large and complex for a single team of academics and rely heavily on interested researchers to spread the story (Shrum et al., 2007). In this case, this was accomplished through a small committee of scientists and medical specialists from the three participating academic institutions. With their shared interest and belief in the value of introducing technically skilled professionals into healthcare practice, the committee members saw the programme's societal relevance and believed it filled an important niche in the medical field. As one of them put it:

> We believe that health care will go through immense changes because of the implementation of technology and that we have to be prepared for these developments. If you start in six years it is already a fact that you are lagging behind.

The first group of professional colleagues involved were already supportive of the programme and at this stage there were few problems. The real challenge came when the idea had to be brought to a wider range of stakeholders – actors potentially sceptical of, or resistant to, the intended change. Several informants emphasize that the programme had to be developed in a very political and competitive environment in which different professional interests were at stake:

> It is a bit of a political minefield out there, because the medical faculties do not recognize this person working in clinical practice. From their perspective, they are a half doctor to whom you really cannot assign authority.

> They are nibbling on vested interests and the finances that go with them.

> The proposed model squeezes out certain groups: one comes; the other has to go.

In addition to being politically sensitive, interviewees mentioned that academic interests also played a role in this process:

> Within the faculty there is also resistance. They always tell you that they already have so many problems. Why should we give money to students who are not from this faculty? Why should the faculty be coordinator of a programme that involves medical centres?

While it is wise from a strategic point of view to communicate extensively with external parties who are not yet part of the network, getting them activated and involved, the intrinsically political and competitive context in which the programme had to be developed at times required a highly diplomatic approach. This meant that from time to time its promoters had to be discreet about developments or very reticent in providing detail. One interviewee explained that, when plans still need to be developed and are not yet fully supported, but the matter is such that it cannot be ignored for long, it is better to first let ideas and visions crystallize before engaging in difficult conversation with others:

> In the beginning you have to arrange so many things and it is very difficult to get everyone on board. [...] If you do not have an idea or vision and

everyone is already informed, the process can degenerate into nothing or will be bogged down. Moving on and involving everyone is a delicate decision.

Other interviewees also noted that medical interest groups known to be critical of the programme's inclusion into the BIG-register were approached very carefully. One interviewee explained how carefully opening up conversations with vested medical interest groups by simply sending them a note, ran into a concrete wall:

> I sent the professional association a short note in which I asked if they could let me know if they considered the bachelor's a good development. From the seven professional associations that were notified, I received a response that they could not say anything about it. [...] There seem to be some signs of cold feet.

While parties often make their resistance heard, they are not necessarily willing to engage in dialogue, and instead protect and guard their boundaries and voice their resistance behind closed doors. This may make it undesirable to start up the conversation with them again.

In addition, plans have to be communicated carefully to academic colleagues who are not directly involved in the process but still need to be informed, such as those participating in joint interdisciplinary clusters or platforms. A support staff member told us about working in the medical cluster in which the participants developing the programme were involved:

> In my perception there is a lot of politics that causes people to restrict communication or not communicate certain issues at all. I have experienced this first hand, for instance when an initiative starts and you ask for input and do not receive any feedback. Meanwhile, colleagues are passing you by left and right and trying to pull work towards themselves. This makes people protect information.

Information is thus treated carefully in exchanges with others. In addition, not much information was disclosed about the ongoing developments at the institutional levels either. While an informant from the technical university explained that he informed colleagues by giving a presentation about it, and that the programme was announced in the university's weekly bulletin, the medical institutions were less eager to be open about the programme, a choice that was not always clearly understood by informants from the technical university:

> They could not go public before everything was official. [...] That process remains very unclear to me. Everybody said that it was a good idea, but then everyone has to have his say in the issue, and this and that. Those in

charge are clearly not in a position to pick it up simply because it is a good idea.

From the perspective of the medical interviewees, announcing the programme was a question of waiting for the right moment to provide openness about developments. This decision was entangled with issues like the recognition of politically sensitive topics, the complexity of the policy environment, and the institutional cultures determining communication and sharing and reputational damage.

To sum up, while the programme was discussed and negotiated openly with many parties in the early stages of the process, not everything was disclosed to everyone at all times. The findings are suggestive of the trade-off that was necessary between activating and enrolling certain actors by telling and informing them about developments, and intentionally staying silent to keep others at a distance and allow room for manoeuvring in order to develop the programme in a meaningfully way.

Sidestepping issues

In addition to the external complexity, to get the programme up and running it had to go through a course-approval process. Learning outcomes had to be determined and assessed and standards of evaluation established. Naturally this required a very complex decision-making process across interdisciplinary and institutional boundaries, involving many different actors with varying levels of authority and with different mandates (such as senior scientists, deans and support staff).

Interactional processes as complex as these can be immensely difficult to manage. In our case study this was exacerbated because there was no single locus of authority, such as a director of educational programmes. Interviewees mentioned that actors were constantly changing and new people always had to be informed about the project from scratch. An interviewee mentioned that participants were constantly bringing new issues to the table and this swallowed up available time. Particular actors also perceived issues and problems and their solutions in different ways, changing their minds during the process even when issues appeared to be settled. For many reasons, working across three institutions thus made it difficult to align multiple constituents during meetings and it became a great challenge to establish plans in a timely manner. A member of the support staff explained how this feature often caused actors to settle issues quickly while everybody was present, leading to short-term decision-making, with issues not being adequately discussed so that they suddenly resurfaced later on:

> People are generally very focussed on short-term results and want to make agreements the same day. In retrospect it is often apparent that things were not that easy and have to be presented 'ex machina' to take it

a few steps further. This costs a lot of extra time to pull things together [...] an extra administrative round becomes necessary, which can lead to new perspectives and approaches. If you have already discussed things with people on the work floor or at other levels, this is a nuisance because everything then once again has to be changed and done in a different way.

In addition to quick decision-making taking place during meetings, actors discussed and resolved things unofficially when particular people were not present. Waiting to settle things in meetings where everyone is present often means more discussion and can result in cumbersome questions that slow up the process. A key informant explained that it is sometimes just easier to reach agreements on small things informally, which then become a matter of 'ticking off' during formal decision-making. This does not mean that other actors or parties are not allowed to participate during meetings – the value of openness and transparent decision-making is usually endorsed – but it is seen as beneficial to the overall course of the process if unexpected and unanticipated contingencies can be resolved 'backstage' and the process can get back on track.

In addition to resolving issues backstage, actors may also intentionally limit the scope of subjects that are open to formal discussions. Complex inter-institutional collaborations naturally involve people with divergent professional perspectives and visions. In our case, not everyone necessarily agreed about the programme's long-term profile and positioning. Perspectives varied between seeing the bachelor's course as a stepping-stone towards a full master's degree or as vocational training that would lead towards a medical degree. Although parties did not adhere to the same long-term vision, this was not necessarily a problem. Difficult subjects like the programme's long-term profile and students' career perspectives were sidestepped or not discussed in any detailed manner during meetings. An interviewee from one of the medical academic institutions explained how the important issue of the programme's final requirements was avoided during committee interaction:

> The subject was raised several times, but in the end it was simply avoided. At one moment we even discussed it for one and a half hours, but after that we moved on to the order of that day, because we want to start with the programme in 2014. That is the overriding goal. The choice has been made that the programme will be realized. [...] That is why we incidentally sidestepped the subject and simply moved on. I have seen that happen on several occasions.

For the process to keep its momentum, certain difficult issues are thus not tabled. They literally must be pushed out of the room to allow for constructive debate on issues and problems that require immediate attention. Constantly raking up contentious issues and seeking the last word would clearly hamper the process.

While silence allows for parties to collaborate and work towards their goals, it can also mean that others do not know about particular communications. Two support staff members explained that meetings were often preceded by a process of 'pre-discussion' that often made the course of the process very unclear to them:

> I frequently did not have a clear view on what was going on because much of it was taking place behind the scenes. At one moment there was some fuss about the unbalanced division of the specialities, but to my surprise it was not discussed during the meeting. It simply never came up. [...] Obviously issues had already been settled between them, as they were never brought up again during the subsequent meetings. Instead we discussed other things I had not anticipated and prepared for.
>
> High-level scientists and decision-makers have contact among themselves. That is not a problem. But it does mean that information streams remain completely unclear to me.

When issues are not clear for actors it is not always easy to address them and get additional information. Because inter-institutional interactional processes are very unstructured by their very nature, it is often uncertain which actors will be present during which meetings. An interviewee explained how he could not find a good time to ask questions and get clarity about the programme's position:

> At one meeting I said out loud: 'Tell me what the benefit of the programmes is!' But it turns out that the right people are not present and you do not get an answer. [...] The result is that the whole discussion is derailed and you have to return to it at another moment. At that point there is always a slight translation of what was discussed. What you said comes across in a different way. Sometimes forceful, and sometimes not so forceful, but never how you initially intended it to be.

It is thus difficult to give internal feedback on the course of the process. When people do manage to do this, there is a chance that they will be wrongly interpreted depending on who is present at the table.

This section has described the friction between professional ambitions and available time. Everyone wants the programme to start within a short time frame and extensive debate would take up too much time. There is not much point discussing the programme's future if people disagree on this. This will come to the table at a later stage, but in the meantime some decisions on more pressing matters will have to be taken. Participants thus focus on things that are feasible in the short term and leave long-term 'dream' projects for later discussion. Visions can coexist side by side because they are not articulated: a silence is maintained. This indicates that silence has benefits for the collaboration; it helps short-term decision-making, but also causes things to become unclear for those who are not actively involved in these decision-making contexts. As we will see more clearly in the next section, silence had unintended consequences for those working in other contexts within the network and phases of the process.

Quick fixes and working backwards

In the decision-making context, the programme still exists mainly on paper. To become a workable programme, effective curriculum design and implementation is needed. This required the involvement and support of new allies in the form of departmental heads and teachers from the participating academic institutions. For this task, a group of domain experts from six technical and medical domains were called in to spread the word about the programme and to inform, motivate and involve departmental heads and teachers across the three institutions.

For the disciplinary domains where collaboration between engineers and medical specialists was already common practice, achieving a working programme posed few difficulties. However, involving people from the clinical workplace was a more significant challenge. Departmental heads and teachers were often weary and morale was lower. Before committing themselves, they wanted to know more about the exact implications of the programme for the medical 'shop floor' or for students' career prospects; issues that until now had mainly remained in the background. As two interviewees explained:

> Medical specialists see the programme as a threat; not enough attention was given to involve them. [...] There is some resistance from the clinical side because people first want to know: what exactly is this programme? What is this person going to do? How will he be deployed? People did not communicate enough about these issues.

> When I ask my colleagues whether they want to teach a class in the programme, they listen and seem interested. But then there is the 'yes, but'. That's the way these conversations go.

The domain experts also explained that they often could not respond to these questions, for example because there were no documents intended for internal discussion and deliberation. The paperwork that might construct strong and appealing stories connecting all the actors in the network did not exist (Krippendorff, 2008). According to one interviewee, this made it impossible to illustrate the programme's relevance to actors with strong opinions:

> Of course the specifics have all been thoroughly written down at a certain time, because that is officially required. [...] But what you really need is a condensed form [...] that can get the message across unambiguously under teachers and departmental heads. That it how things begin to resonate and how you create widespread support.

More significant than the absence of documents was the lack of visibility and formal support from the medical institutions. Several interviewees explained how the lack of publicity and information about the programme on the university's website complicated discussions in the workplace. Despite initially being very positive about the programme, one interviewee stated that without visible support from academic leaders it was impossible to create a space conducive to open and active dialogue with departmental heads and teachers and to ensure the programme was widely popular:

> Without the institution's top-down support and openness it is nearly impossible to talk to, let alone convince, teachers. [...] Senior people feel like you are stepping on their toes, with or without good reason. Others feel passed by. [...] When asked for their cooperation, they respond by saying that they have heard nothing about the programme or that there already exists a similar programme elsewhere. Their response was generally that we should first go talk to their superiors. [...] This is not the right starting point to excite and involve people.

Without an official story it becomes difficult to contextualize plans and fit them into the frames of reference of other relevant actors. It also means that domain experts have to invest a lot of energy in finding quick fixes to solve more or less randomly occurring problems, much like a plumber who runs from one leaking pipe to another. One interviewee explained this as follows:

> The whole design of the programme should have been better positioned within the three institutions. Because it was not adequately anchored, everybody is now keeping the ball in the air because we want it to be a success. As a result, you constantly have to attend to and organize things afterwards that should already have been taken care of. [...] The biggest problem with the programme is that we are continuously working backwards.

Or as other informants from both the medical and technical universities commented:

> How shall I put it, it was a protracted process. [...] There was nobody behind it, no structure. Everybody just messed about.

> Well, you can see it for yourself; it's all half-baked solutions, somewhat amateurish.

It is widely pointed out that quick fixes can cause anxiety (Dorner, 1996; Henriksen and Dayton, 2006; Tucker and Edmondson, 2003). While all supported the process, several interviewees blamed its arduous progress for the disappointment or disenfranchisement they felt. Sometimes they were critical of those who had designed the programme:

It sort of feels that you have been passed over.

Not enough time was spent at the drawing board to develop a workable plan.

The design was simply thrown over the wall. You should not do that. You should support the process and satisfactorily transfer it to others.

As such comments illustrate, although people had good reasons to remain silent in one interactional context, this created uncertainty and became problematic when the programme needed to be implemented in another context. In particular, it created a situation in which the programme's strategic intermediaries, despite their initial interest and very positive attitude to the programme, could not immediately respond to some of the demands that confronted them. They had to invest their time in solving problems they did not anticipate, causing anxiety and decreasing motivation. As a result of this, the programme became framed in terms of challenges rather than possibilities and opportunities.

Silence as a strategy for facilitating collaboration

We started this chapter by showing that meaningful innovation is usually seen as a process dependent on changes in discourses, representations and storylines, and shifts mobilized by multiple and constantly interacting actors and parties (Leeuwis and Aarts, 2011; Ford, 1999). Assuming that silence is also significant in this regard, we used the case of the inter-institutional bachelor's programme to illustrate that, throughout the whole process of developing and implementing the programme, silence was as performative as verbal communication. Strategic and intentional forms of silence are often conceptualized as a tool to secure personal gain, as in the denial, distortion or neglect of the facts (Oreskes and Conway, 2010). However, our study shows that whether specific strategic silences are intended positively or negatively depends on the situation and may be subject to contrasting interpretations within a situation. The presented study demonstrates that in the specific context of developing and implementing educational innovation – fundamentally a question of creating allies and partners (Akrich et al., 2002) – silence served a performative function in establishing, strengthening and managing interdependencies and encouraging the collaboration needed for realizing desired change.

Silence was relevant at two distinct strategic levels: for creating an external stakeholder environment beneficial to developing the programme, as well as for guiding the internal decision-making process. First, silence is used purposefully when the risks of communicating verbally about an issue outweigh the benefits of its resolution or when scientists do not want their plans to leak out prematurely. Exchanging knowledge and information, and engaging in dialogue with powerful parties who seek to influence or pull projects towards themselves, are generally seen as defining aspects of innovation processes (Akrich et al., 2002; Leeuwis and Aarts, 2011). Yet, despite the value of verbal communication, talking with others may not always have the desired effect; sometimes innovations are simply too complicated, controversial or underdeveloped to share with, or sell to, specific parties. Openly addressing issues or confronting a conflict would politicize plans or cause them to be misunderstood and to fail. Silence can take attention away from an issue, protecting and developing plans that are prone to the turbulent culture of bargaining and that still need to gain legitimacy (Bok, 1989). In the meantime, visions can be sold to enough relevant actors to build momentum and ensure that resistance can no longer appropriate or sink ideas or projects.

In addition, silence plays an important role in guiding the internal decision-making process between actors who already recognize and underwrite the value of proposed innovations. Although people purposely choose to collaborate (Knight, 2002), this creates its own specific challenges and dilemmas: people have to meet and discuss ideas with others they may never have worked with before and with whom they may disagree professionally, or whom they do not yet fully trust (Shrum et al., 2007). In order to co-construct the process of innovation, they constantly have to make compromises that justify everybody's views and interests. This requires silence, as when one chooses to communicate a certain topic while avoiding many others in order to leave room for alternative ideas but also to obstruct further questioning of complicated issues and to make it possible to work out short-term agreements.

As Shrum et al. (2007) have observed in relation to multi-organizational scientific collaborations, scientists manage interdependencies by circumventing potential disagreements, abandoning arguments and confrontations until later. As the authors write: 'They define regions of silence, about which they will not talk. They emphasize the shared goal and not the differences in preferred means' (ibid.: p. 175). In our study, actors constantly had to balance day-to-day versus longer-term goals, turning away from sensitive issues and debates that touched upon professional differences in order to get the programme up and running. As such, silence made it possible for diverse ambitions, perspectives and interests to work towards and exist next to each other (Nussbaum, 2006).

The above indicates that silence serves an important strategic function in different contexts of the innovation process. However, we have also pointed to its unintended consequences for the course of interaction, both at other sites in the network and in the long term. First, silence substantially influences how the meaning of innovation is co-constructed in different sites and networks. Scholars have noted that what is talked about in conversations produces further communications, for example because participants take what has transpired in one context and discuss it in another context (Krippendorf, 2008; Ford, 1999). Extending this argument, we suggest that what is not said in one interactional setting is also likely to affect how people talk – or don't talk – about things in another setting, frequently in unexpected and undesirable ways. Keeping the programme out of the spotlight and leaving long-term issues off the table and unresolved, while productive in the decision-making content, caused plans to remain largely invisible and unaccounted for on the work-floor level. This resulted in problems and dilemmas for domain experts who lacked the authority to frame dialogue in such a way to convince departmental heads and teachers of the programme's importance, at times igniting undesirable resistance (Ford, 1999). This indicates that silence in one part of the network can shift the problem elsewhere, causing previously silenced interpretations and voices to emerge and influence the course of the process in undesirable ways (Van Assche and Costaglioli, 2012).

This also has consequences for the co-creation of innovations in the long term. Uncertainty about the programme's meaning led to problematic working routines for those who were trying to make others see its value from their perspective. Previous studies pointed out how the failure to address certain issues can lead to operational problems and behaviour for those who already have a very high workload (Henriksen and Dayton, 2006; Tucker and Edmondson, 2003; Dorner, 1996). In their qualitative study of the internal supply chains of hospitals, Tucker and Edmondson (2003) found that nurses were reluctant to speak up about perceived problems or to ask for help, and instead quietly adjusted and corrected mistakes without addressing the core of the problem itself. This resulted in complex working routines that took valuable time away from patient care. In our study, silence also led to time-consuming and counterproductive activities for domain experts, who had to work with a poorly designed programme and had to engage in uncomfortable communicative situations. From this, we are able to understand how people become less motivated to commit themselves to innovation over time. Failure to talk about changes in a productive manner causes people to become anxious and cynical about proposed changes (Morrison and Milliken, 2000), resulting in a decreased feeling of attachment towards it.

This chapter has offered insights into the role of silence in an academic educational innovation, arguing that fully grasping such processes requires understanding of both silence and verbal communication and their dynamic interaction across various interactional contexts. Conceptualizing silence as an active part of the educational innovation process helps show why it is difficult to connect actors around innovation (Akrich et al., 2002; Van Bommel et al., 2011). For innovation to become meaningful, it must be developed and implemented in different 'micro' settings, but at the same time it needs to become contextualized and anchored within a wider network (Ford, 1999; Akrich et al., 2002; Nicolini, 2010). In such conflicting contexts, silence can provide productive solutions for many context-specific contradictions and tensions, but will unavoidably create new problems in the bigger picture – sometimes very directly, sometimes indirectly, but always preventing structural changes that are needed for innovation to become accepted in the wider network. Silence is thus simultaneously a precondition for, and a limit to, this interaction (Luhmann, 1994). This, it seems to us, is also indicative of the consistency of silence with the paradox of innovation (Fonseca, 2002). The more people rely on silence to carry out context-dependent plans and actions,

Silence in educational innovation 109

the more their silent behaviours will cause that interaction to become even more complex and uncertain.

While we have offered initial insights into these complex issues of silence in the context of educational innovation, further empirical research is needed to deepen the insights provided by our single case study. In particular, our exploratory study serves as an invitation for other researchers to identify and distinguish more subtypes of silence and their consequences for developing and implementing academic innovation. With academic institutions increasingly looking for scientific innovation, this seems to be a productive avenue for generating the insights needed to create and sustain meaningful interdisciplinary projects around important socially relevant themes.

As for its practical implications, our study provides a powerful framework for thinking about the steering of processes of innovation. In designing for innovation, scientists, academic institutions and policy makers need to take into account that change is constructed in and through communication (Ford, 1999) and that not saying things – remaining silent – is a crucial aspect of this process. As we have pointed out, silence has to do with uncertain and constantly changing interactions and interdependencies, which makes it difficult to predict in advance where and when silence will occur, and how and where it will affect connections in the process. This means that silence must be appreciated as a 'natural', inescapable and sometimes necessary part of the messy and unpredictable course of innovation processes. That said, we do think that innovation processes will benefit from involving people who can adequately monitor and address moments of silence and talk and their dynamic interrelation across disciplinary and/or institutional boundaries. We clarify this point by returning once more to our case study.

Despite the bumps along the road, the programme was delivered on time in September 2014. Naturally many things contributed to this, among them the hard work and persistence of some of those involved, positive assessments and the accreditation of the programme. In addition, we would also like to emphasize the importance of putting a domain expert from the technical university formally in charge of the delivery of the programme. Making someone responsible for the process served as a catalyst for encouraging interaction and communication between diverse actors and parties. Under this person's guidance a vision document was written, participants from the three institutions were more actively approached and informed, and workshops were organized in which outsiders were invited to share experiences about developing similar programmes. This all helped to pull the network together. As one of our informants told us:

> You really need someone who keeps control over the process, even when there are bumps in the road. Someone who can give direction. [...] All those people are inclined to make decisions from their own field. Protecting the lines is crucial. If you do not do that you have a wheelbarrow full of frogs that jump in all directions.

It goes without saying that a single person can never control or determine the success or failure of complex interdisciplinary projects such as described in our study. Complex innovations are dependent on the interplay between different actors or interest groups within a network, taking into account their competing interests, wishes and dreams. Nevertheless, from a conversational management perspective (Ford, 1999), we would argue that the process still benefits from including a person who can monitor the quality of conversations needed to connect relevant contexts during all stages of the process and can effectively move between these different contexts. This requires someone who understands the rhythm, melody and harmony of the interplay between silence and talk in institutional environments; someone who can 'spread the story' by clarifying or defending ideas where and when necessary, encouraging people to openly discuss developments and concerns in a productive way while being attentive to their status and interests, but who also recognizes the limits of conversation and knows when it is necessary to remain in the background, and thus, stay silent.

References

Akrich, M., Callon, M. and Latour, B., 2002. The key to success in innovation. Part II: The art of choosing good spokespersons. *International Journal of Innovation Management*, 6(2), pp. 207–225.

Bijker, W.E., Bal, R. and Hendriks, R., 2009. *The Paradox of Scientific Authority: The Role of Scientific Advice in Democracies*. Cambridge, MA: MIT Press.

Bok, S., 1989. *Secrets: On the Ethics of Concealment and Relevation*. New York: Vintage Books.

Carter, C., Clegg, S.R. and Kornberger, M., 2008. Strategy as practice? *Strategic Organization*, 6(1), pp. 83–99.

Casey, B.A., 1994. The administration and governance of interdisciplinary programs. In: Klein, J.T. and Doty, W.G. (eds), *Interdisciplinary Studies Today. New Directions for Teaching and Learning*. San Francisco, CA: Jossey-Bass.

Coleman, K., Jasperse, M., Herst, P. and Yielder, J., 2014. Establishing radiation therapy advanced practice in New Zealand. *Journal of Medical Radiation Sciences*, 61, pp. 38–44.

Czarniawska, B. and Sevón, G., 1996. *Translating Organizational Change*. Berlin: Walter de Gruyter.

Dam-Mieras, R. van, Lansu, A., Rieckmann, M. and Michelsen, G., 2008. Development of an interdisciplinary, intercultural master's program on sustainability: learning from the richness of diversity. *Innovative Higher Education*, 32(5), pp. 251–264.

De Bruijn, H. and ten Heuvelhof, E., 2008. *Management in Networks: On Multi-actor Decision making*. London: Routledge.

Dewulf, A., Gray, B., Putnam, L., Lewicki, R., Aarts, N., Bouwen, R. and Van Woerkum, C., 2009. Disentangling approaches to framing in conflict and negotiation research: a meta-paradigmatic perspective. *Human Relations*, 62(2), pp. 155–193.

Dorner, D., 1996. *The Logic of Failure: Why Things Go Wrong and What We Can Do To Make Them Right*. New York: M Books.

Ephratt, M., 2008. The functions of silence. *Journal of Pragmatics*, 40(11), pp. 1909–1938.

Flyvbjerg, B., 2006. Five misunderstandings about case-study research. *Qualitative Inquiry*, 12(2), pp. 219–245.

Fonseca, J., 2002. *Complexity and Innovation in Organizations*. London: Routledge.

Ford, J.D., 1999. Organizational change as shifting conversations. *Journal of Organizational Change Management*, 12(6), pp. 480–500.

Gladwell, M., 2000. *The Tipping Point: How Little Things Can Make a Big Difference*. Boston: Little, Brown.

Goffman, E., 1959. *The Presentation of Self in Everyday Life*. New York: Anchor Books.

Henriksen, K. and Dayton, E., 2006. Organizational silence and hidden threats to patient safety. *Health Sservices Research*, 41(4), pp. 1539–1554.

Holley, K.A., 2009. Interdisciplinary strategies as transformative change in higher education. *Innovative Higher Education*, 34, pp. 331–344.

Knight, L., 2002. Network learning: exploring learning by interorganizational networks. *Human Relations*, 55(4), pp. 427–454.

Jaworski, A., 2005. Introduction: silence in institutional and intercultural contexts. *Multilingua: Journal of Cross-Cultural and Interlanguage Communication*, 24(8), pp. 1–6.

Krippendorff, K., 2008. Social organizations as reconstitutable networks of conversation. *Cybernetics and Human Knowing*, 15(3–4), pp. 149–161.

Leeuwis, C. and Aarts, N., 2011. Rethinking communication in innovation processes: creating space for change in complex systems. *Journal of Agricultural Education and Extension*, 17(1), pp. 21–36.

Luhmann, N., 1994. Speaking and silence. *New German Critique*, 61, pp. 24–37.

Maturana, H.R. and Varela, F.G., 1988. *The Tree of Knowledge*. Boston and London: Shambhala.

McFadden, K.L.Chen, S.-J., Munroe, D.J., Naftzger, J.R. and Selinger, E.M., 2010. Creating an innovative interdisciplinary graduate certificate program. *Innovative Higher Education*, 36(3), pp. 161–176.

Morrison, E.W. and Milliken, F.J., 2000. Organizational silence: a barrier to change and development in a pluralistic world. *Academy of Management Review*, 25(4), pp. 706–725.

Nicolini, D., 2010. Medical innovation as a process of translation: a case from the field of telemedicine. *British Journal of Management*, 21, pp. 1011–1026.

Nussbaum, M.C., 2006. *Frontiers of Justice. Disability, Nationality, Species Membership*. Cambridge, MA: Harvard University Press.

Oreskes, N. and Conway, E.M., 2010. *Merchants of Doubt: How a Handful of Scientists Obscured the Truth on Issues from Tobacco Smoke to Global Warming*. New York: Bloomsbury Press.

Panteli, N. and Fineman, S., 2005. The sound of silence: the case of virtual team organising. *Behaviour and Information Technology*, 24(5), pp. 347–352.

Pretty, J.N., 1995. Participatory learning for sustainable agriculture. *World Development*, 23(8), pp. 1247–1263.

Rappert, B., 2010. Revealing and concealing secrets in research: the potential for the absent. *Qualitative Research*, 10(5), pp. 571–587.

Rogers, E.M., 1962. *Diffusion of Innovations*. New York: Free Press.

Shrum, W., Genuth, J. and Chompalov, I., 2007. *Structures of Scientific Collaboration*. Cambridge, MA: MIT Press.

Stilgoe, J., Owen, R. and Macnaghten, P., 2013. Developing a framework for responsible innovation. *Research Policy*, 429(9), pp. 1568–1580.

Stone, D., Patton, B. and Heen, S., 1999. *Difficult Conversations: How to Discuss What Matters Most*. New York: Penguin.

Thorp, H. and Goldstein, B., 2010. *Engines of Innovation: The Entrepreneurial University in the Twenty-First Century*. Chapel Hill: University of North Carolina Press.

Tucker, A., and Edmondson, A., 2003. Why hospitals don't learn from failures: organizational and psychological dynamics that inhibit system change. *California Management Review*, 45, pp. 55–72.

Van Assche, K. and Costaglioli, F., 2012. Silent places, silent plans: silent signification and the study of place transformation. *Planning Theory*, 11(2), pp. 128–147.

Van Bommel, S., Aarts, N. and Klerkx, L., 2011. *Zelforganisatie in innovatienetwerken: verhalen van succesenfalen*. Wageningen, the Netherlands: Wageningen University.

Vermeir, K. and Margócsy, D., 2012. States of secrecy: an introduction. *British Journal for the History of Science*, 45(2), pp. 153–164.

Xu, H. and Morris, L.V., 2007. Collaborative course development for online courses. *Innovative Higher Education*, 32(1), pp. 25–47.

Part II
Cultures of silence

5 Talking about secrets

The Hanford nuclear facility and news reporting of silence, 1945–1989

Daniele Macuglia

On 20 May 1984, Reverend William Harper Houff, a Unitarian minister in the city of Spokane in Washington State in the USA, delivered a sermon on the 'Silent Holocaust'. The holocaust of which Houff warned was, he claimed, taking place just 150 miles southwest of Spokane, at the Hanford nuclear facility, the huge industrial complex where plutonium for US nuclear weapons had been manufactured since World War II. Houff accused US officials of having 'uttered almost no words of alarm' about the site, despite copious evidence of health problems caused by the radiation released there. Not only were officials silent themselves, Houff complained, but they had 'frequently taken extravagant measures to silence those who do sound a warning' (Ratliff and Salvador, 1994: pp. 4–5). Houff's sermon was to spark a local campaign which eventually contributed to the release, in February 1986, of about 19,000 pages of classified information about the early history of the site and the radioactive and chemical wastes released during its operation (Gerber, 1992: p. 2; Zwinger and Smith, 2004: pp. 49–50). The facility was decommissioned shortly afterwards.

Today, governments and military bureaucracies often still withhold sensitive information from the public in their attempt to protect law enforcement or to uphold national security (Glazer and Glazer, 1998: p. xiii; see also the Introduction to this volume). In an open society, however, there is also an expectation that people should be able to access information pertaining to possible health hazards. This chapter focuses on the contrasting imperatives of openness and secrecy at Hanford in the decades following the Second World War.[1] By examining the media coverage of the various nuclear accidents that occurred at the facility, I show that the secrecy of Hanford was not the outcome of absolute silence; rather, the silences that surrounded Hanford were partial and context dependent. Newspaper reports reveal that journalists occupied an ambiguous role in their construction: initially consolidating the culture of secrecy through the dissemination of partial silences, but later working with local activists to break the silence. By the early 1980s, I suggest, local journalists writing about Hanford had recast their role from that of local drivers of secrecy to drivers of publicity.

Located in southeast Washington State, about 200 miles from Seattle, Hanford was the plutonium production complex that fuelled the atomic bomb employed in the world's first nuclear explosion, tested in 1945 at the Trinity Site in Alamogordo, New Mexico. It also produced the plutonium for the bomb dropped on Nagasaki, Japan in August 1945. The facility was built in March 1943 on a 640-square-mile desert site and was bordered on the east side by the Columbia River, which flows southeast into the Tri-Cities, a mid-sized metropolitan area composed of Richland, Kennewick, and Pasco (Gerber, 1992: p. 2). Operating in great secrecy throughout its active period, Hanford was created to address the threat represented by Nazi expansion and, later, to counter the Soviet Union's nuclear weapons and military programmes during the Cold War. The site produced more than 67 metric tons of plutonium and supplied most of the 60,000 nuclear weapons produced in the United States from 1945 to 1987 (Gephart, 2003b: p. 5; Anon., 1987).

For more than four decades, Hanford had a significant impact on people and ecosystems and polluted a large portion of the Pacific Northwest (Steele, 1988: p. 17).[2] In the late 1940s and early 1950s, high levels of radioactive waste were released into the air, into the water of the Columbia River and into the ground. Contaminants reached the local human population through environmental pathways such as inhalation, ingestion of wildlife and drinking of contaminated water. The heaviest releases occurred shortly after World War II, but contamination continued all through the Cold War era, possibly causing significant long-term health effects for people living downwind and downstream of the production reactors (Steele, 1988: p. 17).

The public was not directly warned about the offsite contamination and records of contamination incidents were kept secret in classified documents. It was only with the declassification of these documents in 1986 that the world learned about the immense radioactive and chemical discharges from the facility. It now became apparent that decades of continuous manufacturing had left behind 177 single and double-shelled tanks containing about 200,000 m^3 of high-level, radioactive, hazardous waste and 710,000 m^3 of solid radioactive waste, in addition to 520 km^2 of contaminated groundwater beneath the site that is migrating to the Columbia River – all of which has had, and continues to have, a huge impact on US political, economic and ecological programmes.[3] As Congressman Ron Wyden (Democrat) of Oregon remarked in 1988, Hanford was 'the largest, most ultra-hazardous industry of its kind in the world' (Steele, 1988: p. 17).

In some of the most detailed analyses of the history of the Hanford Site, it is not uncommon to read that the government 'kept the public in the dark' about what was going on at the site, that 'major Hanford decisions were made without public awareness or involvement', and that 'Hanford was busy and noisy and industrious, and yet, to the outside world, it was silent' (Gerber, 1992: p. 53).[4] We know that the Manhattan Project maintained a top-secret classification about Hanford's activity and its operational history; even after the end of World War II, less than 1 per cent of Hanford's workers knew they were employed in a nuclear weapons project (Hanford Cultural Resources Program, 2002: p. 1.22; see also Findlay, 2004; Hughes, 2002; Hevly and

Findlay, 1998; Macuglia, 2013b). 'We made certain that each member of the project thoroughly understood his part in the total effort; that, and nothing more', declared Manhattan Project director General Leslie R. Groves (1896–1970) in his memoirs, published in 1962 (Groves, 1983: p. 15). In addition to this, as Michele Gerber noted in 1992, until recent years area residents were not informed of the discharges nor warned of any potential dangers, even when releases far exceeded the tolerable limits and 'maximum permissible concentrations' defined as safe at the time. In fact, Hanford scientists and managers, on numerous occasions throughout the first four decades of operation, specifically told the public that the plant's workings and wastes were all controlled and harmless (Gerber, 1992: p. 3).

We should note, however, that it is one thing to say that people were not directly informed about major Hanford policies and decisions; it is another to say that the facility was silent to the outside world and that the public was kept in the dark about what was going on inside it. Connections between radioactivity and certain forms of cancer were already known by the late 1920s and it seems that some citizens long suspected that something was wrong with safety conditions at the Hanford Site (see, for example, Belton, 2010: p. 74; Mullner, 1999). As discussed by J. Samuel Walker (2009) in his book *The Road to Yucca Mountain*, there were many stories about accidents and waste issues at Hanford and most of the US nuclear facilities operated in an intermediate state of secrecy and publicity. Yet how precisely this ambiguous status was attained and what it actually meant at different times during the Cold War remains unclear (Macuglia, 2013a). How were secrecy and silence experienced by those involved – including employees and local residents – through the Cold War era?

A consistent body of scholarly work on the public understanding of the Hanford Site has focused on the twisted relationships of trust, uncertainty and economic reliance. In addition, studies of lay expertise have examined the dissimilarities, hierarchies and various intersections involved in differently situated types of knowledge. John Findlay and Bruce Hevly, for example, have drawn on local newspapers such as the *Columbia Basin Herald, The Villager*, the *Sage Sentinel* and the *Tri-City Herald*, as well as larger regional newspapers such as the *Seattle Post-Intelligencer*, the *Oregonian*, the *Spokesman-Review* and the *Seattle Times*, to explore Hanford's role as an industrial – as opposed to a scientific – facility located in a specific regional context (Findlay and Hevly, 2011; see also Pope, 2008). In what follows, I present an in-depth analysis of coverage in the *Seattle Times* in the period from the end of the Second World War to the end of the Cold War, to demonstrate the ways in which the Hanford Site existed in a communicative context of partial silence whose qualities and meanings altered over time.

The *Seattle Times* is a privileged vantage point for such an investigation. As the largest and most important newspaper in Washington State, it offers an appropriate perspective from which to shed light on the local dynamics within Washington State, where readers were directly affected by nuclear

118 *Cultures of silence*

issues. Having looked at hundreds of articles and letters to the editors from the early 1940s to the early 1980s, I focus on those pieces that clearly addressed issues of potential danger posed to local communities showing that, despite Hanford's institutional silence, there was a localized, discontinuous dissemination of partial information about radioactive threats posed by the nuclear facility to the local people and ecosystems. This murky local knowledge created a peculiar state of secrecy, a sort of middle ground between 'knowing' and 'not knowing' influenced by instances of personal interests and various protocols of national security.

Problematizing Hanford in the *Seattle Times*

Before the end of World War II, news about Hanford was completely censored. Newspaper editors were explicitly told to cooperate by not speculating or asking questions about the site (Gerber, 1992: p. 45). At this time, journalists, together with the federal government and the military, might be defined as 'drivers of secrecy', inasmuch as they did not promote any disclosure of relevant and unambiguous information.

Secrecy remained an extremely important issue at the end of the war, but the role of the facility in producing the first atomic bomb was now revealed. News reports praised the Tri-Cities community's contribution to national defence. 'Hanford made material for atomic bomb that hit Japs', reported the *Seattle Times* on Wednesday, 8 August 1945 (Cunningham, 1945). The region's victory celebrations were reported throughout the national media. As Gerber put it, the small city of Richland, where Hanford employees lived, 'basked in the praise of the entire nation' (Gerber, 1992: p. 59).

In the years that followed, both national and regional newspapers remained enthusiastic about the nuclear research carried on at Hanford. The *New York Times*, for instance, published lengthy articles about the 'city that plutonium built' and Richland was seen as an oasis in a stretch of the desert where the job of producing some of the most important components of the atomic bomb was carried out with increasing efficiency (Davies, 1949). Atomic discoveries were seen as useful to medicine and industry, and atomic energy was praised for its terrific bursts of power. Hanford was often compared to Grand Coulee Dam, the largest electric power-producing facility in the United States (Mooers and Van De Water, 1945). At that time, Hanford produced as a by-product the equivalent of 1,500,000 kilowatts of electricity for each kilogram of plutonium. This was indeed an impressive result considering that the maximum capacity of the hydroelectric plants at Grand Coulee was estimated to be 2,000,000 kilowatts (Mooers and Van De Water, 1945).

As the Cold War took hold, the feeling of pride in Hanford's activities was accompanied by an increasing sense of responsibility. Journalists continued to receive only partial information about Hanford's operations and faced the delicate problem of needing to be careful in writing their newspaper reports so as not to interfere with governmental plans (Gordin, 2009: p. 91). There was also the obvious risk that enemy agents might read US newspapers, helping them to detect the weakest spots of the US nuclear weapons production system (Kaiser, 2005; Macrakis, 2010).

At the same time, it was clear that the energy produced by the nuclear plant was being injected into the Columbia River in the form of heat and that some of the waste by-products resulting from the production of plutonium, including poisonous and intensely radioactive substances, were released in the form of gases (Mooers and Van De Water, 1945). The general trend, however, was to consider these substances as precious by-products whose employment in future medical and scientific research, after sequestration, could turn out to be extremely useful (ibid.).

Thus the site was celebrated as the place where the United States had begun the largest peacetime construction project in the nation's history. Newspaper reports appearing in this period tended to emphasize the great safety standards adopted by the facility, as in one article that noted that: 'Workers within the Hanford operating areas must wear "pencils" and badges which measure the radioactivity to which they may be exposed' (Anon., 1948). Working conditions there were said to be even safer than in one's home. 'In July', reported the *Seattle Times* in 1949:

> the plant was a less perilous place to be than the average business office or home. The 7,500 workers set a new safety record last month with the only 'serious' injuries consisting of one broken toe and one sprained back. There were 282 'minor injuries'.
>
> (Associated Press, 1949)

Newspaper reports such as this one suggested that employees at Hanford were protected by 'extraordinary precautions' and the nuclear facility maintained remarkable safety and health records (Anon., 1948). We can even read about an employee, a certain Maxine Huntley from Foster, Oregon, who worked 'with complex laboratory equipment' and who wore a special uniform to work in her laboratory, comprising 'coveralls, exposure pencils and badge, and safety glasses' (Anon., 1948).

The first critical, though still muted, articles began to appear in the late 1940s. Some of these referred to surveys of industrial health and safety measures at Hanford conducted by members of the advisory board of the Atomic Energy Commission; others referred to a series of precautions, taken by a group of public health consultants, to prevent contamination of the Columbia River by the Hanford atomic plants (Anon., 1947; United Press, 1949). More censorious articles began to appear in the early 1950s. In 1954, newspaper articles reported that minor emissions of radioactive particles from one of the Hanford chemical-plant stacks occurred on several occasions and that steps had to be taken to control such emissions. These articles referred only to deposits of radioactive particles on the ground, whose concentration outside the site was negligible, and stated that no precautionary steps were necessary (Associated Press, 1954).

In 1955, however, yolks of duck eggs in the area were found to contain about 200 times as much radioactive phosphorous as is found in the bodies of adult river ducks and geese. An article in August 1955 reported that, 'Eggs laid by ducks in the vicinity of the Hanford Atomic Works at Richland are fairly popping with radio-activity [...]' and that, 'ducks in the Hanford area absorb radioactive phosphorous which the big plutonium plant releases as waste into the Columbia River' (Associated Press and United Press, 1955). One day later, the same news was reported in the *New York Times*, which announced that surveys of the air, the earth and the water around the nuclear plant indicated that radioactive chemicals were escaping from the facility (Hillaby, 1955). 'Escaping radioactivity invades wild life around Hanford plant' reads the headline of the lengthy article by John Hillaby, who reported that surveys showed that pollution of the Columbia River by atomic chemicals had affected insects, birds, fish, and other animals. Despite the alarming tone, however, Hillaby presented the situation as under control and reported that no evidence had been found of any clear damage to plants, animals or people.

However, in an article in the *Seattle Times* the following year, Harold F. Osborne reported that, 'Concentrations of a "hot" element, radiophosphorous, up to 1,500,000 times that in nearby river water, have been found in the yolks of eggs laid by ducks nesting near the Hanford plant'. This finding was significant, Osborne suggested, in light of international concern over possible genetic damage to future generations of humans from radioactive fallout (Osborne, 1956). Osborne described a possible contamination route through which radioactive elements produced by the nuclear facility could enter the aquatic food chain. 'Radiophosporous', he wrote,

> survives the cooling period and enters the river. It is absorbed by plankton – small aquatic plants and animals – which is eaten in concentrated form by larger organisms in the river, such as fish. Aquatic insects also become carriers of radioactivity. These in turn are eaten by insect-eating birds such as swallows. Waterfowl which feed on small fish or on crustaceans or vegetation from the river also absorb radioactive materials.
> (Osborne, 1956)

Despite the disquieting news disclosed in some of these articles, the situation was said to be tolerable, especially when compared to worse situations, such as working conditions in uranium mining (Associated Press and United Press, 1955). Osborne himself went on to stress that the high concentrations of radioactive elements in the river were below dangerous levels and that no injurious effect had been found in the water used for cooling purposes at Hanford. This technique of calling the public's attention to obviously dangerous situations and then immediately reassuring them that the situation was actually under control emerges as a standard approach adopted by journalists throughout the Cold War era when reporting on the Hanford Site. It was an approach that mirrored that of the Atomic Energy Commission, which supervised the management of the facility at that time and which responded to any concerns by assuring local communities that they were in good hands.

Beginning in the 1960s, Oregon, Washington, and the United States Public Health Service all joined in a large-scale assessment of the Columbia River to determine how much radiation pollution it contained from the Hanford Atomic Energy Works (Wells, 1961). The programme began with an analysis of marine life and silt at the mouth of the river; at that time, the Columbia River was already highly contaminated. Given the size of the study, it is likely that it would have taken months or years before its findings were published, making it difficult to trace in the newspaper archives. However, it is interesting to note that, despite the announcement of such an extensive investigation, the final results do not seem to have been published in the *Seattle Times*.

By the end of the 1960s, journalists were speculating that Hanford's environmental impact might have turned the public against nuclear power. In 1967 journalist Hill Williams wrote about the important effects that the hot water used for cooling the nuclear reactor at Hanford might have when discharged into the Columbia River:

> Nuclear plants need vast amounts of water for cooling. When the water is returned to a river, it raises the temperature of the river. The temperature increase may only be a degree or two [...] but this could kill off some forms of aquatic life, or it could increase the population of another form. Either result could upset the complex ecology of a body of water.
> (Williams, 1967)

Williams went on to make strong assertions about radioactive releases through discharges of contaminated water in the river. 'Hanford's discharge has made the Columbia the recipient of more radioactivity than any other known river in the world. But the level is considered well-within the safety standards.' Nevertheless, Williams underlined, 'public knowledge of the contamination caused by Hanford undoubtedly will make more difficult the acceptance of nuclear power plants scattered across the state' (Williams, 1967; see also Williams, 2011).

During the following years, press coverage of the dangers of radioactivity intensified thanks both to the rapid growth of the commercial and nuclear industry and the expansion of the media during those years. In 1973, the Hanford Site was subject to the scrutiny of the media when it suffered three major tank leaks, the largest of which took place in June and released 115,000 gallons containing about 40,000 curies of caesium-137 (Gerber, 1992: p. 166). 'Hanford springs another "hot leak"', declared a *Seattle Times* article (Anon., 1973a).

> The Atomic Energy Commission has reported another leak of highly radioactive waste from an underground tank at Hanford, the second in less than a month. The most recent leak [...] spilled 1,500 gallons of radioactive waste into the soil beneath the tank.

The escaped liquid remained 'within "a few feet" of the tank bottom, still at least 150 feet above ground water'. The article continued by noting that an Atomic Energy Commission investigation 'of a much larger leak reported on June 12 is nearing completion. That leak poured 115,000 gallons of radioactive waste into the ground.' With a significant risk of ground water contamination, the Commission's waste-management practices were, the article implied, inadequate.

This incident also received national coverage in several articles in the *New York Times* (United Press International, 1973; Anon., 1973c; Anon., 1974). One piece, headlined 'Leak of radioactive waste', reported that:

> A second leakage of high-level radioactive waste from a storage tank at the Hanford Atomic Energy Commission facility was reported today by the A.E.C. The agency said that about 1,500 gallons of radioactive waste was found missing from a tank. This is in addition to the 115,000 gallons reported to have been lost from the tanks last month.
>
> (United Press International, 1973)

However, even though it noted that this was the second such leak, the *New York Times* still reassured readers that no danger was posed to human life and that the radioactive liquid had not reached the drinking water (Associated Press, 1973). The same conclusion was drawn by Williams in the *Seattle Times* (Williams, 1973).

It is interesting to note that, despite claiming a minimal risk to human health, newspaper articles at this time were sometimes exceptionally detailed in disclosing technical information to the public. Reporting on the 1973 leak of waste into the soil, for example, the *Seattle Times* reconstructed step-by-step the events that led to the discovery of the radioactive contaminating event and specified that the leaking solution included: 'cerium 144-praseodymium 144, cesium 137, europium 155, cesium 134, antimony 125, strontium 89-strontium 90, ruthenium 106-rhodium 106, plutonium 239, plutonium 240, and americium 241, all radioactive' – an impressive variety of highly poisonous substances (Anon., 1973b). The *New York Times*, meanwhile, was highly critical of waste management at Hanford and talked about a nationwide furore over the discovery of the 115,000-gallon leak (Anon., 1973c).

Readers were also reminded that Hanford was the site where about 75 per cent of all high-level radioactive waste in the United States was stored. The Atomic Energy Commission was accused of 'failing to file an environmental impact statement as required by the National Environmental Policy Act of 1969 and of violating sections of the Atomic Energy Act of 1954 for containment of radioactive wastes' (Anon., 1973c). It was emphasized that, since 1958, 'radioactive leaks from storage tanks or from the pipes connecting the tanks totalled about 530,000 gallons' (Anon., 1973c). Despite the attempts to reassure readers that no hazard was posed, it was clear that something was wrong with the plutonium production complex.

Things got worse in August 1976, when a chemical explosion at Hanford contaminated ten workers (Williams, 1976). One of them, Harold McCluskey of Prosser, Washington, received such a heavy dose of radiation that he became known as 'The Atomic Man' (Associated Press, 1983). In the early morning of 30 August 1976, reported Hill Williams, McCluskey was 'peppered with radioactive americium when a 2-gallon vat of nitric-acid solution exploded. The acid also caused burns' (Williams, 1976). The explosion tore off the protective mask he was wearing; his eyes and skin were spattered with acid and radioactive fragments (115 in his face alone), he breathed the radioactive fumes that were filling the room, and he received the most severe internal burns that anyone at the time was aware of.

Doctors who had treated McCluskey at the Kadlec Hospital's emergency decontamination facility thought that he had absorbed about 500 times the maximum allowable lifetime dose of radioactivity. According to the *Seattle Times*, McCluskey had made 'medical history', having absorbed a dose of americium larger than any human being at the time had ever received (Associated Press, 1983). McCluskey became so radioactive that he was kept in a steel and concrete isolation tank for five and a half months to avoid exposing others and doctors approached him wearing special protective clothing (Haitch, 1980).[5]

This event inevitably attracted public attention in the state of Washington, as can be ascertained from the articles on the tragedy published between 1976 and 1983 (see, for example, News Services, 1977; Associated Press, 1979; Associated Press, 1983). McCluskey never developed cancer or radiation sickness, but his eyes remained extremely sensitive to bright light. The government gave him $275,000, plus free medical care, for the rest of his life. He died of coronary artery disease in August 1987. Despite the accident, he always claimed to be pro-nuclear energy. His story was reported in many national publications and he was annoyed by those authors who tried to make him into an anti-nuclear symbol. 'Just forget about me being antinuclear, because I am not', he said in one interview, 'we need nuclear energy. It's about all we've got left to get power from' (Associated Press, 1983).

Hanford attracted further public attention when the federal government announced in 1979 that it was considering selecting the site as the major commercial radioactive waste repository for the nation. The political and environmental activist Ralph Nader declared that Hanford faced the possibility of becoming the 'radioactive waste dump of the Western world' (King, 1979). Nader's national prominence ensured that his critical assessment attracted press coverage (King, 1979).

Further problems affecting the site also found their way into the press, such as another spill of radioactive material in 1982 and the contamination of seven workers in two incidents involving rips in devices used to handle radioactive material (United Press International, 1984). Hanford's mismanagement and safety violations were again emphasized in 1984 when, as the *Bulletin of the Atomic Scientists* put it a few years later, 'a scientist working

124 *Cultures of silence*

inside the super-secret plant went public with fears that sloppy procedures would lead to a serious accident'. The scientist, Casey Ruud, approached the *Seattle Times* with his concerns that plutonium had gone missing at the plant (Steele, 1988: p. 17). In the first of a long series of articles based on the information leaked by Ruud, the paper reported that 'between 10 to 13 kilograms of plutonium are unaccounted for at the super-secret plutonium processing plant at the Hanford Nuclear Reservation' (Associated Press, 1984b). Although officials were reported as claiming this was just a record-keeping problem, the newspaper construed the matter as extremely serious and another confirmation of Hanford's problems with the safety management of intensely toxic and long-lived radioactive substances. As the article made clear, the 'unaccounted-for material' was enough to build an atomic bomb like the one that had devastated Nagasaki.

By this time, Reverend Houff had delivered his 'Silent Holocaust' sermon in Spokane, alerting his congregants to the dangers of exposure to radiation. As a result, a group of activists in the city formed the Hanford Education Action League (HEAL) to conduct research and educate the public, an initiative that sociologist Myron Peretz Glazer and historian Penina Migdal Glazer have identified as 'an important step in breaking the code of silence' (Glazer and Glazer, 1998: p. 31). HEAL would go on to bring about the public release of Hanford's environmental records. At the same time, farmers near Hanford also began to speak out about the unusual number of deformities in their animals – and in local human beings as well. As later recorded by journalist David Proctor, 'some lambs near Hanford were born without eyes, mouths or legs. Some had two sets of sex organs, others had none' (Proctor, 2001).

The damage to human beings was also severe. One woman living in the area named Juanita Andrewjeski had suffered three miscarriages. Together with her husband Leon, she kept a 'death map' of their neighbourhood. 'On it', Proctor recalled,

> were 35 crosses for heart attacks and 32 circles for cancer. One girl was born without eyes. Another couple had eight miscarriages and adopted all their children; two children were born without hipbones; one farm wife killed her baby and herself after her husband died of cancer.
>
> (Proctor, 2001)[6]

'It seemed like every man we knew was having heart problems', said Andrewjeski. 'We knew something was wrong, real wrong, because they were young men' (Coates, 1990).

Interventions like the Andrewjeski death map, in identifying apparent clusters of death and disease, provided local residents with the evidence they needed to make their voices heard (Taylor, 2008: p. 95). Calling themselves the Downwinders of Hanford, they began to attract the attention of the press. In 1984, for instance, the *Seattle Times* reported that a group of Downwinders worried about 'danger to Spokane from radiation emissions at the Hanford nuclear reservation briefed the City Council about their concerns and asked the council to take action to protect the city' (Associated Press, 1984a). Meanwhile, concerns also continued to circulate about Hanford becoming the 'radioactive waste dump of the Western world' if it were selected as the US's commercial waste repository. According to Karen Dorn Steele, a local journalist who first reported on the ecological impacts of Hanford, it was this circumstance that finally led to the declassification of Hanford's previously secret environmental records as the Department of Energy came under pressure to release environmental data on the impact of the site, both from state officials who were now able to invoke the Nuclear Waste Policy Act of 1982 and from local activists and journalists invoking the Freedom of Information Act (Steele, 1988: pp. 18–19).

Accounting for the dynamics of silence

Different levels of 'silence', 'consciousness', and 'spin' become apparent as one works through a systematic analysis of the articles published in the *Seattle Times*. Public knowledge of accidents and radioactive waste was not just a binary issue of 'knowing' or 'not knowing', but rather a matter of partial and fragmenting silences adhering to the Hanford Site. What sorts of things were known? And what sorts were ignored? And why did certain 'secrets' suddenly become public and penetrate people's awareness even though everyone always 'kind of' knew them?

The events reported in the *Seattle Times* did not actually coincide with the most dramatic contaminations that we now know to have taken place at the site, something that local communities could not easily detect partly because connected with areas of restricted access. Instead, minor but still very troubling events were reported, a series of environmental mishaps that people and public agencies could detect without the employment of sophisticated equipment and which therefore could not easily be hidden by the managers of the Hanford Site. This was the case, for example, with easily detectable leaks of radioactive material, variations in the Columbia Basin animal and plant populations, absorption of radioactivity by wildlife, increases in the Columbia River temperature, accidents involving Hanford employees, and the presence of chemical waste resulting from the production of plutonium. Such events, significant though they were, were relatively minor when compared to the overall amount of pollution produced by the facility. For this reason, it is not possible to assert that people were clearly conscious about what was going on at the nuclear facility or about the dangers that they were facing.

Articles like the ones written by Harold Osborne in 1956 and by Hill Williams in 1967 are representative, but also very peculiar in their nature inasmuch as they simultaneously conveyed information about potentially serious radioactive dangers and implied that the situation was always under control. Their writing technique promoted a sense of ambiguity, resulting in an inefficient capability to clarify controversial aspects related to radioactive contamination. Newspaper articles such as these offered a flow of partial and fragmented pieces of information, generating what might be called a middle ground between knowing and not knowing. This peculiar feature could be due in part to the journalists themselves receiving incomplete information, but also to the possible direct or indirect influence that the federal government had on the press (Gordin, 2007: p. 91). Despite not being subject to overt censorship, journalists and others complied with a nuclear consensus for most of the Cold War. With the first Soviet nuclear explosion in August 1949, it was clear that the American atomic monopoly was over. Dramatic indications of the alarming tension between the USA and the Soviet Union were published in newspapers almost every day. A bigger and more sophisticated atomic arsenal seemed to be the only way that the USA could face the terrible threat posed by the Soviet Union (Gordin, 2009: pp. 247–284). For this reason, scientists and workers at the Hanford Site, along with the people living in the Tri-Cities, considered themselves nothing less than front-line soldiers (Findlay and Hevly, 2011: p. 100; D'Antonio, 1993: p. 11). The security of the whole nation rested on their shoulders, and they had to commit themselves to doing their best to build efficient reactors and to produce the best isotopes to fuel the US atomic arsenal (D'Antonio, 1993: p. 11).

Patriotism and pride in the importance of being part of the 'Atomic City' meant the local people were unlikely to speak out against Hanford, but silence was also institutionally imposed by the Hanford employers. Workers were required to sign agreements that obligated them not to discuss with anyone, not even their families, what they saw and heard while at work and they were allowed to receive information only in relation to the specific, and often limited, tasks they were required to complete (Gerber, 1992: p. 47). In order to keep their jobs and avoid disciplinary action, Hanford workers avoided talking about what they did all day. As Gerber noted, 'Richland became a town where people simply never talked about their work' (ibid.). Even family members seemed to know nothing about the work of the nuclear site; yet, at the same time, they realized that 'speaking up or asking questions was forbidden and dangerous' (Glazer and Glazer, 1998: p. 30). Town shops near Hanford commonly featured posters displaying sentences like 'Don't be caught with your mouth open', 'For security "freeze up" this winter' or 'Don't talk shop', which directed people to maintain a strict secrecy about the nuclear facility (Gerber, 1992: p. 48). In addition, FBI agents 'periodically questioned Richlanders about their neighbor, searching for weakness, such as alcoholism or infidelity, that might make Hanford workers targets for blackmail' (ibid.).

In this context, environmental safeguards were a secondary concern. Even if someone suspected that Hanford's management of the radioactive waste was not perfectly safe, it was not quite possible to imagine anything like a serious environmental catastrophe. Radiobiological standards were different then compared to today (Walker, 2000). In any case, some of the best atomic scientists in the world were working at Hanford. It would have been natural to assume that if there were a problem, this brilliant team of scientists would have realized it and found an effective solution.

Thanks to a combination of genuine consent, personal interests and outside pressures, the Tri-Cities community had come to identify with Hanford's mission. The community was almost entirely dependent on Hanford's payroll. At the time when Hanford began to be constructed in the sagebrush, only about 250 people had lived in Richland. 'Today', reported the *New York Times* in 1962, 'it has about 25,000 and the biggest payroll is at the Hanford Works' (Turner, 1962). The nuclear facility was the only real business in the area, the economic engine of that arid and isolated place, and some of the local people may have traded potential environmental safety for military, political and economic benefits.

There were rumours about possible hazards caused by the site, but it was not until the early 1980s that these rumours and stories penetrated people's awareness enough for them to decide that it was time to do something. If the government and the Hanford management had hoped for a state of absolute silence, this was eventually disrupted by a series of accidental mishaps that called public attention to the problem and required a series of ad-hoc coverups to maintain social quietude – cover-ups that did not always look totally convincing.

Beginning in the 1980s, the disruptive potential of these Hanford accidents was amplified by a series of unexpected and surprising events at other nuclear production complexes around the United States (Amundson, 2002; Bartimus and McCartney, 1991; Del Sesto, 1974; Limerick, 1987; Lindee, 1994; Ringholz, 1989; Wellock, 1998). These accidents at other facilities affected the way that the public perceived Hanford and may have accounted for the change of tone in newspaper articles at this time. In 1983, Oak Ridge National Laboratory, near Knoxville, Tennessee, came under the scrutiny of the press after the publication of documents describing the discharge of 2.4 million pounds of mercury into the East Fork Poplar Creek between 1950 and 1977. In December 1984, the Fernald Feed Materials Production Center, a uranium processing facility located about 20 miles northwest of Cincinnati, was the focus of health and safety difficulties, accompanied by a series of public complaints, after the release of records documenting nuclear contamination. Fernald released millions of pounds of oxides and uranium dust into the atmosphere for almost thirty years. The site was also responsible for the leak of 230 tons of radioactive material into the Greater Miami River, along with other serious contamination and safety events (Gerber, 1992: p. 6).

Also in 1984, a class action suit was won by twenty-four plaintiffs representing 1,200 people who were deceased or living victims of radiation caused by nuclear atmospheric testing at the Nevada Test Site.[7] The following year, Arthur Dexter, a physicist working for DuPont Corporation, made public a memorandum describing previously classified instances of misconduct that took place at the Savannah River Site, a nuclear reservation located in South Carolina (Carothers, 1988). In December of that year, further problems came

128 *Cultures of silence*

to light at Savannah River, including cracks in the reactors, broken monitoring instruments and disconnected safety systems (Gerber, 1992: p. 5). Such incidents acted as a catalyst for popular awareness and mounting concerns about nuclear policies. They also suggest that the non-absoluteness of secrecy that characterizes the Hanford Site was shared by many other nuclear facilities in the United States.

Journalists reporting on Hanford had by this time shifted from their original role as drivers of secrecy to drivers of openness and publicity. The emergence of local activist groups helped consolidate this transformation by providing local journalists with alternative stories and a range of critical voices on which they could draw. Around the same time as Houff's sermon and the formation of HEAL, two other major voices joined the debate, both of which highlight the way in which members of the public started taking an active role in shaping the operational history of the Hanford Site, in part by working with local journalists (Glazer and Glazer, 1998: p. 31).

The first was Tom Bailie, a farmer affected by thyroid disease and sterility who grew up near the Hanford Site at the peak of its Cold War emissions. Even before the official public document release in 1986, Bailie was an activist who looked with growing suspicion at the inordinately high rates of cancer and disease suffered by the local people (Glazer and Glazer, 1998: p. 31). In 1985, he persuaded Karen Dorn Steele, a reporter for the *Spokesman-Review*, to release information about deformities in local animals and the high incidence of particular diseases and deaths among the human population (Glazer and Glazer, 1998: p. 31). In an interview from 1991, Bailie illustrates the way in which local residents occupied a middle state between knowing and not knowing prior to the official release of Hanford's classified documentation.

> I was born in 1947. My first memories as a child are of space-men-like people walking by my house and waving. I remember seeing deformed animals and watching my parents dispose of them. [...] Men in suits and ties would come into our school and pass Geiger counters around the classroom and over the children. We would wonder why one day all the jackrabbits would be lying around kicking and dying in the sagebrush, when just the day before they were alright. We always believed it was because we had done something wrong. We accepted a lot of guilt. Grandma always explained it was *God's will*. Early on in life one of my jobs was to shoot the deformed kittens and calves – because my father didn't like to do that. Seeing all the sick people and the handicapped children around us, I guess deep inside I knew something wasn't right, but I didn't have anything else to compare it with.
>
> (Glazer and Glazer 1998: pp. 29–30)

The second voice is that of Casey Ruud, the Hanford scientist who, in 1984, complained of the missing plutonium at the nuclear facility and was ignored by his superiors.[8] Ruud risked his job when he contacted the *Seattle Times* investigative reporter Eric Nalder in 1986 to disclose information about the safety violations at Hanford and discharges of nuclear waste into the Columbia River. Ruud's revelations were the focus of months of extensive press coverage by Nalder and other journalists at the *Seattle Times*, stories that, as Glazer and Glazer put it, 'sparked a congressional furor' (Glazer and Glazer, 1998: pp. 29–30).

The culmination of public efforts to disseminate information about Hanford came when HEAL and other activists forced the release of classified information about Hanford (Thomas, 2000: p. 41). 'When Hanford's managers told us not to worry, that there was no proof that anyone had ever been harmed by Hanford's past operations', HEAL member James P. Thomas recalled some years later, 'we grew more suspicious and began demanding documentation to support the official contention that everything was right. We felt that the government had lied in other cases, and we wanted to find out if it was lying about Hanford' (Thomas, 2000: pp. 40–41).

HEAL contacted an employee of the Environmental Policy Institute, Robert Alvarez, who, together with his colleague Bernd Franke, compiled a detailed request for official documentation and filed it under the Freedom of Information Act (FOIA) (Thomas, 2000: pp. 40–41). The request was signed by HEAL and other groups concerned about Hanford's operating history and was submitted in January 1986. In February that year, Hanford released about 19,000 pages of reports about its history from 1944 to 1985, including the revelation that a 1949 experiment had spread radioactivity across neighbouring states.[9] A few months later, another FOIA request was filed which resulted, in April 1987, in the release of more than 20,000 more pages.

Meanwhile, public concerns were significantly increased after the explosion of the Chernobyl reactor in Ukraine, in 1986. According to Steele, the Chernobyl nuclear accident, in a graphite-moderated reactor, turned national attention to the aging 4,000-megawatt N-reactor at Hanford, which was also graphite-moderated and which was shut down in 1987 (Steele, 1988: p. 22). In 1989, the Department of Energy together with the Washington State Department of Ecology and the United States Environmental Protection Agency signed the Hanford Federal Facility Agreement and Consent Order, more commonly known as the Tri-Party Agreement, a document that outlined 'legally enforceable milestones for Hanford cleanup over the next several decades'.[10] Between 1987 and 1993, a further 150 FOIA requests were filed. Several million pages on Hanford's history are now available to the public.

Conclusions

On a scale ranging from total secrecy to complete openness, the Hanford case is located somewhere between these two extremes. Various drivers of secrecy have shaped the history of the nuclear facility. Among them were military, political and economic agents that acted according to specific interests. The military was mostly concerned about the possibility that enemy agencies could have received important information regarding the health of the US nuclear weapons industry, whereas politicians and economists were interested in maintaining social order and in protecting the value of funds invested in the project. Public awareness would have meant the possibility of interference, fear and protests potentially capable of affecting these military, economic and political interests.

In addition to these general drivers of secrecy there were also local drivers of secrecy, social groups which acted at a local level and which had no interest in speaking out. Among these were the Hanford employees who knew about the various safety violations (those, for instance, who wrote the reports eventually disclosed in 1986), but also local journalists who were careful in writing about the nuclear facility in order to comply with the security sensitivities of the nuclear consensus.

We could say that Hanford was 'institutionally silent' in that there was no institutional flow of information from the authorities to the people. There was, however, a localized dissemination of partial information whose general trend is depicted in the articles analysed in this chapter. By the early 1980s, local communities and journalists, responding to a range of environmental, legal and cultural changes, started to break the institutional and precautionary silences, collaborating with each other to uncover the state of secrecy which had been imposed on the military and scientific activities of the nuclear site. By this time, local journalists had reinterpreted their role from drivers of secrecy to drivers of publicity, producing articles that amplified social concerns and disseminating the information that contributed to the eventual closure of the facility in 1987.

Notes

1 An early version of this chapter appeared in Macuglia (2013a; 2013b).
2 For a general overview of the US nuclear weapons industry, see: Hales (1997); Hewlett and Anderson (1962); Hewlett and Duncan (1969); Hewlett and Holl (1989); Rhodes (1986; 2007).
3 See: 'Hanford Quick Facts', State of Washington Department of Ecology website, web.archive.org/web/20080624232748/http://www.ecy.wa.gov/features/hanford/hanfordfacts.html; 'Hanford Facts', Physicians for Social Responsibility website, www.psr.org/chapters/washington/hanford/hanford-facts.html.
4 See also Hanford Cultural Resources Program (2002: p. 1.12) and Gephart (2003a: p. 6.1). On the general problem between Hanford and its local public, see Macuglia (2013a).
5 It is important to note that another victim of the same accident was Marvin E. Klunds, 43, of Richland, but his case received much less media coverage. McCluskey was treated with calcium DTPA and with the experimental drug zinc DTPA by Dr Bryce Breitenstein (News Services, 1977).
6 Also see the interviews and data collected in D'Antonio (1993).
7 The legal case was called *Irene Allen v. United States of America* because Irene Allen was the first on an alphabetically ordered list of twenty-four test cases.
8 For attempts to monitor employees of nuclear plants, see Hacker (1987; 1992; 1994).
9 In 1949 alone, between 7,000 and 12,000 curies of radioiodine – and even greater amounts of radioactive xenon-133 – were secretly released to the air on 2–3

December in an intentional experiment known as the Green Run (Associated Press, 1986).
10 'Hanford Overview', Department of Energy Hanford website: www.webcitation.org/68BrRtsd7.

References

Amundson, M., 2002. *Yellowcake Towns: Uranium Mining Communities in the American West*. Boulder: University Press of Colorado.
Anon., 1947. Hanford safety, health studied. *Seattle Times*, 22 Sep., p. 5.
Anon., 1948. Supervision, security at plutonium works. *Seattle Sunday Times Rotogravure*, 20 Jun., p. 2.
Anon., 1973a. Hanford springs another 'hot' leak. *Seattle Times*, 11 Jul., p. A14.
Anon., 1973b. Reports of nuclear leak unread, says A.E.C. *Seattle Times*, 31 Jul., p. A8.
Anon., 1973c. Company criticized by A.E.C. on leak of radioactive waste; analysis and report. *New York Times*, 5 Aug., p. 36.
Anon., 1974. Radioactive leak found. *New York Times*, 18 May.
Anon., 1987. Science watch: growing nuclear arsenal. *New York Times*, 28 April.
Associated Press, 1949. Atomic-energy plant is safer than your home. *Seattle Times*, 13 Aug., p. 4.
Associated Press, 1954. Hanford project to be made 'Radiation Area'. *Seattle Times*, 15 Oct., p. 19.
Associated Press and United Press, 1955. Duck eggs near Hanford radioactive. *Seattle Times*, 17 Aug., p. 2.
Associated Press, 1973. Leak of radioactive waste held no peril to humans. *New York Times*, 14 Jun.
Associated Press, 1979. Radiation casualty hopes he can help. *Seattle Times*, 15 Nov., p. A16.
Associated Press, 1982. Radioactive liquid spills at Hanford. *Seattle Times*, 6 Apr., p. D7.
Associated Press, 1983. Nuclear survivor 'Atomic man' makes medical history. *Seattle Times*, 18 Aug., p. A22.
Associated Press, 1984a. Group is worried about radiation. *Seattle Times*, 24 Oct., p. E7.
Associated Press, 1984b. Plutonium missing at Hanford. *Seattle Times*, 3 Dec., p. B1.
Associated Press, 1986. 1949 test linked to radiation in northwest. *New York Times*, 9 Mar.
Bartimus, T. and McCartney, S., 1991. *Trinity's Children: Living Along America's Nuclear Highway*. New York: Harcourt Brace Jovanovich.
Belton, T.J., 2010. *Protecting New Jersey's Environment: From Cancer Alley to the New Garden State*. New Brunswick, NJ: Rutgers University Press.
Carothers, A., 1988. Plutonium politics: the poisoning of South Carolina. *Southern Changes*, 10(4), pp. 4–5, 8–11.
Coates, J., 1990. Victims used 'death map' to connect cancers, plant. *Chicago Tribune*, 23 Jul., p. 4.
Cunningham, R., 1945. Hanford made material for atomic bomb that hit Japan. *Seattle Times*, 8 Aug., p. 1.
D'Antonio, M., 1993. *Atomic Harvest: Hanford and the Lethal Toll of America's Nuclear Arsenal*. New York: Crown Publishing Group.

Davies, L.E., 1949. Atomic plant set for big expansion; Hanford Works on threshold of the second phase of 400 million post-war program. *New York Times*, 4 Nov., p. 5.
Del Sesto, D.L., 1974. *Science, Politics and Controversy: Civilian Nuclear Power in the United States, 1946–1962*. Chicago: University of Chicago Press.
Findlay, J.M., 2004. The nuclear West: national programs and regional continuity since 1942. *Journal of Land, Resources, and Environmental Law*, 24(1), pp. 1–15.
Findlay, J.M. and Hevly, B., 2011. *Atomic Frontier Days: Hanford and the American West*. Seattle: University of Washington Press.
Gephart, R.E., 2003a. *Hanford: A Conversation About Nuclear Waste and Cleanup*. Columbus, OH: Battelle Press.
Gephart, R.E., 2003b. A short history of Hanford waste generation, storage, and release. Pacific Northwest National Laboratory. Available at: www.pnl.gov/main/publications/external/technical_reports/PNNL-13605rev4.pdf
Gerber, M.S., 1992. *On the Home Front. The Cold War Legacy of the Hanford Nuclear Site*. Lincoln: University of Nebraska Press.
Glazer, P.M. and Glazer, M.P., 1998. *The Environmental Crusaders: Confronting Disasters and Mobilizing Community*. University Park: Pennsylvania University Press.
Gordin, M.D., 2007. *Five Days in August: How World War II Became a Nuclear War*. Princeton, NJ: Princeton University Press.
Gordin, M.D., 2009. *Red Cloud at Dawn: Truman, Stalin, and the End of the Atomic Monopoly*. New York: Farrar, Straus and Giroux.
Groves, L.R., 1983. *Now It Can Be Told: The Story of the Manhattan Project*. New York: Da Capo Press.
Hacker, B.C., 1987. *The Dragon's Tail: Radiation Safety in the Manhattan Project, 1942–1946*. Berkeley: University of California Press.
Hacker, B.C., 1992. Radiation safety, the AEC and nuclear weapons testing: writing the history of a controversial program. *The Public Historian*, 14, pp. 31–35.
Hacker, B.C., 1994. *Elements of Controversy: The Atomic Energy Commission and Radiation Safety in Nuclear Weapons Testing, 1947–1974*. Berkeley: University of California Press.
Haitch, R., 1980. Follow-up on the news. Atomic man. *New York Times*, 10 Aug., p. 49.
Hales, P., 1997. *Atomic Spaces: Living on the Manhattan Project*. Urbana: University of Illinois Press.
Hanford Cultural Resources Program, 2002. *U.S. Department of Energy, Hanford Site Historic District: History of the Plutonium Production Facilities, 1943–1990*. Columbus, OH: Battelle Press.
Hewlett, R.G. and Anderson, O.E., 1962. *The New World, 1939–1946*. University Park: Pennsylvania State University Press.
Hevly, B. and Findlay, J.M., 1998. *The Atomic West*. Seattle: University of Washington Press.
Hewlett, R.G. and Duncan, F., 1969. *Atomic Shield, 1947–1952*. University Park: Pennsylvania State University Press.
Hewlett, R.G. and Holl, J.M., 1989. *Atoms for Peace and War, 1953–1961: Eisenhower and the Atomic Energy Commission*. Berkeley: University of California Press.
Hillaby, J., 1955. Escaping radioactivity invades wild life around Hanford Plant, survey shows pollution of the Columbia River by atomic chemicals – insects, birds, fish, other animals affected. *New York Times*, 18 Aug., p. 2.
Hughes, J., 2002. *The Manhattan Project: Big Science and the Atomic Bomb*. New York: Columbia University Press.

Kaiser, D., 2005. The atomic secret in red hands? American suspicions of theoretical physicists during the early Cold War. *Representations*, 90, pp. 28–60.

King, W., 1979. Nader hits 'cruel' bill, nuclear waste. *Seattle Times*, 18 May, p. A14.

Limerick, P. N., 1987. *The Legacy of Conquest: The Unbroken Past of the American West*. New York: Norton.

Lindee, M.S., 1994. *Suffering Made Real: American Science and the Survivors of Hiroshima*. Chicago: University of Chicago Press.

Macrakis, K., 2010. Technophilic hubris and espionage styles during the Cold War. *Isis*, 101, pp. 378–385.

Macuglia, D., 2013a. Die nuklearen Anlagen von Hanford (1943–1987): Eine Fallstudie über die Schnittstellen von Physik, Biologie und die US-amerikanische Gesellschaft zur Zeit des Kalten Krieges. In: Forstner C. and Hoffmann D. (eds), *Physik im Kalten Krieg: Beiträgezur Physik geschichte während des Ost-West-Konflikts*. Berlin: Springer Spektrum, pp. 77–87.

Macuglia, D., 2013b. Hanford and the middle ground between 'knowing' and 'not knowing'. *Bulletin of the Atomic Scientists*, 'Voices of Tomorrow', 31 Oct. Available at: http://thebulletin.org/hanford-and-middle-ground-between-knowing-and-not-knowing

Mooers, C. and Van De Water, M., 1945. Atomic discoveries useful to medicine and industries. *Seattle Times*, 23 Sep., p. 3.

Mullner, R.N., 1999. *Deadly Glow: The Radium Dial Worker Tragedy*. Washington, DC: American Public Health Association.

News Services, 1977. Radiation victim amazed at recovery. *Seattle Times*, 16 Mar., p. A2.

Osborne, H.F., 1956. Columbia River is their laboratory. *Seattle Times*, 7 Oct., p. 8.

Pope, D., 2008. *Nuclear Implosions: The Rise and Fall of the Washington Public Power Supply System*. Cambridge: Cambridge University Press.

Proctor, D., 2001. Nuclear murder: America's atomic war against its citizens and why it's not over yet. *Boise Weekly*, 27 Jun.–3 Jul.

Ratliff, J. and Salvador, M., 1994. Building nuclear communities: the Hanford Education Action League. Paper presented at the Annual Meeting of the Speech Communication Association. Available at: http://files.eric.ed.gov/fulltext/ED379721.pdf. Accessed Jul. 2015.

Rhodes, R., 1986. *The Making of the Atomic Bomb*. New York: Simon & Schuster.

Rhodes, R., 2007. *Arsenal of Folly: The Making of the Nuclear Arms Race*. New York: Knopf.

Ringholz, R., 1989. *Uranium Frenzy: Boom and Bust on the Colorado Plateau*. New York: Norton.

Steele, K.D., 1985. Cancer deaths cited in arguments against DOE policing Hanford. *Spokane Chronicle*, 15 Aug., p. A3.

Steele, K.D., 1988. Hanford's bitter legacy. *Bulletin of The Atomic Scientists*, 44(1), pp. 17–23.

Taylor, B.C., 2008. *Nuclear Legacies: Communication, Controversy, and the U.S. Nuclear Weapons Complex*. Lanham, MD: Lexington Books.

Thomas, J.P., 2000. 150 requests. *Bulletin of the Atomic Scientists*, 56(6), p. 41.

Turner, W., 1962. Plutonium City calm; residents doubt Russia would bomb site that produces warhead 'punch'. *New York Times*, 6 Nov., p. 18.

United Press, 1949. Pollution to be studied. *Seattle Times*, 13 Nov., p. 15.

United Press International, 1973. Leak of radioactive waste. *New York Times*, 11 Jul.

United Press International, 1984. 7 workers contaminated at Hanford nuclear site. *Seattle Times*, 15 Dec., p. A14.

Walker, J.S., 2000. *Permissible Dose: A History of Radiation Protection in the Twentieth Century*. Berkeley: University of California Press.

Walker, J.S., 2009. *The Road to Yucca Mountain*. Berkeley: University of California Press.

Wellock, T.R., 1998. *Critical Masses: Opposition to Nuclear Power in California, 1958–1978*. Madison: University of Wisconsin Press.

Wells, J., 1961. Maritime news around the world. *Seattle Times*, 7 Sep., p. 14.

Williams, H., 1967. B.P.A. attacks fear – nuclear plant obstacle. *Seattle Times*, 17 Jul., p. 23.

Williams, H., 1973. Is Hanford waste really a threat? *Seattle Times*, 15 Jul., p. C5.

Williams, H., 1976. 'Serious' health problem for Hanford blast victim. *Seattle Times*, 2 Sep., p. B7.

Williams, H., 2011. *Made in Hanford: The Bomb That Changed the World*. Pullman, WA: Washington State University Press.

Zwinger, S. and Smith, S.D., 2004. *The Hanford Reach. A Land of Contrasts*. Tucson, AZ: University of Arizona Press.

6 Silence and selection

The 'trick cyclist' at the War Office Selection Boards

Alice White

In March 1939, psychiatrists wrote to the War Office of Britain to offer up their services in the likely event of war. The response? A resounding silence. This unpromising start marked the first words (and the first silence) in a discussion of psychological science that would span the war.

The psychiatrist, or 'trick cyclist' in soldiers' slang, was a controversial figure during World War II. At War Office Selection Boards, psychiatrists sought a voice to speak not only of the deviant populations that they conventionally studied, but also to discuss normal and even 'superior' members of society. Winston Churchill, amongst others, was not at all sure about this, noting the 'immense amount of harm' they might do. Suspicions of 'these gentlemen' and their interest in taboos such as sex resulted in a number of enquiries into, and limitations upon, their work at selection boards during the war (Churchill, 2010: p. 815). The most contentious site of such negotiations was the psychiatric interview, where psychiatrists assessed soldiers put forward for commission.

This chapter analyses the technique of the psychiatric interview at the War Office Selection Board as the point of intersection between Army authorities, soldiers and psychiatrists: those commissioning science, those subject to the gaze of science and those practising science. Silences in and around the interview punctuated larger discussions around democracy and authority, and who might speak on whose behalf. I begin with a brief summary of how the problems of selecting Army officers were raised and how official silence on what was expected of psychiatrists provided them with the opportunity to experiment in this new field. The use of unofficial spaces for conversations about their work enabled psychiatrists and their supporters to develop their methods until they were sufficiently robust to be accepted by military colleagues.

I then discuss the silences in psychiatric interviews themselves. These interviews were a key method for selecting officers, and were both literally and metaphorically a place where voices might speak or be silent, through choice or coercion. They thus offered interviewees opportunities for advancement in the Army but also potential for regulation. Finally, I examine who wanted to silence the psychiatrists and limit their interviews, and why. The 'tightest hand' was increasingly kept over them, with suspicious Army senior leadership censoring what psychiatrists were permitted to ask and even attempting to have them completely removed from officer selection work. Psychiatrists attempted to resist such efforts to control their voice. Informal networks, off-the-record conversations and manipulations of omissions in orders and instructions were key to eluding limitations and keeping psychiatrists in the battle for the borders of their discipline.

Speaking up on selection

It is far from obvious how psychiatrists came to be involved in the work of officer selection. A long and complicated battle had to be fought to get from the first letters sent by psychiatrists (and ignored by the Army authorities) to the stage where psychiatrists were actively involved in choosing Army officers. However, by early 1939, J. R. Rees, a psychiatrist from the Tavistock Clinic in London, had decided that he and his colleagues should be conducting work with the Army, and Rees was a determined man.[1] Speaking on behalf of his colleagues in the psychological sciences, he sent the letter to the War Office to which there was no reply. This would not be the first time that the psychiatrists failed to have their voices acknowledged. The problem, however, was solvable: though the psychiatrists themselves were not considered worthy of a reply at this time, there were others whose status did require a response who could ask on their behalf. From this point forward, the psychiatrists used proxies wherever possible, putting the case for their work to the highest military authorities via military men of established rank and respect.

Following Rees's unsuccessful letter, the psychiatrists arranged to have questions asked in Parliament about the psychiatric provision for war (Trist and Murray, 1990: p. 1). The questions implied that soldiers' mental health was at risk from the lack of experts to safeguard it, and suggested that the British Army was in danger of repeating the mistakes of the past by failing to utilise available expertise. Harold Boyce, Member of Parliament for Gloucester, sought reassurance that the Secretary of State for War, Oliver Stanley, would 'ensure that these men are not lost, as many of our most distinguished specialists were in the last war, in doing general services?' Stanley had to admit that only nine officers were commissioned for employment in psychiatric work (House of Commons, 1940: cc. 1892–3). One of those who had been commissioned was Rees, who had managed to get himself appointed as a consulting psychiatrist to the British Army in 1939, meaning that he was in charge of any Army psychiatrists who were in Britain. As Stanley had admitted, this wasn't very many men to begin with; however, the parliamentary questioning meant that the government was unable to stay silent on the matter and was forced to address the issue. Shortly afterwards, Rees was given the opportunity to appoint a number of Command Psychiatrists.

The 'officer problem' was becoming difficult for the Army to avoid. There was a chronic shortage of officers for the Army, and those who were obtaining commissions were mocked as relics of an older world of 'polished cross-straps, swagger canes, long haircuts, and Mayfair moustaches' (Field, 2011: p. 264). Between September 1939 and mid-1942, when nearly 2 million men entered the Army, the method of selecting officers was very simple. Men were marked as potential officer material by commanders of training units and then interviewed, often for only around twenty minutes, by a commanding officer. There was a level of official silence about what constituted a good officer; no training was given to those selecting the men, and often 'they tended to base their conclusions on personal hunch or preconceptions that there was an "officer-producing class"' (Broad, 2013: p. 102). For example, during the little time that officer candidates were given to speak about themselves, they were reportedly often asked: 'What school did you go to? What is your father? What private income have you? ... Did you go to a public school?' (House of Commons, 1942: cc. 1954–5). Men's backgrounds were seen to speak louder than their personalities, and it was believed that good candidates were missing out on commissions because blinkered senior officers could not recognise their talents, instead promoting those with the right 'old school tie'.

The resultant image of the out-of-touch gentleman officer was blamed for poor morale and battlefield failures, and mocked relentlessly in the press, particularly in the figure of Colonel Blimp, the creation of cartoonist David Low. There were a few attempts to defend the 'natural leadership' of the 'old school tie men', such as a notorious letter to *The Times* by Lieutenant-Colonel Bingham, which resulted in his sacking (Bingham, 1941). However, most of the voices in the press were outspokenly scathing about the poor state of officer selection. It was seen as hurting the morale of the Army and stunting the supply of officers, as men lacking 'old school ties' refused to put themselves forward if they were to inevitably fail. The existing selection methods were also seen as working against British principles: 'the simple liberal proposition that men should have equal opportunities is one of the things we surely are fighting for against the German idea of the master class and the master race' (Jones, 1941). By overlooking the middle and working classes simply because of dialects or educational background, the Army were seen as ignoring good candidates to an extent that was seen as immoral.

In the early years of the war, officer selection was thus a problem in need of a solution, one which psychiatrists felt they could supply. Psychiatrists by this point had been dispatched to work with different local Army commands. These men lacked any specific orders, and there was complete silence from the Army hierarchy as to what their role should entail. Command psychiatrists thus set about investigating what they felt were the pertinent problems of their local commanding officers. Top of the list for several was the matter of officer selection. The psychiatrists were supported in their work through connections and unofficial discussions with local military figures, who sought their advice on problems and took interest in their experiments.

The first large-scale experiments in new methods to choose officers were prompted by Lieutenant-General Andrew Thorne of Scottish Command, who had seen new 'scientific' selection methods in Germany when working there as military attaché before the war. When Thorne had previously attempted to trial new methods in Britain, he was accused of being the 'bloody Freud of the British Army!' (Rees, 1945: p. 53). It was only when Thorne was promoted to General Officer Commanding-in-Chief and transferred to Scotland (a move lamented in his biography as having 'side-lined' him from true military work) that he was able to start selection experiments. This supposedly peripheral role, where little was officially expected from him, enabled him to pursue innovations that had long interested him and that would later be described by some as making 'revolutionary' changes to the Army (Bidwell, 1973: p. 121).[2]

Other experiments soon followed those initiated by Thorne, were seen as great successes and received endorsement from the local military commanders. Sir Ronald Adam, who took over Northern Command in June 1940, was one of those won over by the psychiatrists' work.[3] In June 1941, Adam was appointed Adjutant General and given the power to implement reform to selection procedures. In December 1941, the director of personnel selection Brigadier K. G. McLean and Scottish Command held a conference in Edinburgh on officer recruitment. There, the psychiatrists and their supporters argued that the combination of intelligence tests and psychiatric interview was the only efficient method for selecting officers. One month later an experimental War Office Selection Board (WOSB) was formed. By March, techniques for selection had been refined and the boards were ready to be implemented across Britain and beyond. Explaining the new scheme in the House of Commons at this point, Edward Grigg noted that:

> There are two questions which, the House will no doubt be interested to know, the Adjutant-General has ruled out of order. One is, 'How much money have you got?' and the other is 'How much money has your father got?' Those two have been completely ruled out.
> (House of Commons, 1942: cc. 1993)

This would be the first mandate about what could and could not be said at WOSBs, but not the last.

Strategic and problematic silences at selection boards

Talking and silence were strategically alternated at the selection boards. When a new batch of candidates arrived, the president of the board addressed them with the aim of reassuring them that the WOSB 'was interested neither in the status of our family nor in the place of our education' (Fleming, 1944). The president thus addressed concerns linked with the perceived unfairness of the old selection procedure before they could be voiced. Following this, the psychiatrist was introduced and asked 'to say a few words'. These few words fulfilled a key purpose for the psychiatrists, as men were 'told that if their notion of a psychiatrist is simply to get mental defectives out of the Army and neurotics into and out of hospital, then they have something new to learn'

(Rickman, 2003: p. 150). Within the space of a few minutes, then, meeting one group of officer-candidates at a time, psychiatrists were attempting to push back the culturally constructed boundaries of their profession that had limited them to the role of alienist working with strange behaviour in asylums. These few words, aimed to dispel notions of what a psychiatrist was not, were an attempt to legitimise the idea that their subject included 'normal' men. However, they were silent on what a psychiatrist *was*, leaving candidates to form their own ideas during the three days that the WOSB took place.

Not only did psychiatrists attempt to broaden what their discipline could advise upon, they aimed also to extend psychological thinking to other members of the WOSBs too. The military men who headed the WOSBs as presidents underwent a process of orientation whereby they were trained to view situations from a psychological viewpoint. The president was to be trained by his psychological colleagues because he was 'better ... the more he has been trained in first principles. How else can he make successful adaptations to changing conditions?' (Smith, n.d.). This training was not only intended to increase the psychological voice and make practices more generalisable, it was also intended to prevent 'the temptation to superimpose unchecked judgements instead of evaluating the evidence, and to "interfere"' (Harris, 1949: p. 212). In training others in their principles, psychological staff hoped to silence what they saw as unscientific opinions.

Psychiatrists also aimed to initiate a dialogue with the candidates. Men were told that 'the WOSB's tests are a new experience and they are invited to participate in them as observers ... and to feel free to ask questions' (Rickman, 2003: pp. 150–151). Whilst this appeared to give officer candidates a voice in a way that sharply contrasted with the older interviews that the WOSBs replaced, it was also a method of gaining the co-operation of the psychiatrists' subjects. By inviting him to participate in the process, a candidate at the WOSBs was thus converted from a passive subject 'into a proto-psychologist and encouraged to adopt a scientific gaze' (De Vos, 2010: p. 161). Men were not simply encouraged to give their opinions. They were encouraged to participate critically in a psychological procedure. In a subtle and indirect way, the opinions and questions of the candidates became reconstructed as the valuable reasoning of involved 'observers'.

The 'conversation' between officer candidates and psychiatrists began before the WOSB interviews themselves. Before the interview, there was often an 'informal "cocktail party" type of meeting' to establish initial contact between individual candidates and the psychiatrist before the interview (Trist et al., n.d.: p. 6). In addition, psychological pointers, questionnaires and group tests were all conducted in advance of the interviews so that the psychiatrists would have a bank of information from which to draw. This enabled the psychiatrists to strategically allocate their precious time:

> Where the interviewer felt he knew the candidate from his previous contact he would devote a shorter time to him in interview. Conversely,

where a candidate presented a complex picture he could be given more time.

(Trist et al., n.d.: p. 5)

The officer candidate's identity was thus established as much through unspoken revelations on paper and in a group, as it was in conversation at the interview itself. In some cases this silent communication of personality was even given more weight than the interview. The more the candidate was believed to have 'said' about himself in these tests, the less time was considered necessary for an interview. The psychiatrists noted that 'the increasing amount of discussion amongst the observer team of candidates on the basis of ... preliminary contacts added to [the] tendency to make the interview directed to the elucidation of specific aspects of the candidate' (Trist et al., n.d.: p. 5). The more that others had already discussed a candidate, the less he was required to divulge one-to-one. In some cases, the candidates were expected to be totally unaware of the tests' aims in order to reveal their personality through the anxiety this produced. In others the men were asked to look within themselves, attempt to view themselves objectively and record their findings for the boards. Candidates were then sometimes rendered almost silent by the expert voice, which interpreted test data and implied that little more could be added by the man himself. In the revelation of their character, therefore, the would-be officers were sometimes passive and sometimes active.

The psychiatric interview was structured such that it proceeded from the present day and led back to the candidate's past. In this sense it was seen as therapeutic because it resulted in the candidate 'confronting himself' and 'achieving some piece of insight however slight into some of the implications of one's traits and behaviour' (Anon, 1945: p. 6). Despite enrolling candidates as proto-psychologists, the theoretical underpinnings of the WOSBs' procedure were concealed from candidates. The psychiatrists believed that their silence about their intentions produced a more valid picture of the 'real' man, including the bubbling to the surface of his anxieties. For instance, the psychiatric interview was revealing and challenging for the candidates because:

> the boundaries of the psychiatric investigation are not clearly defined in the candidate's mind. If he is uncertain of himself he is liable to imagine the widest and wildest variety of probes and investigations into his personality, and to fear that his weaknesses will surely be found out!

(Rickman, 2003: p. 155)

The opening talk mocked the idea of the psychiatrist dealing only with madmen, but it was a talk characterised by absence – the psychiatrists capitalised on the unresolved question of what it was they actually did, and how, by leaving boundaries ill-defined. Psychologists had long employed silent deception as a method to produce what they saw as more valid results (Pettit, 2013), but the psychiatrists at WOSBs took this further. Silence-induced ambivalence about the intent of the psychiatric interview was seen as a useful method, not only because it passively prevented bias, but because it actively developed anxiety and thus revealed the true inner character of candidates.

There were fears that in some particular forms of the WOSB a lack of understanding of the modern role of the psychiatrist might prevent candidates from speaking up. Because of this the psychiatric interview was disguised as a part of the medical check-up at boards where schoolboys were selected for pre-officer training university courses. Boys learned that a man was a doctor, but his speciality of practice was not mentioned (Anon., 1945: p. 3). Again, this reflected previous use of deception in the psychological sciences to achieve results, including the disguise of the person of the psychological investigator (Pettit, 2013). Despite these precautions, word usually got out: 'the majority of candidates were quite familiar with the role of the Army Psychiatrist about which they had heard from their brothers and friends or the Press' (Anon., 1945: p. 3).

The analysis of these revelations claimed to be one of three things central to the method of the psychiatrist's interview. This silent interpretation was so integral that psychiatrists noted that their interview could not 'in fact be distinguished from that by the President in terms of content and scope or even in terms of the general method of proceeding' (Anon., 1945: p. 6). The interview itself was unremarkable, but the unspoken element (both in terms of what the candidate revealed and in the 'technical interpretation' the psychiatrist's training brought to the revelation) was what 'distinguished the interview by the psychiatrist' (Trist et al., n.d.: p. 12).

However, psychiatrists also considered themselves particularly adept at conducting interviews due to their insight into the interviewee's psychology. Psychiatrist John Rickman was particularly interested in the manipulation of the balance of words and silence at officer selection. In his wartime essay on psychological theory and practice, 'The psychiatric interview in the social setting of a War Office Selection Board', he argued that in the ideal discussion 'the talk is fluid and balanced by informed statement from my side, not at all a string of questions' (Rickman, 2003: p. 151). So fundamental was this idea of productive dialogue to the WOSBs' procedure that it was even articulated in parliamentary discussion of the new boards, where Edward Grigg explained that:

> The object of the questions ... is to try and get the candidate to talk and to express himself and his own personality ... It is no use cross-examining the candidate. You get nothing by that means. You have to try and find out what the man is like, and the way to do it is to get him to talk himself.
>
> (House of Commons, 1942: cc. 1999)

The psychiatrists believed that not only should the psychiatrist abstain from asking too many questions and making too many interpretations, but they should also actively employ silences as 'a powerful diagnostic technique, whose method is inconspicuous' (Rickman, 2003: p. 154). They considered that the psychiatrist's training in the therapeutic context provided him with the ability to 'judge how far he should probe, when he should leave things untouched or when he should bring them out, so that the candidate was benefited and not necessarily disturbed' (Trist et al., n.d.: p. 14). Even the point at which the conversation had apparently concluded, as the candidate was heading out of the door, might be valuable to the psychiatrist, as 'he may betray a cynical or contemptuous attitude which he could control when sitting during what he regarded as the interview proper' (Rickman, 2003: p. 153). Such skilful deployment of silence was claimed to be distinctive to the psychiatrist.

Psychiatrists suggested that their expertise was what enabled them to employ 'oblique evaluation', or an indirect approach to questioning. Oblique evaluation was seen as required precisely because there were limits on what psychiatrists could ask due to 'a social norm which expressed what candidates and the authorities alike accepted as relevant enquiry' (Trist et al., n.d.: p. 16). For instance, a psychiatrist could not productively ask 'Was (or is) your home-life happy?' without producing a 'defensive and stereotyped' response. Such questions had to therefore remain unasked. In order to get around these limits, psychiatrists evolved oblique questions such as: 'Do you think you take after one or other of your parents in your own temperament?' Some topics were particularly taboo, and where detected, certain 'emotionally charged trends ... e.g. a tendency to homosexuality, had to be conducted with great sensitivity' (Trist et al., n.d.: p. 18). These tactful omissions and silences were vital to the acceptability of the psychiatrists' interviews.

Rickman argued that silence was particularly effective when dealing with the talkative candidate 'who blusters or bluffs his way through difficulties, or tries to' because those who 'try "to work up an effect" find this bland silence particularly trying'. The silent psychiatrist was essentially a blank canvas onto which the candidate projected information about his assumptions and attitudes to others. The psychiatrist 'may turn into a persecuting figure in his mind, or into a friendly one, or into a fool to be kicked out of the way' (Rickman, 2003: p. 155). By staying silent, the psychiatrist also invited the candidate to step into the role of psychiatrist himself, and he was credited with having 'insight' if he seemed aware of his own behaviour. The psychological orienting of officer candidates was thus seen as a positive, desirable, 'officer' quality. Once the candidate had poured into the silence 'the diagnostic information [the psychiatrist] wants, [the psychiatrist] breaks the silence' (ibid.: p. 155). Silence at the psychiatric interview therefore acted as a space for the officer candidate to fill with information about themselves. There were two layers to the information thus acquired: the superficial information as provided by the words explicitly written or spoken, and the underlying signs of the candidate's unconscious personality traits and his assumptions about the world and its human dynamics. However, although the psychiatric silence was described as useful, it was also a potentially dangerous technique. In deploying it with blustering candidates, it posed a danger to the psychiatrist's morals lest they derive a degree of pleasure from the process, and Rickman advised that the psychiatrist must '*watch himself*' to see whether he is getting an irrelevant satisfaction from the effect'. Furthermore, the analysis could cause harm to the candidate if prolonged or if not sufficiently concluded.

Though risky, silence was for the psychiatrist, 'a powerful diagnostic technique' that could be used to get a man to talk about himself. However, candidates who were not forthcoming enough were seen as problematic, both to those dealing with them and to themselves (Rickman, 2003: p. 154). WOSB memoranda noted that the psychiatrist 'is on the watch for points at which on the one hand things become sticky and the candidate closes up, and on the other for points at which the candidate becomes more communicative' (Anon., 1945: p. 6). This watching for silences indicates that the interviewer was not simply interested in what was said, but also in the way in which information followed (or failed to follow), revealing how a candidate dealt with people and therefore whether he was a potential officer. 'Sticky' silences where a man 'closed up', it was argued, could be read as reflective of a man struggling to make human contact. Thus for those who lacked a voice, the psychiatric interview was strategically planned to include therapeutic techniques in order to get shy men to say more about themselves. After a gentle, general beginning to the conversation, the psychiatrist was to pose the question of when the candidate really began to get over their shyness. With the usual response being that he still had not, the psychiatrist would respond along the lines that 'be that as it may, the situation is no doubt better now … we can see in what way you can help yourself still further in this matter in the future' (Rickman, 2003: p. 154). Silence on the part of the candidate was not only an impediment to the psychiatrist in measuring his character, it was also seen to be a problem for the man himself which could be overcome with expert guidance. Silence was almost deemed to be a curse which limited a man's opportunities, and the psychiatric interview therefore was a 'privilege' for the man because in a 'kind of *quid pro quo* … he gives the psychiatrist rather embarrassing details of his shyness, in return he gets a method of dealing with his difficulty' (Rickman, 2003: p. 154).

This idea of a balanced interview as a fair exchange with the candidate was echoed in the idea of therapeutic closure, the final feature that the psychiatrists argued characterised the psychiatric interview. At the end of the session, a 'sealing-off' technique was employed. This was designed to fulfil the implied promise from the introductory talk that the process would be interactive – the candidate was asked if he had any questions for the psychiatrist and was given the opportunity to voice his thoughts and opinions about the process. Under the old selection process, there had been no opportunity for candidates to voice their thoughts, and the feedback which leaked out in the form of rumours about an unfair system was seen to result in low morale. The chance for the candidate to speak at the psychiatric interview provided a less harmful outlet for these thoughts: 'the emotional significance for the candidate of the question being raised was the important matter' (Trist et al., n.d.).

However, this sealing-off was not only intended to provide for emotional release, it was also a continuation of the assessment of character. The candidate's questions were a demonstration of 'his capacity to think freely ... and to criticise [the board's] work in a constructive (or destructive) way'. When the candidate did ask questions, they were met with a strategic lack of answers on the part of the psychiatrist, who turned the questions back for him to answer, 'the amount of help he needs being a measure of his grasp of the situation' (Rickman, 2003: p. 156). The amount that a man could say by himself without the need for the psychiatrist to guide his speech was a measure of his capability; the psychiatrist could remain silent because the process of psychologisation had occurred to such an extent that the candidate could make his interpretations himself. As such, the men seen as most capable by the psychiatrists were those who were, or had become, like the psychiatrists themselves.

Attempts to silence the 'trick cyclist'

Although psychiatrists thought deeply about when to speak and when to be silent at the WOSBs, there were many who considered that the psychiatric profession should be entirely silent on the matter of officer selection. By and large, WOSBs had proved hugely successful with the vast majority of candidates. They were 'almost universally approved by those who go through them whether they pass or fail' (Cripps et al., 1942: p. 3). Aversion to the psychiatrists' work came from those in the higher reaches of military authority, for example General Sir Bernard Paget and Winston Churchill, who felt their authority and the traditional system of the Army was being threatened. In response to the statement that some disapproved of the psychiatrists, the MP Geoffrey Cooper pointedly asked 'Was that on the part of senior officers or other ranks?', knowing full well that it was senior officers (House of Commons, 1946: cc. 960). Some public school candidates (who felt cheated of a position that once would have been guaranteed) also complained (Crang, 2000: p. 34).

Psychiatrists deployed silence in their interviews to expose those who tried to 'blind interviewers with science' and trick their way into a commission by spouting terms that made them sound knowledgeable. Ironically, concerns that psychiatrists were doing just the same themselves led to attempts to silence them. General Paget was dismayed by psychiatrists' involvement in selection boards, being particularly concerned by the psychiatric interview and the presence of the psychiatrist at the final meeting where conclusions were collectively drawn about candidates. He felt that 'psychiatrists were dominating the selection procedure by virtue of their technical knowledge and the ability to present evidence' (Crang, 2000: p. 35). The voice of the psychiatrist compared with the military men was seen by Paget to be problematic, bamboozling them with jargon and forcing them into silence. Psychiatrists themselves admitted that it was 'much easier' to validate the opinions of the psychiatrist at the board conference because 'where the psychiatrist's opinion differed substantially from the president's he could usually furnish evidence that explained the difference' (Trist et al., n.d.: p. 2). They saw this as the opportunity both to begin a discussion about a man and to educate others in their science. Conversely, men such as Paget considered that psychiatric judgement closed down discussion because its objective manner unfairly trumped the more relevant experience of military men at the boards.

In addition to concerns about what the psychiatrists said at the conference, there were worries too about the probings of the psychiatric interview. Because of fears about the psychiatrists' descent into disturbing realms, 'grievances arising from a sense of such exploitation [of what could acceptably be asked] were quick to rebound against WOSBs' (Trist et al., n.d.: p. 15). Winston Churchill himself wrote to John Anderson, the Lord President of the Council, in 1942 to express his concerns:

> I am sure it would be sensible to restrict as much as possible the work of these gentlemen [psychologists and psychiatrists], who are capable of doing an immense amount of harm with what may very easily degenerate into charlatanry. The tightest hand should be kept over them ... it is very wrong to disturb large numbers of healthy normal men and women by asking the kind of odd questions in which the psychiatrists specialize.
> (Churchill, 2010: p. 815)

Although the psychiatrists had anticipated 'a social norm which expressed what candidates and the authorities alike accepted as relevant enquiry', they were unable to avoid the problem of the negative connotations that their discipline carried in the minds of those such as Churchill (Trist et al., n.d.: p. 16). Psychiatry was often conflated with psychoanalysis and thus was a subject almost unspeakable to some. During the interwar period, few doctors (let alone politicians and generals) would mention an interest in psychoanalysis 'without the verbal equivalent of spitting three times over the left shoulder, and even to speak about the revival of war memories carried the risk of being accused of advocating free fornication for everyone' (Culpin, 1952: p. 71). The strength of opposition to psychiatry resulted in attempts to prevent the psychiatrists' ideas from being voiced in relation to officer selection.

Because of his suspicions about their 'odd questions', Churchill initiated an Expert Committee on the Work of Psychologists and Psychiatrists in the Services to investigate, and ideally prevent, what they were doing. Though they were ostensibly examining all psychological work, the Committee had been particularly invited by the War Cabinet 'to consider the question whether there might be a tendency to use the psycho-analytical technique too extensively, and whether, if unwisely handled, it might encourage the very tendencies it was hoped to combat' (Cripps et al., 1942: p. 1). This clearly related to psychiatrists' interest in 'odd questions', and expressed the suspicion that the mere presence of psychiatrists might have a damaging effect and 'put ideas of fear and mental instability into the minds of otherwise healthy men'. However, it also related to concerns that psychiatrists threatened traditional military beliefs, such as the idea 'that unintelligent types ... would make good fighting soldiers' (Cripps et al., 1942: p. 8). The Expert Committee therefore embodied concerns both about what psychiatrists might say to candidates and also the idea that psychiatrists' beliefs subverted the authority of senior military leadership. The Committee was a focus for efforts to silence the psychiatrists on both fronts.

The protocols by which the Expert Committee should work had not been stated and the psychiatrists capitalised on this omission to their advantage. Ronald Adam, the military commander who supported the new methods, once again assisted the psychiatrists by intervening to ensure that Stafford Cripps, the Lord Privy Seal, investigated. Had Adam not intervened, there was a chance that a 'prosecutor' like Churchill's physician Lord Moran might become judge, jury and executioner (Thalassis, 2004: p. 95). Instead, the psychiatrists came under the scrutiny of Cripps, a man 'fascinated' by modernisation, something that the rhetoric around the boards had always emphasised. In addition to discussions of modernisation, the psychiatrists shrewdly also provided unspoken expressions that they and Cripps were like-minded: they laid on a vegetarian 'feast' to cater to his preferences with 'special quantities of carrots ... provided for his meals' (H. Vinden, quoted in Shephard, 2001: p. 195). Cripps subsequently supported the psychiatrists, writing in his report that there was 'no substance of the criticisms' which had been levelled at them (Cripps, 1942). He also explicitly stated that the psychiatry used in officer selection was not the same as the controversial psychoanalysis he had been charged with rooting out, explaining that 'conditions would not permit of [psychoanalysis] being used in war-time even if those responsible in the service admitted its value' (Cripps et al., 1942: p. 7).

However, the attempts to silence the psychiatrists did not end with Cripps's investigation. In 1943, instructions were issued that psychiatrists should interview intimately no more than half of the candidates and that no questions on sex or religion were permitted. These instructions produced great frustration amongst the psychiatrists, who felt that the authorities had failed once again to understand their method and its fundamental concepts. Psychiatrist Robert Ahrenfeldt noted that it should have been clear:

> to all but the most prejudiced, that it was of the greatest importance ... to enquire in appropriate cases into so significant an aspect of the human mind, behaviour and social adaptation as sexual adjustment. Similarly, it should have been obvious that [where the question of religion arose] they were attempting a fundamentally sociological evaluation of a man's attitude to established authority.
>
> (Ahrenfeldt, 1958: p. 64)

Yet it was evidently far from obvious. Panicked concerns that psychiatrists might be hunting out complexes and having inappropriate and damaging conversations drowned out the explanations that the psychiatrists provided.

As well as their frustrations about the limits imposed on their questioning, psychiatrists were concerned that the limitations on the number of interviews would return them to their previous status as alienists who dealt only with problem cases. They worried at possible 'anxiety aroused in candidates where certain candidates only were interviewed ... candidates would wonder why they had been so selected' (Anon., 1945: p. 2). When universally delivered, the psychiatrists' opinions were welcomed by the candidates. When they were silent on most cases and only called in to deal with 'problem' cases, the psychiatrists were seen as a worrying unspoken indicator of a man's deficiencies before judgement had been officially passed on him. In such a scenario the candidate was tainted by the mere association with the psychiatrists and so in return resented them and became uncommunicative in an attempt to hide from view complexes he felt they were hunting.

To get around the seemingly conclusive ruling limiting their work, the psychiatrists carefully manipulated the unspoken elements around the instructions. They noted that 'the wording of the rule had a certain ambiguity', and so they managed to find space for their work in the space between 'no interview' and 'intimate interview'. Psychiatrists 'interpreted the ruling ... in terms of "intimate interviewing" as the essential point' and 'strove to give ... short interviews which they could not feel in any way to be intimate' (Anon., 1945: p. 2). A memorandum was circulated outlining the 'correct interpretation' of what constituted an 'intimate' interview:

> An interview is *intimate* when it entails a direct confrontation of certain difficulties and problems of the candidate. This entails an extension of direct questioning beyond the usual limits and for this reason must be followed by a full application of therapeutic closure. Intimate interviews are necessary only in the exceptional case and are conducted solely by the Psychiatrist.
>
> *Non-intimate* interviews include all other interviews whether by the President, Psychiatrist or Psychologist.
>
> (Anon., 1945: p. 9)

In this way, the psychiatrists tactically obeyed the order to the letter, whilst resisting efforts to silence their influence over the WOSB.

The psychiatrists felt that there was a delicate balancing act between the said and the unsaid when it came to matters of selection:

> If on the one hand there were limits in the selection setting to what (except under special circumstances) might be asked by direct questioning, there were equally limits as to what might be omitted. For example, the 'atrocity' stories which arose from feelings that psychiatrists had transgressed one limit were paralleled by 'atrocity' stories regarding the scant and superficial questions of the old Interview Boards.
>
> (Trist et al., n.d.: p. 17)

They hinted that the danger of silencing the psychiatrist was to return to the 'old school tie' method of superficial selection of the familiar, rather than the sometimes intrusive programme of scientific selection of the ablest men. In February 1945, the Expert Committee fully vindicated psychiatric expertise in its report to the War Cabinet. However, this report would not be published for more than a year. In addition:

> Despite the expert committee's acknowledgement of the contribution of psychiatry and psychology to the selection of officers, in September 1946 a War Office Committee headed by Lieutenant-General Sir John Crocker recommended that both psychologists and psychiatrists should be withdrawn as permanent members of the WOSBs.
>
> (Crang, 2000: p. 38)

Sir Ronald Adam was no longer in a position to support the psychiatrists, and Lieutenant-General Sir Richard O'Connor, the replacement for Adam as Adjutant-General, supported this recommendation. The Chief Psychologist to the War Office, Bernard Ungerson, noted that this decision was made 'in spite of the contrary advice from all the very senior psychologists and psychiatrists who advise the War Office' (Ungerson, 1950).

Conclusion: was the 'trick cyclist' silenced?

Was it then the case that psychiatrists were back to square one, with their advice being issued and duly ignored by the War Office, as had been the case with Rees' 1939 letter? Ben Morris, in reply to Bernard Ungerson, argued that it was not unjustifiably optimistic to consider that progress had been made, since 'the Army continues to use the advice of psychological technicians regarding the selection and training of members of WOSBs' (Morris, 1950). As this chapter has discussed, though the boards themselves no longer had a psychiatrists' voice present, that voice was being imparted to others via psychologisation. Other board members and even the men who passed through the boards themselves were being trained in psychological thinking and evaluation of personality. These 'proto-psychologists' thus spoke with a voice inflected by psychiatric thought even without the psychiatrists being physically present at boards. Furthermore, Morris also argued that whilst his preference would be for psychological staff to be present, 'the existence of such an adviser is not an essential component of the *methods* themselves' (Morris, 1950). There were many tests in which psychological information about the candidates was drawn out before they reached the interview. Many of these psychological tests remained integral parts of the boards, thus giving voice to psychiatric theories and approaches in a hidden, less objectionable fashion through the appearance of 'objective' data on a Hollerith computer tabulated page. Though psychiatrists were no longer active in selecting officers, their presence remained in the British Army.

In addition to this, the popular acceptance of the psychiatrists' work by those subject to their gaze had opened up numerous opportunities to them. After the war, Rockefeller and Medical Research Council funding flowed in the direction of the Army psychiatrists in part because their methods had been proven acceptable to large populations by work such as the WOSBs. Private companies such as Unilever employed individuals on the basis of their work at WOSBs, and the National Fire Service, Civil Service, the India Office, the American Office of Strategic Services (a precursor of the CIA) and the Palestine Police all adopted variants of the scheme and called on psychiatrists to advise them on its implementation. Though the psychiatrists had been silenced on the matter of officer selection, they were asked for their opinions on who would make good managers and foremen in industry.

Finally, the psychiatrists had learnt valuable lessons about the opportunities that could be located in what was unsaid. They could omit a full explanation of their role, as they did particularly with the schoolboys; they could disguise psychiatric methods as being military in appearance; and they could work around instructions and act in the spaces of what was not explicitly ruled out to find a space for their practices, as they did with the 50 per cent rule limiting the number of intimate interviews. This sort of manipulation of silence, and the construction of techniques and approaches in the spaces of the undefined, would prove vital in Army psychiatrists' next project, to rehabilitate returning prisoners of war. Here, if the men themselves were resistant to being tainted with psychological diagnoses, the government was even more reluctant to accept the existence of psychological problems lest they be required to provide compensation. Civil Resettlement Units have their psychological roots so silently embedded that they are often not recognised even today as psychological constructions. The roots of this later work can be found in the silences of the WOSB.

Notes

1 Rees had worked with ambulance units and the Royal Army Medical Corps during the First World War and then helped Hugh Crichton-Miller to establish the Tavistock Clinic to treat 'functional nerve cases' in returned servicemen, so he had a long history of work in military psychiatry.
2 There remains historiographical debate over how much the backgrounds of officers changed as a result of changes to selection. However most historians agree that a cultural change did occur in how the matter was approached by the Army (Crang, 2000).
3 Adam had taken one of the intelligence tests and not only completed it in a fraction of the time allotted, but also scored at the very top of the scale. It is possible that a degree of his support for the psychiatrists stemmed from the reinforcement at this moment of the idea that he himself was ideal leadership material.

References

Ahrenfeldt, R.H., 1958. *Psychiatry in the British Army in the Second World War.* London: Routledge and Kegan Paul.

Anon., 1945. *Interviews by the Psychiatrist and the Psychologist at WOSBs*. Box 205802222. London: Tavistock Institute Archives.
Bidwell, S., 1973. *Modern Warfare: A Study of Men, Weapons and Theories*. London: Allen Lane.
Bingham, R.C., 1941. Man management. *The Times*, 15 Jan., p. 5.
Broad, R., 2013. *The Radical General: Sir Ronald Adam and the New Model Army 1941–46*. Stroud: The History Press.
Churchill, W.S., 2010. *The Hinge of Fate: The Second World War*. New York: Rosetta Books.
Crang, J.A., 2000. *The British Army and the People's War, 1939–1945*. Manchester: Manchester University Press.
Cripps, S., 1942. *Use of Psychologists and Psychiatrists in the Services – Enquiry by Lord Privy Seal*. WO 32/11972. London: The National Archives.
Cripps, S., Alexander, A.V., Grigg, P.J. and Sinclair, A.H.M., 1942. *The Use of Psychologists and Psychiatrists in the Services*. CAB 98/26/10. London: The National Archives.
Culpin, M., 1952. A criticism of modern trends in the treatment of psychoneurosis. *Medical Press*, 228, pp. 71–73.
Daston, L. and Galison, P., 1992. The image of objectivity. *Representations*, 40, pp. 81–128.
De Vos, J., 2010. From Milgram to Zimbardo: the double birth of postwar psychology/psychologization. *History of the Human Sciences*, 23(5), pp. 156–175.
Dicks, H., 2009. John Rawlings Rees. In: *Munk's Roll: Lives of the Fellows*, online edition. Royal College of Physicians, p. 387. Available at: http://munksroll.rcplondon.ac.uk/Biography/Details/3726. Accessed 14 May 2014.
Field, G.G., 2011. *Blood, Sweat, and Toil: Remaking the British Working Class, 1939–1945*. Oxford: Oxford University Press.
Fleming, U., 1944. Men into officers. *Spectator*, 6 Jul., p. 9.
Harris, H., 1949. *The Group Approach to Leadership-testing*. London: Routledge and Kegan Paul.
House of Commons, 1940. *Psychological Specialists*. Vol. 357, cc. 1892–1893.
House of Commons, 1942. *Captain Margesson's Statement*. Vol. 377, cc. 1924–1999.
House of Commons, 1946. *Psychologists and Psychiatrists*. Vol. 431, cc. 959–961.
Jones, E., 1941. Isn't this what we're fighting for? *Daily Mail*, 24 Jan., p. 2.
Morris, B., 1950. A reply to Colonel Ungerson. *Occupational Psychology*, 24, p. 59.
Pettit, M., 2013. *The Science of Deception: Psychology and Commerce in America*. Chicago: University of Chicago Press.
Rees, J.R., 1945. *The Shaping of Psychiatry by War*. New York: W.W. Norton.
Rickman, J., 2003. The psychiatric interview in the social setting of a War Office Selection Board (1943). In: King, P. (ed.), *No Ordinary Psychoanalyst: The Exceptional Contributions of John Rickman*. London: Karnac Books.
Shephard, B., 2001. *A War of Nerves: Soldiers and Psychiatrists in the Twentieth Century*. Cambridge, MA: Harvard University Press.
Smith, B., n.d. *History of WOSB: Handwritten Notes*. Box 205802222. London: Tavistock Institute Archives.
Thalassis, N., 2004. Treating and preventing trauma: British military psychiatry during the Second World War. Ph.D. thesis. University of Salford.
Trist, E. and Murray, H., 1990. Historical overview: the foundation and development of the Tavistock Institute. In: Trist, E. and Murray, H. (eds), *The Social Engagement*

of *Social Science: The Socio-Psychological Perspective*. London: Free Association Books, pp. 1–34.

Trist, E., Sutherland, J.D. and Morris, B., n.d. *WOSBs Write-Up Manuscript*. Box 205802225. London: Tavistock Institute Archives.

Ungerson, B., 1950. Mr Morris on officer selection. *Occupational Psychology*, 24, p. 55.

7 The silenced subject
Oral history and the experience of cancer research

Catriona Gilmour Hamilton[1]

Introduction

This chapter examines the silences of cancer research in Britain from 1970 to the present. It considers the process of communication and decision making in the consultation room, an experience that we might assume to have moved from a shadowy and silent past toward a more transparent and outspoken present. However, as I will show, silences remain, albeit differently configured and charged with different meanings.

The period begins with almost complete silence on the subject of cancer and the conduct of research. From the perspective of 2015, when individuals can expect to be empowered and autonomous participants in research into their illnesses, it is difficult to grasp just how different things were in 1970. I will begin by briefly mapping out this period of considerable change.

Information – or an absence of silence – is the unit of currency in the process of patient empowerment, autonomy and choice, whether we speak of it in terms of enhancing understanding of illness and treatment, as a tool to enable parity with health professionals, or simply as the expectation that one's questions will be cogently and honestly answered (Tomes, 2006; 2007; Mold, 2010; 2013). At the beginning of the 1970s, there were no accessible sources of information for the person with cancer. It was a diagnosis engulfed in silence. Libraries held basic medical text books for those with the literacy to comprehend them. Women's magazines, and the occasional newspaper article, offered scant and often sensational coverage. The journalist and well-known agony aunt Claire Rayner observed in 1980 that cancer storylines evoked considerable interest from readers but editors remained reluctant to cover such material, fearful that it would compromise sales. For information about what to expect of the diagnosis or its treatment, individuals had to rely instead on the people caring for them.

This reliance was confounded by the fact that it was common practice among doctors in Britain to withhold information about a cancer diagnosis. The prevailing paternalism ensured that patients were prevented from exposure to the unvarnished truth for fear of a catastrophic loss of hope. The merits or otherwise of disclosure attracted heated debate throughout the 1970s and were widely debated in medical journals and sociological analyses (McIntosh, 1977; Novack et al., 1979).

This situation resulted in corresponding silences in the conduct of cancer research. How could consent to participation in cancer research be sought if the individual did not know they had cancer? When medical practitioners began to reconsider the question of cancer disclosure, they were influenced by an increasing demand for enhanced scrutiny of research practice. From the 1970s until the turn of the twenty-first century, scrutiny of research moved from internal professional self-governance to external independent review, culminating in the mandating of external regulation by impartial research ethics committees. These committees became increasingly insistent that researchers should seek the informed consent of research subjects, a process that necessitated the provision of information about trial rationale, discussion of potential risks, opportunities to ask questions and the ability to grant consent free from coercive influence.

Of course, the process was contested. On the subject of informed consent, for example, although enshrined in good practice since the 1964 Declaration of Helsinki, there is evidence to indicate that people taking part in cancer clinical trials were not routinely asked to give consent until well into the 1980s. But by degrees the silences associated with research conduct were progressively challenged (Gilmour Hamilton, 2013; Hedgecoe, 2009; Hazelgrove, 2002).

This, coupled with the rise of cancer information provision in the voluntary sector and the advent of the world wide web, meant that by the 2010s the individual with cancer might expect to feel suitably empowered and respected as an equal partner, equipped with the information to facilitate autonomous choice.

To rely on that big picture, however, risks an assumption: that current patients benefit from an improved regulatory environment such that the 'research partners' of today are reliably protected from the silences of the past. Yet the ubiquity of cancer information and the official recognition of the patient voice might be misleading. To avoid this risk of complacency, it is necessary to attend to the hitherto silent voice of the individual research participant.

A case for subjectivity

Although the patient voice is now potent in the politics of research conduct, it is conspicuously silent in the historiography. Medical historians generally, under the influence of the biographical and linguistic turns in the humanities, have used life stories and individual narratives as the appropriate methodological and interpretative response to the challenge of building a patient-centred history (Chamberlayne et al., 2000; Condrau, 2007). Yet, with a handful of exceptions, the history of cancer research tends to marginalise patient subjectivity, much as the methodology of the randomised controlled trial rigorously obscures individual identity.

To maintain this silence is to overlook the rich potential of individual subjectivity. This chapter examines the experiences of silence as recalled in individuals' memories, using oral narratives of individuals treated for lymphoma in Britain since 1970 who were either subject to innovative treatments or were participants in clinical trials.[2] These silences are the manifestations of power and powerlessness: the silence of what is not said; the silence of not being heard; the metaphorical silences of what cannot be known. Individual accounts suggest that silences are not necessarily confined to the past, but instead they have undergone a process of redistribution in recent years.

Perhaps most significantly, oral narratives undermine the legacies of the recent history of research governance, the origins of which were largely reliant on exploitation of the silenced, including prisoners in Nazi concentration camps, mentally incapacitated children and the African-American men of Tuskegee (Weindling, 2004; Lederer, 1995; Jones, 1993). These histories cast silence as a malign entity, exemplar of brutal medical hubris or, at best, a misplaced paternalism. Individual oral narratives suggest more nuanced definitions of silence. They question the simplistic dichotomy that assumes that silence is malign and that information is empowering. In fact, in the artful exchange of good communication, in the fraught process of becoming informed, silence can be a resource, skilfully applied and strategically appropriated.

'It wasn't spoken about'

The silences that characterised the British experience of cancer in the 1970s are illustrated by Margaret's story.

In 1971, Margaret was aged 20 and living with her family in County Durham. She had an office job and led an active life, going on walks with her youth club across the Pennines and working towards a Duke of Edinburgh award. Her activities were gradually curtailed by unexplained fatigue and a chronic cough. She had started to sweat profusely at night and was mortified at the thought that she might be wetting the bed. After a protracted period of repeat visits to her GP, she was eventually referred for investigations in Newcastle and was diagnosed with advanced Hodgkin's disease just before her twenty-first birthday.

Hodgkin's lymphoma is a cancer of the lymphatic system which is diagnosed in around 1,400 people in the UK each year.[3] It is now among the most curable of adult cancers, but at the time of Margaret's diagnosis advanced Hodgkin's disease was considered incurable. Some progress had been suggested by the work of a team from the US National Institutes for Health, who had tested a combination of chemotherapy drugs during the 1960s, the results of which were published in the *Annals of Internal Medicine* in 1970. The combination was abbreviated to MOPP – mustine, oncovin, procarbazine and prednisolone – and had demonstrated unprecedented success in people with advanced Hodgkin's lymphoma (DeVita et al., 1970).

Margaret was treated for two years with repeated courses of radiotherapy before her consultant decided to try the chemotherapy treatment he had learned of during his stay in America. Margaret was an early UK recipient of MOPP, in late 1973. It was successful, and she went into a permanent remission.

In a manner typical of British practice at the time, Margaret – although an adult – was told nothing of her diagnosis:

> Well I don't think cancer was mentioned then. Obviously my parents and sister knew but my mother didn't want me to know.
> *So they told your parents?*
> Yes they told my parents. They actually told them I only had four weeks to live. Er, I think my mother just didn't want … you know. And I suppose at that time cancer was, like, it wasn't spoken about. She didn't want me to know. I mean, I was old enough but I suppose they wanted to protect us in a way.

Margaret remembers hearing other women on the hospital ward whispering to each other at night and mentioning 'cancer'. It was not a word that people said out loud. She only heard of 'Hodgkin's disease' when she read a nurse's notes on a pathology form two years after her diagnosis. When she summoned up the courage to ask her doctor about it, he refused to discuss it:

> *When did you learn what was wrong with you?*
> When I had my blood taken and I saw them write it down, upside down! And then I saw my consultant and I said, you know, 'Have I got …?' And he *still* didn't elaborate. He didn't give me any bumph and say, 'Go home and read that', he said, 'Go and … look it up.' Well the only place I could find was the library and I found one article that said, 'Invariably the patient dies.' So when you're 23, 24 you don't [pause] I mean now there's the internet and everything, the Hodgkin's Society, but then, nothing like that.

Margaret's experience was one of total disempowerment and isolation. Even after a distance of forty years, she speaks of her experience with an emotional immediacy that belies that length of time. Her account is marked by the silence of her healthcare team and by the silencing of her attempts to find out what was going on. For Margaret there have been prolonged legacies of that silence: bitterness, a mistrust of health professionals and sadness at her prolonged isolation. Under such circumstances any notion of consent, autonomy or choice is rendered meaningless. This kind of silence is all encompassing.

'The decision has to be yourself'

Compared to Margaret's story, Lina's account of her experience in 2003 might suggest an interval well in excess of a mere thirty years. It takes place at a point in history at which silence might be assumed to be something of the past.

Lina is in her early 40s and lives in Buckinghamshire. She is a hairdresser and works from a small purpose-built conservatory-cum-salon to the rear of her large and immaculate kitchen. In 1997 her husband Darren, a painter and decorator, developed a lump in his neck, which was dismissed as insignificant by his GP. However, by 1998 the lump had grown and a biopsy in February of that year led to a diagnosis of advanced follicular lymphoma. He was 33 years old.

Follicular lymphoma is a subtype of non-Hodgkin's lymphoma. It is the most common indolent lymphoma, with an annual UK incidence of around 2,500. It is typically a disease of later life, so Darren was young for such a diagnosis. Advanced follicular lymphoma, in which disease is present in multiple sites in the body, is a disease that has long frustrated efforts to cure it. It behaves as a chronic illness with periods of remission followed by relapse, the duration between relapses typically becoming progressively shorter. The disease can develop resistance to prior therapies and can transform, meaning that changes in cell characteristics make the disease more aggressive. At the time of Darren's diagnosis, median survival for advanced follicular lymphoma was 8–10 years.

Because Darren had no symptoms and was otherwise well, he was not treated and had regular check-ups for the first five years. But in 2003 he became acutely unwell and a large tumour was found near his kidney. The disease had transformed to become aggressive, a situation associated with a poor prognosis. By this time the couple had three children, the youngest just a baby.

Darren and Lina were offered treatment with a stem cell transplant. Already established in the treatment of relapsed aggressive lymphomas, doctors were increasingly looking at stem cell transplants for follicular lymphoma, on the basis that it might secure that elusive prolonged remission. Stem cell transplant therapies can involve a transfusion of one's own cells (autologous) or the transfusion of donor cells (allogeneic). Autologous stem cell transplants rely on very high dose chemotherapy to deliver a comprehensive blow to the malignant disease. The consequent life-threatening obliteration of the bone marrow, without which no blood cells are produced, necessitates a transfusion – or 'transplant' – of haematopoietic stem cells to re-establish the bone marrow. Allogeneic stem cell transplants work on the premise that the donor cells will replace the immune system of the recipient, and that this new immune system will kill malignant disease: the graft-versus-lymphoma effect. However, a significant risk of these donor transplants, aside from the prolonged recovery period during which blood cell counts are very low, is the graft-versus-*host* effect, in which the new immune system attacks the organs of the recipient's body. At the time of writing, donor transplants are considered a potential cure for follicular lymphoma, but at the time of Darren's treatment, the use of donor transplants was in its comparative infancy and each individual treatment was framed as 'experimental.'

Darren had a course of chemotherapy following which the couple had to choose between autologous and allogeneic transplant:

They then decided that he would benefit from a stem cell transplant, either be it from himself or from a donor. And I think at that time I know that they used to say that the mortality rate was pretty high when you were having a donor transplant, you know, they were doing the stem cells. So, er, anyway his brother was a perfect match so we ... and he'd agreed, you know, he'd be happy to be the donor if we decided to go down that avenue.

[...]

So as he had then finished the CHOP [cyclophosphamide, hydroxydaunorubicin, oncovin (vincristine), prednisolone] which he had had done locally we were then in sort of a bit of a dilemma as to whether he should go and try and get cured and go for the donor from his brother or go for a stem cell from himself. Erm but like I said the mortality rate at that point, at that time, was quite high. Erm and they did say if it did relapse you did have that option of then being able to, in time, if you needed it, to go for the donor. So after much sort of – oh it was awful, really, because at the end of the day the decision, although the doctors can advise you, the decision has to be yourself, and my husband kept saying 'what should I do, what should I have?' And I just, you know, I sort of [sighs] you want them to be cured but you don't want them to die [pause]. So in the end we said we'll go for the stem cells from himself.

The choice was between two uncertainties: a highly toxic treatment with less chance of cure versus one that might cure but with significant mortality risk. It was a choice based entirely on the information given to them at the hospital and what Lina found online:

I mean we were given a lot of information and I constantly, constantly went on the internet and read about it so yeah we were pretty informed I think.

Was it written stuff or was it mainly conversations?

No written stuff. I read a lot of stuff. And I mean Darren's, you know, like with anything, Darren's not interested, he doesn't wanna know, that's how he dealt with his illness. I would have to [pause] and then I'd read to him and that's what we'd discuss [pause] the decision [pause] you know, but in the end to make that decision [sighs] oh I can't remember. But basically it was he said, 'No, I'll go for my own stem cells.'

At no point in the interview does Lina suggest resentment at having to make the decision – she accepts that 'the decision has to be yourself'. But being informed did not make that decision any easier. Indeed, the trauma of that original choice remains in Lina's telling of it a decade later. She speaks in an anxious tone, her voice trembling slightly, interspersed with heavy sighs. She appears to struggle with the strong feelings these memories evoke. It is as if she still doesn't quite comprehend how they reached their decision, in spite of 'constantly, constantly' being on the internet. Her repetition of 'constantly' suggests that her memory is of ongoing attempts to gain control without ever feeling *in* control.

In spite of full disclosure and access to information, Lina's account is suggestive of silences. Darren's doctors are silent in her account. Lina tells her story with scant reference to their words. Compared to other interviewees, who offer detailed accounts of meetings at the hospital, conversations with individual personalities and direct quotations of what was said, Lina refers to these people with only the occasional 'they'. She makes no mention of meetings or conversations even though they undoubtedly took place: in her recollection, in what she remembers of how it felt, she casts herself and Darren as isolated protagonists.

The most significant relationship in terms of decision making is the one Lina had with her computer. By 2003 the internet had become a substitute for what would once have been a conversation. Clearly, that conversation would have been vulnerable to the persuasive bias of the clinician. The clinical view would have been subject to what we are now, in the era of the evidence hierarchy, encouraged to be suspicious of: experiential knowledge, anecdote, prejudice. But the conversation would have constituted a human interaction that is curiously absent from Lina's account.

She suggests a decision-making environment that is simultaneously a cacophony and a silence. She consults the infinity of the internet, with its multiple voices, the experts, the survivors, the quacks. But the web is mute in response to what she is most desperate to know. Although Darren survived the transplant, and will be one of those who eventually contribute to growing confidence in donor stem-cell therapies, his own case must be decided in the evidence-free silence of the unknown.

Silence is not necessarily an entity of the past, but rather it is something that has changed shape. The imperatives of individual autonomy have silenced coercive influence but there is a silence that persists in the isolation of individual decision making. Ostensibly, the comparison between 1973 and 2003 suggests considerable progress: Margaret's doctors made her decisions for her, Darren's doctors are silent on the subject of what his choice should be; she had no information, he had access to infinite quantities. But these superficial markers of empowerment are too simplistic and risk complacency. Without the addition of the emotions and subjectivities of individual oral history, historical understanding is insufficiently nuanced.

'I'll do this one for you'

Unlike Lina and Darren, Carol was reluctant to use the internet to help her with decision making. Her treatment choices relied on the exchanges of the consultation room. Carol's experience has considerable chronological scope, from her diagnosis in 1984 up to the early 2000s. It suggests still more permutations of silence in research conduct: that silences persist in the securing of consent in spite of the efforts of research governance; and that individual voices are silenced by those with greater power.

Carol is a retired nurse living in North Yorkshire. In 1984, when she was aged 36, she was diagnosed with advanced follicular lymphoma. She had two school-aged boys at the time and was working night duty. After six months treatment with oral chemotherapy she was monitored with watchful waiting until her disease relapsed in 1994 and again 1999, when she was treated with intravenous chemotherapy.

Following the second relapse, her consultant invited her to take part in a clinical trial of a new drug, rituximab. Rituximab had been licensed for use in follicular lymphoma in 1997. In combination with chemotherapy it would eventually come to be considered a significant innovation in follicular lymphoma therapy, helping to prolong and improve disease remission. At the time of Carol's treatment it was tested as a single agent. Here she describes her decision to take part in that trial:

> But that's when I went on this trial. I'm not too sure why I went, they just said 'Well will you go, it just suits you.' And I said, 'Well OK if that's what you think.'
>
> And the guy at the time that asked me he was very into research and things and was very keen at that time on I think stem cell transplants were just beginning to emerge in lymphoma treatment and he was suggesting that maybe I'd like to do this. And I said 'Maybe I wouldn't, thank you. I'll do this trial for you.' And he said 'Well you know it's £6,000 and you have four doses of this thing.' So I said 'Right OK.' So off I went to do this thing. And that was the only information we kind of got about it!

By 1999, NHS research ethics committees were supposedly monitoring clinical research to ensure adequately informed consent. Yet Carol's experience suggests enduring silence, her doctor reluctant to say more than 'It just suits you.' The rhetoric around consent, choice and communication, about provision of information and protection from coercion, had scant impact on this particular negotiation. Note Carol's reference to being told what the treatment costs, her saying 'I'll do this one for you' and 'If that's what you think.' She describes acting on her doctor's imperatives, her decision to participate in the trial offered as an appeasement for her refusal to have a stem cell transplant. This exchange is suggestive of a stand-off: the doctor's reluctance to engage in a conversation and the silencing of that patient's attempts to broaden discussion.

By 2004, Carol had reluctantly agreed to go ahead with a stem cell transplant:

> also at that time they were still going on 'You need to have a stem cell transplant.' And I was thinking 'No I don't.' I didn't understand them, I didn't really want this thing.

[...]
And I said 'I don't think so. What are you going to do, are you going to make me well and in 18 months' time you want me to come back and go through some more chemotherapy to knock out my bone marrow and give me some cells back?' I said 'Why? Why would you want to do that?' I was doing a degree at the time I said, 'No, I don't, No, I'm working, I'm doing a degree, I don't want all this performance thank you very much.' I said 'I'll just have the chemotherapy now and I'll just get on with the rest of my life.' I couldn't work this out. They were on and on and on about this from sort of '93 onwards. And I finally succumbed in 2004 and said 'Well, fine, if that's what you want to do then fine, we'll do it.'
[...]
And yet for me I wasn't that sure. Even then, you know, I wasn't that sure that it was the right process for me but I thought 'Oh, I've put them off all this time, maybe they do know a bit more than me.'

The emotional register of Carol's description suggests an adversarial relationship in which her point of view is silenced. Her repetition of her own objections indicates that her memory is of feeling unheard. She speaks of being persuaded, not of coming to a mutually agreed decision. Carol 'succumbed' because she had 'put *them* off all this time'. She does not feel that she was given adequate information about what she should expect. For reasons that are obscure, her team decided to maintain silence on the serious side effects of high dose therapy, which came as a profound shock to her.

The treatment did not have the outcome she had hoped for. After an experience that involved very great physical and psychological distress, one that she felt she might not survive, the outcome was a disappointment:

and within twelve months the lumps had appeared again, although it was another three years before I needed treatment. So I suppose I got four years and *they* think that was good [laughs a little] *they* think that there was some success in it although I think I probably felt that there should have been a lot more years free and it wasn't.
[...]
And like I say they still think it was the right way forward. And I've seen the consultant at Jimmy's since – she did the radiotherapy – and I said to her 'I wish I'd never had that stem cell' and she said 'You can wish all you like but …' she said 'there was success, Carol.' So they're still not going to say it wasn't successful.
[...]
The only consolation was that I did collect an awful lot of stem cells! They've still got some of mine, they said 'Well, you're top of the league.' I said 'That's *really* good news, that's great …' you know … And then they say 'Well you were the first person in less than the tenth day the cells reappeared,' and they said, 'And in the Unit, in Leeds, that hasn't

happened before' and I said 'That's probably because I'm praying like hell to get out of here, I really don't want to be here!' If you look at it in that way you collect a load of stem cells and the cells returned very quickly but it was still a *long* road to recovery, which we didn't know half of the problems, didn't know that I wouldn't be able to swallow for *days* on end and that I would have to have diamorphine for *days* on end and that it would take an awful long time to recover. We didn't know any of that, you know, that I'd end up being so thin because it was just too sore to swallow.

(Speaker's emphasis)

Definitions of success or failure are subjective and determined not by the individual who is treated but by the people observing her. Carol's subjective interpretation of her experience, her regret, is silenced by the assessments of her doctors who are determined to assert its success on their own terms: she yielded copious stem cells and she was the first person in her hospital whose blood count recovered within ten days, something that they seemed pleased about for the institutional record. But for Carol, 'success' is compromised by memories of physical and emotional difficulty and a remission too short to justify the trauma. For her, the treatment was a setback.

Written transcription fails to capture the emotional qualities of Carol's narration. Her tone as she says 'That's *really* good news' is sardonic. She sounds bitter that the clinicians were insisting on one interpretation and failing to hear her own. The trauma of the treatment is exacerbated by this failure to hear what she is trying to say, and the silencing of her point of view feels like betrayal.

Carol's experience suggests actual silences that evade historical context. The process of withholding information, the use of coercive language and the silencing of what should be half of a conversation persist into the twenty-first century. This example suggests that individual subjective assessment can be silenced in the appraisals of experimental and high risk treatments. Such an account does more than undermine historical grand narrative: silences such as Carol describes risk compromising trust and undermining potential partnerships with patients. They thus threaten the relationships on which future research participation depends.

'Tell me what it is in English'

If Margaret, Lina and Carol suggest some malign silences of recent history, then Tim's experience indicates that silence might be something to be usefully negotiated.

Tim is an Anglican cleric and former BBC journalist living in his parish in Norfolk. He was 23 years old in 1973 and working as a technician on a film project interspersed with some painting and decorating work. He had been experiencing inexplicable fatigue and went to discuss it with his GP. Almost in passing, he pointed out a lump in his neck that had grown quite large, obscured by his long hair and profuse beard. He was referred to the Middlesex Hospital in London and was soon diagnosed with advanced Hodgkin's lymphoma.

Given the gravity of his illness and the paucity of treatment options, Tim's treatment was from the outset an exercise in hopeful experimentation. Like Margaret, Tim was one of the earliest UK recipients of MOPP chemotherapy and he went into remission. But his remission was short-lived and within twelve months his disease had come back. There was no known treatment for someone with relapsed disease, but his doctors suggested an experimental new drug called CCNU. Tim was one of just eleven people with relapsed Hodgkin's disease to be treated with it. The CCNU worked – in fact he was the only one of the eleven to survive – but he was given eight times the amount that would now be considered the adult dose. As a consequence, he developed leukaemia 11 years later.

Tim's experience was deeply traumatic. That he tells it with such equanimity is a consequence of the communication he had with his doctors and his opportunity to participate in discussions about his care – opportunities so comprehensively denied for Margaret. An articulate man with innate authority and confidence, Tim had insisted on being told what was going on from the outset:

> I remember saying to Professor Semple one day when he was getting the lab reports back and they were all standing around my bed, as they do, and I said 'Well, read it to me', and I said, you know, I said to him 'You know, this is a teaching hospital, and on the subject of my Hodgkin's disease I can pretty well assure you that I'm the best student you've got! So if there's anything I don't understand I'll ask you to explain it to me.' And so that was the kind of relationship that was set up, and so very quickly they told me everything. They would read verbatim lab reports and I would say 'Tell me what it is in English!', so that I understood.

Their openness was not a given. The consultant who diagnosed Tim was initially reluctant to disclose his diagnosis, but Tim's personal qualities combined with the attitudes of younger doctors persuaded him to open up:

> the doctors on his team then told me that there had been a heated debate about whether or not I should be told. It came to a head when the junior doctors became uncomfortable with not answering my direct questions, even having to lie or mislead, because of Prof. Semple's instructions not to tell meant that they were not at liberty to say. […] The junior doctors argued that I should be told and that I 'could take it'. So Prof. Semple did and went on to open up considerably after that.
>
> (Email correspondence, 2013)

The silenced subject in cancer research 163

When asked about his role in facilitating communication, Tim suggests that the process of discussing life-threatening illness involved dual input and dual responsibility. Having established that he wanted honesty, one of the doctors pointed out: 'You can ask any question you like, but if you don't want to know the answer you shouldn't ask the question.'

Silence is not something to be avoided unconditionally but rather it is something that is negotiated: Tim had to choose what he wanted to hear. Tim does indicate that there were times when strategic silences were necessary:

> But there is also a good time to lie.
> [...]
> And I'd been phoned up and they said well we've got the results would you come in and see us in outpatients?
> [...]
> And he sat me down and no holes barred he said something to the effect of 'You've got leukaemia.' And my first reaction was '... well what's the prognosis?' And he just said, 'We are hopeful.' Now I now know, having talked to David about this, of course the one thing he had *no* evidence whatsoever to say 'hopeful' at all, he didn't think I had a hope in hell, but the important thing there is, from my point of view, the ethics is that you don't withdraw hope.
> [...]
> They thought that was curtains for me, but David didn't withdraw hope and I think that's *really, really* important.
>
> (Speaker's emphasis)

Silence is something that when strategically deployed can mean the difference between coping and not coping with life threatening illness. Elsewhere in his interview, Tim gives another description of the strategic negotiation of impossible decision making. Here he describes a conversation that took place in 1984. Tim's doctors were not convinced of the success of his leukaemia treatment, and did not know how best to proceed:

> And they said, 'Look, Tim, it's up to you. If you want to stop treatment now that's fine, but if we give you another course of treatment we have no idea if it will work or not.' I mean he was quite frank, he had no idea, 'We've never been here before.' And Margaret [his wife] was with me and it was quite a shock, I mean, you know, but we obviously took the decision, 'Well OK David we won't hold it against you if it doesn't work, but we would go for a course of chemotherapy.' But David made it quite clear that he had absolutely nothing to go on.
> [...]
> And then David talked about bone marrow transplant. And he said 'We've thought about whether or not we should do a bone marrow transplant with you.' And I think David said that at that time he had

performed more bone marrow transplants than anyone else in this country, and he said 'With what I know now about bone marrow transplants I think the risks that I will kill you in the process are too great and I'm not going to recommend it.'

[...]

And they weren't saying, the attitude wasn't, 'Tim, we've got to the point where we don't know where we're going and therefore we're going to dump all the heavy stuff on you, you make the decision,' it wasn't like that at all. It was actually the way it was put, look ... it was actually a discussion about quality of life and about my involvement in that choice without any sense of which they are trying to offload difficult questions onto the patient.

The question of adequate communication is not one of disclosure versus non-disclosure, or information versus silence. This extract highlights shared responsibility in decision making, with the burdens equally distributed. Tim describes parts of the exchange that his doctor assumes responsibility for – 'I am not going to recommend that' – and parts that are negotiated, but Tim is never left alone with the responsibility for an impossible decision. This is in striking contrast to Lina's experience thirty years later.

Tim's story suggests that to consider silences purely as manifestations of medical paternalism is inadequate. Such a view obscures the fact that sometimes silence is necessary to allow an individual time to take in what is happening and to decide when, and if, they are ready for the truth. It overlooks the active role of the patient in entering into a negotiation, not as a passive recipient of information, but as someone making choices about which silences to respect.

'Neither treatment would let me down'

Ann's story also suggests the merits of strategic silences. Ann is a retired teacher in her 60s living with her husband in Sheffield. She moved to Yorkshire from Cambridge to pursue her career and was at the point where she felt ready to become a head teacher. She had not had a mirror in her bathroom for some months; their house was undergoing a prolonged process of redecoration. When the mirror was finally replaced, Ann was surprised to notice that her collarbone on the right side had disappeared. She was referred for a biopsy and was diagnosed with advanced follicular lymphoma in January 1994 at the age of 44.

After three months of watchful waiting, Ann was offered participation in a clinical trial of an agent called fludarabine. At the time of Ann's treatment, fludarabine had been used with considerable success in chronic lymphocytic leukaemia, a disease very similar to follicular lymphoma.[4] Early phase trials had established that people with follicular lymphoma responded well to fludarabine, with one author suggesting a 60 per cent response rate even in those with multiple prior treatments (Keating, 1994). So there was reason to hope that fludarabine might be an important contributor to treatment. Ann's trial was a phase III trial that would compare fludarabine with the then standard chemotherapy combination of CVP in previously untreated patients with advanced follicular lymphoma. Here she describes the process of giving her consent:

> I was offered treatment in the April which actually came as a surprise. I'd got my head round getting as much done to the house as possible and Brian and I were decorating literally on a Sunday afternoon and I got a phone call from Prof who said he'd been trying to get hold of me for days.
> [...]
> Anyway he finally got hold of me on the Sunday evening and to say could I get into Weston Park on the Monday morning. There was a possibility of starting on a trial treatment and they'd like to talk to me about it on the Monday morning. So obviously I went into Weston Park on the Monday, met with Prof, met with a trial nurse, and had a conversation about the trial. Of course at that time I didn't know which half of the trial I would be in, obviously, but I agreed to take part and then went back to work.
> [...]
> They described it as a new treatment being tested against the best of the old treatments and that it would be a random selection as to which I got and from the information they had neither treatment would let me down – that was the sort of phrase they were using – that they anticipated the new treatment being as good as the best of the old treatments and it might just be better. They talked about side effects of both treatments, erm, the old treatment involving hair loss and a much more ... what was the word they used? ... rigorous ... a much more rigorous effect on my body than the new treatment because the initial trials of that before they went widespread suggested that the side effects were somewhat less. There was certainly minimal hair loss, if any, and not as much sickness that the old treatment tended to produce.
> Asked if I had any questions and I couldn't think of any which I think is often the case [...] yes, you think about it when you get home. But about a week later I got a phone call to say that I had been accepted onto the trial ... and that I would in fact be getting the new treatment, I'd been randomly selected for the new treatment, which I was then told was a drug called fludarabine. And that started after two weeks so it was two weeks after the initial discussion that I was put onto fludarabine as a trial.

Ann's memory of the process of securing her consent suggests subtle applications of silence. There is a simplicity to the explanation she was offered and the suggestion that some information was not provided. Notably, she was not told the name of the drug to be tested until after she had agreed to take part in the trial. She is told only that the side-effects are not as rigorous as those of standard chemotherapy. They have discussed hair loss and sickness, side-effects typically of most concern to patients, but not its significant depression of the immune system which was understood at the time (Wright et al., 1994).

Ann's description also indicates that she did not ask the name of the treatment. Although equipped with a background in biology and the literacy to find out more had she wished to, Ann was content with the explanation she was given. She was provided with a leaflet, but the information it contained was peripheral to a decision made on the basis of her conversation with a trusted doctor. Hence the process was a simple but delicate balance between what her doctor chose to tell her and what she chose to know.

Ann's overall experience was quite ambivalent. Her tumours returned within weeks of finishing treatment, putting her back to square one. She didn't need further treatment for some years, which might suggest she could have been spared trial participation altogether. In fact, single agent fludarabine was soon discredited because of its sustained impact on individual immunity. Ann, in common with a significant number of other patients on that arm of the trial, developed shingles as a result of the treatment and had to take early retirement. She was robbed of a cherished ambition to become a head teacher and the decision had significant financial implications for herself and her husband.

So an experience that might, with hindsight, have been interpreted very differently is told with equanimity. Although ambivalent about the trial outcomes for herself, she has no regret about how her decision was negotiated. At no point did she doubt that her best interests were paramount. She had, she says, complete trust in the professionals. It is notable that the trial description was framed around Ann – it was about what the trial might offer *her*, not what she might offer the trial: 'neither treatment would let *me* down'; 'less rigorous effects on *my body*.'

Presentism would mean we judge the exchange she describes in terms of information withheld and paternalistic silences, or perhaps even the gentle coercive influence of 'her Prof' who telephones his patients at home. But such analysis would be flawed. It would silence Ann's right to interpret her own experience, much in the way that Carol's subjective assessments were silenced. It would condemn Ann's choice to discount information and to seek no more than she was told.

Another silence in Ann's experience, one she relates as if chiding a student who might do better, is the failure to inform her of the trial results, something she learned by chance some years later. She had become a volunteer in medical education, participating in teaching sessions for medical students in her capacity as a former cancer patient. It was at one of those sessions that a student asked about the trial and another doctor explained that single agent fludarabine had been discontinued in the treatment of lymphomas because of its tendency to cause marked suppression of the immune system. She turned to her Professor at that point, chiding him and saying, 'You never told *me* that!' Individuals want to know the outcome of their contribution to research. The failure to inform participants of trial results, a failure that is widespread, is a persistent silence in the conduct of cancer research that comprehensively undermines any notion of equal partnership.

'They must know more than they're saying'

At a talk that I gave at a university seminar, I had an interesting conversation with a recent participant in a large international clinical trial. His experience makes a fitting conclusion to this discussion, exemplifying as it does the silences of the evidence hierarchy.

Connor was 63 in 2013 when he was diagnosed with aggressive prostate cancer and told that his condition was incurable.[5] He was asked whether he would consider taking part in a clinical trial. Connor agreed in principle. As someone with what was effectively a terminal illness, he had expectations that discussion of the trial would take place in the context of his treatment options and his longer-term care. But the researchers engaged in no such conversation, resisting any discussion about the rationale for the trial. When he asked if there was a hypothesis that was being tested, the researcher replied, 'No.' He was told, 'Well the trial may be of benefit, it may make no difference whatsoever, or it may be detrimental in the long run and we don't know.' There was, he was informed, no good evidence to offer him. Connor experienced great frustration with a lack of information and a strong sense that the research team were strenuously avoiding the risk of persuading him one way or the other. But the consequence was that he felt like nothing meaningful was being said.

He asked what doctors had observed thus far about the various arms of the trial. The trial had been going for a couple of years and he felt sure that the investigators must have observed something of the experiences of the different arms of the trial. But to disclose this – to anticipate interim analyses – would be to pre-empt results and betray the trial methodology. The interests of that methodology trumped the enquiries of an individual faced with terminal illness. Failure to answer this question left feelings of mistrust and a conviction that they were not telling him something; 'they must know more than they are saying', he said.

The silencing that Connor describes suggests that the initiatives designed to protect individual participants – a preoccupation with avoiding coercive language, a desire to inform only on the basis of what constitutes scientific evidence – have the potential to limit trust rather than facilitate it (O'Neill, 2002). In particular, Connor sought access to experiential evidence. The observations based on just a handful of men would have been far more meaningful to him than the explanation that there was no evidence on which to base his choice. The emotional and psychological qualities of his experience – the knowledge that a silence was being maintained – bear strong resemblance to those of Margaret in 1973.

There is no reason to assume that the bad experience of one person reflects that of the majority. But Connor's account does suggest new manifestations of silence in contemporary research practice. Preoccupations with the evidence hierarchy – a structure that casts anecdote or individual experience as unreliable – have resulted in the silencing of important aspects of individual decision making. As we have seen in the recollections of Ann and Tim, these conversational and experiential exchanges are often what contribute to long term acquiescence with the events of one's past.

Conclusion

This chapter has made a case for the epistemological value of marginal, subjective, individual memories. The oral narratives I have discussed have the potential to complicate and undermine the assumptions that the big picture implies; in this case, the assumption that the experience of cancer research has moved from a silent and shadowy past to a transparent and informed present. They also suggest more nuanced and more useful definitions of 'silence' in clinical research, definitions that see past the vulnerability of the research subject and acknowledge that person's active participation in the process of becoming informed.

If we are to interpret silence as a manifestation of power – the power to withhold information or the power to silence another – then the interviews I have discussed suggest that it is not safe to assume that the silences that once characterised research conduct are a thing of the past. Silence persists, changes shape, or is redistributed. The silence once typical of medical practice has been replaced by the cacophony of the internet, but the world wide web – an entity that in spite of its multiple voices is ill equipped to advise individuals – must remain mute on the subject of often painful individual choices. In consultations with medical practitioners, experiential, personal, anecdotal human exchange in research decision making has been silenced in favour of purely evidence-based information. But this ostensibly laudable step to protect individuals from coercion might lead patients to feel that nothing meaningful is said or that information is withheld. Individual subjective experience, such as Carol's assertion that her treatment was not sufficiently successful on her own terms, can be silenced by the more powerful and persuasive narrative of the clinical view that assesses outcomes on the basis of collective, comparative analyses. Clearly, collective analysis of survival or relapse is crucial, but failure to hear the subjective assessment of an individual leaves that person feeling at odds with the people caring for her. There is also a persistent silence in the failure to report the outcomes of research to those who take part in it. This failure compromises the fabled 'partnership' with patients and undermines confidence in its existence.

The comparison of individual experiences within historical perspective makes potentially meaningful and enlightening comparisons between the past and the present. Comparing Tim's account of negotiated decision making in the early 1980s with Lina's more isolated decision making of 2003 clearly undermines any assumption that people feel more empowered as a consequence of assumed autonomy and access to information. Similarly, Connor's suspicion that he was not being told the truth in 2013 invites comparison between the present and 1973, when Margaret was also kept in the dark. Such comparisons suggest continuities in experience that we might not expect and highlight aspects of patient experience that require ongoing attention.

In particular, these connections rely on the emotions revealed by oral history that are not otherwise apparent: what silence and its consequences *feel* like. Oral history concerns not only what is said but how it is said, and emotion in narrative performance will reveal the extent to which people experience trust, confidence, isolation or anxiety in a manner not evident in a written transcription. Carol's words – 'that's great', 'they thought it was a success' – might be interpreted variously but for her sardonic tone of voice that reveals feelings of bitterness. It is regrettable that the reader is unable to hear the emotion of the spoken word and must rely instead on the silence of a transcription. Like other oral historians, I trust that the academic publishing of oral history might increasingly embrace the technological potential to liberate the recording from the archive.

The comparative emotions of these differing accounts also suggest a more valuable conceptualisation of silence. 'Silence' in cancer research is a pejorative term, a malign hangover from a past marked by medical hubris and paternalism. Perhaps due to its historical origins in the experience of the silenced and vulnerable, research governance has lost sight of the fact that many participants in research are active agents in the process of becoming informed. The experiences of the people discussed in this chapter indicate that negotiated silences are a critical part of the process of becoming informed. There are times, for example, when withholding information is strategically essential and times when individuals choose not to find out more. Such choices must be interpreted as active appropriations of silence in negotiating a fraught and traumatic experience. Without this broader interpretation of silence we perpetuate assumptions about the passivity of the individual research participant.

This chapter has argued for the epistemological significance of marginal, subjective, individual recollection. The hitherto silenced voice of the research participant reveals the centrality of subjectivity in looking back at the recent past and in looking forward to how best to proceed.

Notes

1 I gratefully acknowledge the contributions of those who shared their experiences of cancer treatment with me. I would also like to acknowledge the valuable comments and continuing support and encouragement of my supervisors, Dr Viviane Quirke and Dr Marius Turda. This chapter is an outcome of Ph.D. research funded by a Wellcome Trust Programme Grant, number 0906580/Z/11/A, on Subject Narratives of Medical Research.

2 Interviews were conducted at the interviewee's home by the author between March 2013 and July 2013. Unless otherwise indicated, quotes come from these interviews. Names are unchanged unless otherwise stated.
3 Hodgkin's lymphoma is the current term for what was known as Hodgkin's disease. Incidence figures have not changed since 1973 (Cancer Research UK, undated).
4 The distinction between the two diseases is often difficult to make. It will be described as leukaemia if malignant cells predominate in the blood and as lymphoma if malignant cells are predominantly in the lymphatic system.
5 Connor is a pseudonym.

References

Cancer Research UK, undated. Hodgkin's lymphoma incidence statistics. Available at: www.cancerresearchuk.org/cancer-info/cancerstats/types/hodgkinslymphoma/incidence/uk-hodgkins-lymphoma-incidence-statistics#trends. Accessed 29 March 2015.

Chamberlayne, P., Bornat, J. and Wengraf, T. (eds), 2000. *The Turn to Biographical Methods in the Social Sciences: Comparative Issues and Examples.* London: Routledge.

Condrau, F., 2007. The patients' view meets the clinical gaze. *Social History of Medicine*, 20, pp. 525–540.

DeVita, V.T., Serpick, A.A. and Carbone, P.P., 1970. Combination chemotherapy in the treatment of advanced Hodgkin's Disease. *Annals of Internal Medicine*, 73, pp. 881–895.

Gilmour Hamilton, C., 2013. Stories from the sharp end: human expectations and experiences of cancer research. *Wellcome History*, Winter, pp. 16–18.

Hazelgrove, J., 2002. The old faith in the new science: the Nuremberg Code and human experimentation ethics in Britain, 1946–1973. *Social History of Medicine*, 15, pp. 109–135.

Hedgecoe, A., 2009. 'A form of practical machinery': the origins of research ethics committees in the UK, 1967–1972. *Medical History*, 53, pp. 331–350.

Jones, J.H., 1993. *Bad Blood: The Tuskegee Syphilis Experiment.* New York: The Free Press.

Keating, M.J., 1994. Introduction. *Leukemia and Lymphoma*, 14, p. 1.

Lederer, S.E., 1995. *Subjected to Science: Human Experimentation in America Before the Second World War.* Baltimore, MD: Johns Hopkins University Press.

McIntosh, J., 1977. *Communication and Awareness in a Cancer Ward.* London: Croom Helm.

Mold, A., 2010. Patient groups and the construction of the patient consumer in Britain: an historical overview. *Journal of Social Policy*, 39, pp. 505–521.

Mold, A., 2013. Repositioning the patient: patient organisations, consumerism and autonomy in Britain during the 1960s and 1970s. *Bulletin of the History of Medicine*, 87, pp. 225–249.

Novack, D.H., Plummer, R. and Smith, R.L., 1979. Changes in physicians' attitudes towards telling the cancer patient. *Journal of the American Medical Association*, 241, pp. 897–900.

O'Neill, O., 2002. *Autonomy and Trust in Bioethics.* Cambridge: Cambridge University Press.

Tomes, N., 2006. The patient as a policy factor: a historical case study of the consumer/survivor movement in mental health. *Health Affairs*, 25, pp. 720–729.

Tomes, N., 2007. Patient empowerment and the dilemmas of late-modern medicalisation. *The Lancet*, 369, pp. 698–700.
Weindling, P., 2004. *Nazi Medicine and the Nuremberg Trials: From Medical War Crimes to Informed Consent*. Basingstoke: Palgrave Macmillan.
Wright, S.J., Robertson, L.E., O'Brien, S., Plunkett, W. and Keating, M.J., 1994. The role of fludarabine in hematological malignancies. *Blood Review*, 8, pp. 125–134.

8 Reconstructing ancient thought
The case of Ancient Egyptian mathematics

Elizabeth Hind

Introduction

Traditionally, writings on ancient Egyptian mathematics have appeared as a prologue to the true start of the history of mathematics: the Greeks. Linear narrative writing and the perceived debt to the Greeks have ensured that in the majority of histories the Egyptians are hardly mentioned, if at all. Writers about the topic tend to be historians of mathematics, not Egyptologists. Their goal is to trace the origins of mathematical thought and to place different cultures in a timeline. It is even rarer to find reference to mathematical texts in works on the ancient Egyptian culture and their achievements than it is to find them mentioned in histories of mathematics. The overall impression is summarised by the Egyptologist Barry Kemp and the historian of mathematics Otto Neugebauer:

> Practised scribes must have developed a degree of mathematical intuition, but the idea of pursuing this as an end in itself – to create the subject of mathematics – did not occur to them.
>
> (Kemp, 1989: p. 117)

> The mathematical requirements for even the most developed economic structures of antiquity can be satisfied with elementary household arithmetic that no mathematician would call mathematics.
>
> (Neugebauer, 1952: p. 71)

There are many silences in the study of ancient Egyptian mathematics; some are intrinsic to studying a culture that has been dead for over 2,000 years, others are artefacts of modern historiographical techniques and contexts. Combined, these silences ensure that the character of ancient Egyptian mathematics is defined by a lack: a lack of abstract reasoning, a lack of proof, a lack of progress, a lack of general theorems, a lack of scientific attitude of mind, even a lack of beauty. Yet it is difficult, if not impossible, to extricate each of these silences in order to study them separately.

Analysis and debate may be discouraged by an overview that sees Egyptian mathematical achievement best described by its deficit, its 'non-achievements'.

Egyptologists, perhaps already disinclined to take part in complex debates on mathematical philosophy, tend not to engage with Egyptian mathematics because they have been told by mathematicians that the texts are not important. However, the goal of analysing Egyptian mathematical texts should be a culturally sensitive, contextualised understanding of the abilities and techniques of the ancient Egyptians, to which Egyptologists are essential. If the silences about Egyptian mathematics can be better understood, then their malign influence on our overall impression of the achievements of the ancient Egyptians can be lessened.

There is a gap between the modern appreciation of the nature of mathematics and what we see in ancient surviving texts. The way in which this overshadows any positive achievements is amply shown by the prologue to *God Created the Integers*, a history of mathematics that begins with Euclid. Defending his choice of starting point, the editor Stephen Hawking explains:

> The Egyptians employed this skill to build the great pyramids and to accomplish other impressive ends, but their computations lacked one quality considered essential to mathematics ever since: rigor. For example, the ancient Egyptians equated the area of a circle to the area of a square whose sides were 8/9 of the diameter of the circle. This method amounts to employing a value of the mathematical constant pi that is equal to 256/81. In one sense this is impressive – it is only about one half of one percent off of the exact answer. But in another sense it is completely wrong. Why worry about an error of one half of one percent? Because the Egyptian approximation overlooks one of the deep and fundamental mathematical properties of the true number, π: that it cannot be written as any fraction, that is a matter of principle, unrelated to any issue of mere quantitative accuracy.
>
> (Hawking, 2007: p. xiv)

This extract demonstrates how despite its impressive feats, Egyptian mathematics does not warrant inclusion in a history of the subject because it did not live up to ideals of mathematics that were only fully realised millennia after the texts were written. Importantly, such a judgement rests on the presumption that the Egyptians did not realise that their method was imprecise, because they did not write this down in a way that we can read about it.

When attempting to reconstruct the mathematical thought of the ancient Egyptians there are many different sources for silence, which interact and overlap. With multiple silences their specific origin can be difficult, if not impossible, to identify. This can lead to accusations that the ancient Egyptians were ignorant, when in reality there may be multiple plausible explanations. For example, if a particular mathematical idea is not in the extant texts, is that because it was known about but not written down, was written down but has now been lost, or was simply never known? Given the paucity of surviving material, each of these explanations is possible.

The non-survival of evidence causes many problems. It must be presumed that a lot of material has not survived. Papyrus, the material most often used for ancient Egyptian mathematical texts, survives in only very particular circumstances. We have only a few texts from a specific period of Egyptian history. There is no way of knowing whether the texts have survived through luck or through agency. Even where the texts do survive, language is a particular barrier in understanding ancient Egyptian mathematics. There are the usual issues of translation, but there are additional problems due to the way that the ancient Egyptian language has had to be deciphered. We may only have one or two examples of a mathematical term, so its decipherment will depend on the mathematical context. The word for the diameter of a circle, for example, is clearer once we see how its value is used in the mathematical working that accompanies a problem because it is the only dimension that would make mathematical sense in the solution of the problem.

Particular silences can magnify each other. For example, if it is presumed that the Egyptians did not have an abstract understanding of mathematics then this can mean that mathematical terminology is translated in a particular, and prejudicial, way. This then leads those who cannot read the original texts to be even more doubtful about the specialist and abstract understanding of the Egyptians. Combined with a traditional history of mathematics that traces its origins to the ancient Greeks, the effect is that Egyptian mathematics remains rarely discussed, even within Egyptology.

Survival of evidence

Differential survival of evidence has always been a problem for Egyptologists. Most of what is known about the culture is derived from objects and texts that are found in tombs. This is because material that is placed in a tomb is sealed and is only at risk from tomb robbers, both ancient and modern. The ancient Egyptians stored a lot of artefacts in the tomb so that they would be available for the tomb owner to use in the afterlife. This quirk of Egyptian funerary belief means that a lot of the evidence that we have for the ancient Egyptian way of life comes from funerary artefacts and paintings, such as those on display to the public in museums. This prevalence of funerary artefacts in public displays of Egyptian culture has led to a popular image of the ancient Egyptians as a people obsessed with death. Settlements were made of mud brick and even the palaces of pharaohs were mostly temporary buildings that do not leave a trace. Another main source of evidence for the worldview of the Egyptians comes from their temples. While Egyptologists understand them to be economic centres as well as religious ones, it again leaves the general observer with an impression of a mystical people, not serious scholars and mathematicians.

Not surprisingly the textual evidence for ancient Egyptian mathematics is very scarce. Papyrus is organic and will break down through hydrolysis, oxidation and the agency of micro-organisms and insects. Most papyri survive because they are funerary texts preserved in the tomb or are part of archives or personal libraries that have been placed in the tomb (Leach and Tait, 2000). Personal libraries will be placed in tombs only under exceptional circumstances and papyri that have survived in settlement contexts are extremely rare. Other writing materials such as ostraca, wooden tablets and animal skins were used but were certainly not normal.

In these circumstances we cannot expect a large number of mathematical texts to survive. While it is normal practice in Egyptology to combine textual, artefactual and archaeological evidence, this is not the case in the history of mathematics, where textual evidence is the only kind that is permitted. While textual evidence may be the only source for certain types of mathematical object, such as the statement of a mathematical rule, when assessing the importance that the Egyptians placed on mathematical style of thought, then other sources must also be considered. For example, mathematical achievement should be placed within the context of Egyptian art, which used a grid system for strict proportional control, and their religious belief in the deity Ma'at, the embodiment of order. We also have textual evidence that the Egyptians valued knowledge. Take this extract from the *Song of the Harper* (Claggett, 1989: p. 220):

> Blessed nobles too are buried in their tombs.
> (Yet) those who built tombs,
> Their places are gone.
> What has become of them?
> I have heard the words of Imhotep and Hardedef,
> Whose sayings are recited whole.

This cultural background is as important to understanding the motives of the Egyptian mathematicians as is an appreciation of taxation practices and styles of building. With this enriched context, apparent silences can be better understood.

Furthermore, the main sources date from the Middle Kingdom and therefore represent only a very small percentage of the total span of ancient Egyptian history. Egyptian history covers a few thousand years and is split into several different periods. It is a significant length of time. The pyramids were built during the Old Kingdom, with the main pyramids being constructed during the 4th Dynasty, dating from c. 2543 to 2120 BCE. Tutankhamun died in approximately 1324 BCE while Cleopatra VII died in 30 BCE. To put all this another way: the death of Cleopatra is closer in time to the Battle of Hastings than it is to the building of the Great Pyramid at Giza.

A huge percentage of our evidence for the mathematical achievements of the ancient Egyptians comes from a relatively short space of time. This is problematic if we are to make pronouncements about the general character of the civilisation: Egyptian civilisation lasted for around 3,000 years. To an untrained eye, Egyptian culture looks homogeneous, but there were of course developments in ideas and culture in this time which we need to be aware of.

176 Cultures of silence

Main sources

One of the greatest problems facing an accurate assessment of the mathematics of the ancient Egyptians is the paucity of surviving material. The two main sources of evidence for ancient Egyptian mathematics are the Rhind Mathematical Papyrus (RMP) and the Moscow Mathematical Papyrus (MMP). Both these texts were produced during the Middle Kingdom, which includes the 11th and 12th dynasties and dates to approximately 1980–1769 BCE (Hornung et al., 2006).

The Rhind Mathematical Papyrus can be dated through the dedication which appears at the front of the text, to: 'year 33, month 4 of [the season] Akhet [under the majesty of] King of [Upper] and Lower Egypt Awserre', which puts it in the reign of the Hyksos King, Apohis, who reigned c. 1585–1542 BCE. The dedication states that the Rhind Mathematical Papyrus is a copy of an older text, dating the earlier text to the reign of Amenemhet III who ruled c. 1844–1797 BCE in the 12th Dynasty. The papyrus was found in a small building near the Ramesseum, on the west bank on the Nile at Thebes.

The Moscow Mathematical Papyrus does not have a dedication to date it exactly. However, linguistic analysis of the text places it at about the same age as the text that the scribe of the Rhind Mathematical Papyrus copied. It would appear that the text was originally found in a tomb not far from the tomb where the Rhind Mathematical Papyrus was discovered (Claggett, 1999).

Other sources include the Kahun Fragments, the Berlin Papyrus, the Mathematical Leather Roll and the Reisner Papyrus. These sources also date to a similar age, around the 12th Dynasty. They mostly found their way into publication through private collections so their provenance is unclear. It is known, however, that the Kahun fragments were found during Flinders Petrie's excavations at the workers' town at Kahun and consist of six small fragments devoted to mathematical problems.

The mathematical papyri generally consist of series of worked problems. Problems solved include the calculation of the volumes and areas of simple geometric shapes, the sum of an arithmetic progression and problems in the proportional distribution of bread and beer. Often, but not always, the problem is introduced by a statement describing what it concerns. If there is arithmetical working in the problems then the working often accompanies the narrative text. One example is shown here to demonstrate how an Egyptian problem was set out and the narrative format of the working out and explanation. It also includes a diagram in the original text which is here shown in a neatened format with the numerals transcribed:

> Example of working out a triangle.
> If it is said to you a triangle of twenty in area;
> that which you put on the length, you put a third and a fifteenth of it on
>   the breadth.
> You are to double the twenty, it becomes forty.

Reconstructing ancient Egyptian mathematics 177

You are to work on a third and a fifteenth in order to find one, it becomes two and a half times.
You are to make forty times two and a half; it becomes one hundred. You are to take its square [root];
it becomes ten. Look, it is ten in length. You are to take a third and a fifteenth
of ten, it becomes 4. Look, it is 4 on the width.
You will find it good.

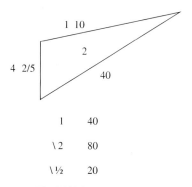

```
     1      40
    \ 2     80
    \ ½     20
```
Total 100 Square root 10
(Moscow Mathematical Papyrus Problem 17)

This problem starts with a simple statement of what the problem is concerned with, the area of a triangle. The problem then states that we are dealing with a triangle with a known area and we are told that the breadth of the triangle is a third and a fifteenth of the length. Note that a third and a fifteenth equals two fifths, but the scribe writes it as the sum of two unit fractions in line with standard Egyptian procedure.

The first step to work out the answer is to double the area, so that it becomes a rectangle. The next step is to work out what we would term as the reciprocal of a third and a fifteenth. The answer, two and a half, is a necessary step in the calculations as the Egyptians would work out a division by multiplying one number until they reach the other. This is equivalent in method to multiplying by the reciprocal. We are then told to multiply the area of the imaginary rectangle by this factor to obtain 100. We can then find the square root of this number to find one length and then use the original ratio to find the other dimension. Translated into modern algebraic notation the problem would read:

$$\frac{x\left(\frac{2}{5}x\right)}{2} = 20$$

$$x\left(\frac{2}{5}x\right) = 40$$

178   *Cultures of silence*

$$x^2 = 40 \times \frac{5}{2}$$

$$x^2 = 100$$

$$x = 10$$

At the end of this problem we are shown a diagram to demonstrate that the answer is correct. We are given a picture of the triangle, inside which the area two is shown.[1] Surrounding the triangle we are given details of the arithmetic necessary to complete the problem. At the bottom, the calculation of two and a half times forty is shown; the left hand column shows the multiplier and the right hand the result of multiplying it by the multiplicand. The tick marks show which terms should be added to give the final answer. In this case we are told this equals one hundred and the square root is quoted as ten. We are also told on the left hand side that 1/3 and 1/15 times ten is equal to four, although we are not shown how this calculation is carried out.

The subject matter appears to be, as several writers have suggested, rooted in practical need, solving problems that the scribe would face in carrying out his duties. These included taxation, surveying and payment of workers. The worked problems in the two main sources include geometrical, algebraic and arithmetical problems. The geometrical problems include working out the areas and volumes of straight edges and curved figures, the volume of a truncated pyramid and the slope of a pyramid. There are several examples of working out an unknown quantity through different means. Arithmetical problems include dividing up a quantity of loaves fairly, arithmetical progressions and working out the quality of bread.

One side of the Rhind Mathematical Papyrus contains what is commonly referred to as the 2/n table. Although it is not strictly a table, it is a list of unit fraction identities for two divided by every odd number up to 101 that is probably intended as a reference when performing calculations with fractions.

It is possible that these texts were intended as a learning aid for a scribe who would need to be able to carry out similar calculations in the course of his duties. Just like a modern school textbook, problems take an inspiration from real world situations, but present them in a way that explores the mathematics, such as using values that make the calculation easier. However, while these problems seem rooted in practical need there are glimpses of mathematics for its own sake. For example, Rhind Mathematical Problem 79 contains the sum of a geometric progression which, while it is couched in everyday terms, resonates with us as a problem for its own sake because of its similarity to a famous rhyme:

An inventory of a household.

1	2801
2	5602
3	11204
Total	19607

7	Houses
49	Cats
343	Mice
2301[2]	Wheat
16807	Hekat[3]

(Rhind Mathematical Papyrus Problem 79)

The problem that is posed here is similar to the nursery rhyme 'As I was going to St Ives'. A geometric progression is to be summed. Its first term is 7, the common ratio is 7 and it contains five terms. The sum is worked out in two ways. The second way is simply to work out the terms of the progression and then add them together.

The first method shown is more remarkable. The formula that we would use to work out the sum is:

$$a\frac{r^n - 1}{r - 1}$$

where a is the first term, r is the common ratio and n is the number of terms.

Replacing the terms of this progression:

$$7\left(\frac{16807 - 1}{7 - 1}\right) = 7\left(\frac{16806}{6}\right) = 7 \times 2801$$

This is exactly the calculation that is performed in the first part of this problem, 2801 is multiplied by 7 to give 19607. The working shows 2801 multiplied by 2 and then 4, which can then be added to give the total of 7 times 2801. This method of working out the sum shows that the scribe who prepared the problem was using a method very similar to our modern method. How the scribe achieved this answer is not explained and as this is the only example of its type it cannot be compared to any other problem from the Rhind Mathematical Papyrus. It may be that the Egyptians only knew how to calculate the sum of a geometrical progression of this type, one in which the first term is also the common ratio. This makes the formula simpler:

$$a\frac{l - 1}{r - 1}$$

where l is the last term.

This simpler formula does not contain the r^n term which makes the arithmetic easier. However, even in this case the method used to solve the problem is not the straightforward addition of each separate term, but a much more sophisticated method that would have involved abstract mathematical reasoning.

Multiplication

Of all of the Egyptian mathematical techniques, it is their method of multiplication that is most often used as evidence for their poor understanding of abstract mathematical reasoning. For example, Neugebauer wrote:

> In general, multiplication is performed by breaking up one factor into a series of duplications. It certainly never entered the minds of the Egyptians to ask whether this process will always work. Fortunately it does; and it is amusing to see that modern computing machines have made use of this principle to exactly the same end.
>
> (Neugebauer, 1952: p. 73)

Here, Neugebauer equates the fact that the existing texts do not contain an explicit discussion about the methods of multiplication, with ignorance on the part of Egyptian mathematicians. Yet the extant texts are full of examples of multiplications of both whole numbers and fractions, and techniques are similar between the different texts. If we consider the many different sources of silence that are at work, explanations for the deficit emerge. The Egyptians might have doubted whether their practice would always work and so passed their reasoning on through an oral tradition. Another possibility is that they may have written their technique down on a text that was not placed in a context where it would survive.

Even writers who are more sympathetic to the mathematical achievements of the ancient Egyptians are prone to over-simplifying the Egyptians' approach, making it appear that they were rigid in their methods. In the case of multiplication it is often said that Egyptian multiplication was performed by repeated addition. For example: 'One of the great merits of the Egyptian method of multiplication or division is that it requires prior knowledge of only addition and the two times tables' (Joseph, 2000: p. 63). Joseph describes the method as multiplication by repeated doubling. In this method two columns are shown, the left hand column starts with 1 and the right hand column one of the numbers to be multiplied. Both columns are doubled to give a new row. This doubling process is repeated until the left hand column shows at least half of the multiplier. The multiplication can be completed by marking with a tick all of the numbers in the left hand column that add up to the multiplier, the corresponding numbers in the right hand column added together will give the correct total. For example if 14 and 56 were multiplied using the method it would be written out thus:

	1	56
	\2	112
	\4	224
	\8	448
Total	14	784

Yet, contrary to Joseph's statement, the scribes who contributed to the existing texts do in fact show flexibility in their approach to multiplication, particularly in their approach to fractions, which are not mentioned by Joseph. Techniques of multiplication in the Rhind Mathematical Papyrus can be classified into eight broad types. From the relative frequency of each type (Table 8.1), it can be seen that repeated doubling is a common method but certainly not the only one used.

Here we deal with a silence manufactured by the way in which research into Egyptian mathematics has been carried out. Only a minority of scholars possess the linguistic skills required to understand the texts in their original form. The emphases they place on particular techniques are likely to be replicated by other scholars, including most historians of mathematics, who lack these skills. Alternative techniques and interpretations are effectively silenced.

Unit fractions

Unit fractions are one of the most distinctive features of Egyptian mathematical texts, and one of the most difficult for modern readers to feel comfortable using. With the exception of 2/3, the ancient Egyptians used only unit fractions (that is a fraction with a numerator of one), in all the calculations in the

Table 8.1 The relative frequency of multiplication techniques in the Rhind Mathematical Papyrus

Multiplication type	Number of examples in the Rhind Mathematical Papyrus	Rhind Mathematical Papyrus problem numbers
A Multiplication by repeat doubling, with no fractions	6	26, 32, 41, 48, 50, 52,
B Multiplication by repeat doubling of fractional multiplicands	7	24, 25, 27, 43, 54, 55, 69
C Multiplication by repeat doubling and additional fractional multipliers	5	35, 42, 53, 69, 70
D Multiplication of fractions with no doubling	19	7–20, 30, 35, 56, 58, 67
E Multiplication of fractions by repeat halving	2	80, 81
F Other methods	8	41–46, 49, 79
G Trivial multiplication in algebraic type problems	7	24, 25, 27, 35, 36, 37, 38
H Multiplication with no working shown	9	40, 61, 62, 72, 73, 74, 75, 78, 82

182 *Cultures of silence*

extant mathematical texts. Other fractions were expressed as a sum of unit fractions, with the largest fraction (smallest denominator) first and then decreasing in size. No fraction could be repeated. For example, in Moscow Mathematical Papyrus Problem 17, translated above, the fraction of two fifths is expressed as one third plus one fifteenth.

The use of unit fractions has intrigued many writers on Egyptian mathematics. It has been described as 'at once the glory and the straitjacket of Egyptian methodology' (Robins and Shute, 1987: pp. 58–59). The 'glory' refers to the technical skill revealed by the texts in the manipulation of the unit fractions. The scribes show great talent in their use, but large parts of the texts are taken over by exercises in the manipulation of fractions which, had a modern notation system been used, would be unnecessary.

A good example of a problem that simultaneously shows the skill of using unit fractions and their apparent awkwardness is Problem 30 of the Rhind Mathematical Papyrus. In the translation below, unit fractions are shown by placing a line over the denominator, in the case of two-thirds it is shown by placing two lines over a three:

If a scribe says to you: 10 has become $\overline{\overline{3}}\,\overline{10}$ of what? Let him hear: Make $\overline{\overline{3}}\,\overline{10}$ in order to find 10

	\1	$\overline{\overline{3}}\,\overline{10}$
	2	1 $\overline{3}\,\overline{5}$
	\4	3 $\overline{15}$
	\8	6 $\overline{10}\,\overline{30}$
Total	13	$\overline{30}$

Making $\overline{30}$ times $\overline{23}$ [sic] to find $\overline{\overline{3}}\,\overline{10}$
Total, the quantity that says it, 13 $\overline{23}$

	1	13 $\overline{23}$
	\$\overline{3}$	8 $\overline{\overline{3}}\,\overline{46}\,\overline{138}$
	\$\overline{10}$	1 $\overline{5}\,\overline{10}\,\overline{230}$	Total 10

(Rhind Mathematical Papyrus Problem 30)

The problem is to find a quantity that when multiplied by $\overline{\overline{3}}\,\overline{10}$ becomes 10. In modern notation the problem to be solved is $x\left(\frac{23}{30}\right) = 10$. The scribe uses a repeated doubling method to experiment with what the answer should be. He notices that 13 times $\overline{\overline{3}}\,\overline{10}$ comes very close, leaving a remainder of $\overline{30}$. Although it is usual in Egyptian texts to mark where a number is a remainder, that mark is omitted here. The next line contains a mistake as 23 should not have been shown as a fraction, this line is meant to show that 23 times $\overline{30}$ is equal to $\overline{\overline{3}}\,\overline{10}$. This equality is left without explanation. The last section of the problem shows that 13 $\overline{23}$ multiplied by $\overline{\overline{3}}\,\overline{10}$ is equal to 10.

This problem is typical of Egyptian mathematical texts as in some places the steps needed to be taken are explained, in other cases results are given without explanation. In the top multiplication of this problem it is shown that 13 times $\overline{\overline{3}}\,\overline{10}$ is $\overline{30}$ short of ten, in the second one the result of 13 $\overline{23}$ multiplied by $\overline{\overline{3}}\,\overline{10}$ is given with few intermediate steps. The intermediate steps themselves are far from obvious results. 13 $\overline{23}$ multiplied by $\overline{\overline{3}}$ and then expressed as unit fractions is a complicated sum, but it appears that the scribe who prepared this papyrus did not think that it was worth showing those steps. Silences are presented in this short problem for which there are few clues about their origin or meaning.

It is a matter of interpretation whether to praise Egyptian mathematicians for their skill in using unit fractions or to criticise them for using a system that appears cumbersome next to our own. The lack of modern notation seems to make the process of calculations harder, but the Egyptian mathematicians appear to have put a lot of effort into practising these techniques until they were able to perform complicated calculations. Are we to presume a lack of imagination on the part of the scribes, or take the effort such calculations entailed as evidence that the use of these fractions had a function outside of the texts within the administrative system of Egypt? The statement, albeit with a mistake, in the centre of this text that 23 multiplied by $\overline{30}$ is $\overline{\overline{3}}\,\overline{10}$ would perhaps show that ancient Egyptian mathematicians were capable of complex mental arithmetic within their system of fractional notation.

We know from the context of the problems that fractions were important in paying groups of people. This is confirmed by the Kahun fragments. In a system that did not have coinage, people were paid a part of the resources available; the more important the individual the larger their part was. The scribe would be required to add up all the parts and divide the total amount by this sum. Each share would then be worked out separately. These types of calculation would have required great skill in fractions. It would also have been necessary to ensure that each of the workers felt that they had been treated fairly. The lack of intermediate steps when calculating a complex sum could be taken as proof of the mathematician's ability to perform complex mental calculations, or it could be that calculations were performed on the scribe's equivalent of a rough book, perhaps on ostraca. There have even been suggestions that an Egyptian scribe would have a series of reference tables at his disposal so that he could perform calculations, including adding integers (Gillings, 1972). Again a silence in the text can produce contradictory interpretations; one that assumes a great skill on the part of the mathematician, and one that supposes the scribe to be an unthinking calculation machine reliant on tables.

The 2/n table

One reference table that does exist in the surviving material is known as the 2/n table and appears in the Rhind Mathematical Papyrus. Because of the Egyptian method of showing fractions as a sum of unit fractions and the common fractions of 2/3, 1/4, 1/2 and 1/3, it was regularly necessary for the scribe to be able to double unit fractions as part of a multiplication. In order to facilitate this, the Rhind Mathematical Papyrus contains a list of divisions of 2 by odd numbers. Only the odd numbers were necessary in this table because 2 divided by an even number is a unit fraction. While too much emphasis may have been placed on repeated doubling in Egyptian multiplication, particularly in the case of fractions, there is certainly a link between the methods of multiplication and the need for a reference source like the 2/n table.

There are thousands of possible decompositions for each of the entries; why does the scribe select one particular decomposition over another? The practical nature of the table is shown by the five precepts that govern which of the decompositions is selected (Gillings, 1972). These precepts are guiding principles that govern which decomposition is selected. They are not generally inflexible rules, but rather principles that are traded off to give the most practically useful decomposition. The rules govern the number of terms in the decomposition and also the nature of the terms, such as a preference for smaller, even numbers.

These precepts show that the Egyptians knew their own mathematical system well and had found ways in which to deal with its limitations. For example, using even numbers is important because if even numbers are used then doubling them is easier because you can halve the number, eliminating the need for future reference to a table. Similarly, keeping the number of terms small would ease calculations, and the end result would be better for the real-world need to pay workers. The need to keep the numbers small could also provide a hint that the use of fractions was linked to the need to physically divide resources as it is far easier to cut a fifth than a twelfth, particularly in something that is prone to generating crumbs, such as bread, where the number of cuts needed could be important. If these precepts show the usefulness and glory of the system then the third precept, which is the only one to be a binding rule, shows the straitjacket. No number could be repeated.

Together the precepts show that Egyptian mathematics was a working system. It suggests that the table has been honed over years of use and passed on, implying a long tradition and a community of mathematicians who shared results. If this is the case then it shows that the Egyptians were invested in their system of writing fractional quantities. Given what we know of the administrative and taxation systems of the ancient Egyptians and the strength that these systems gave to their civilisation, it is hard to believe that their system of notation arose through a lack in the governmental and administrative systems. The conclusion that, despite the difficulties they posed for performing calculations, unit fractions had a purpose is compelling, even if hard evidence is short.

The last entry in the table is a mathematical curiosity and shows that the 2/n table was a working document, but hints at mathematical insight on the part of the scribe who produced the table. The entry is for 101, presumably because the need to look up entries beyond this were rare, and gives the following decomposition:

$$2 \div 101 = \overline{101} + \overline{202} + \overline{303} + \overline{606}$$

At first sight this entry appears strange because it breaks several of the precepts that govern the other entries in the table. The reason that this entry is remarkable is that it gives a complete solution to the problem. Six is a perfect number, that is a number that is the sum of its own positive divisors. This property means that the following equation is always true:

$$2 \div n = \bar{n} + \overline{2n} + \overline{3n} + \overline{6n}$$

Egyptian mathematics is criticised because it is presumed that it lacked rigour, but this feature of the table shows a much more theoretical aspect to Egyptian mathematics. It could be argued that the scribe was just responding to a practical need and did not necessarily have a theoretical understanding of perfect numbers and their properties, but the production of a complete solution to a problem – especially if it is not practically optimal as is the case here with the violation of the usual precepts – is a feature of abstract mathematics.

Translation problems

Egyptian hieroglyphs were famously deciphered by Champollion when he realised that the three parts of the Rosetta stone were the same text written in different languages using different scripts. Phonetic equivalence could be established by comparing the hieroglyphs contained in cartouches with the Greek and Demotic names of people such as Cleopatra and Ptolemy. Establishing the meanings of words is more problematic because even if a phonetic value for the word is obtained, this does not give any clue to the meaning of the word. Words from the ancient Egyptian language have to be ascertained through concordances and careful detective work. Translating a text, even translating between living languages, requires the translator to select the words in the target language that most closely represent the meaning of the word in the original language.

There are two particular problems when attempting a translation of an ancient Egyptian mathematical text: decipherment of rare words and the precise rendering into modern mathematical terms where these terms have very precise meanings. Egyptian mathematical texts contain words that have a specific meaning within the context of a mathematical problem, for example the words for dimensions of geometrical shapes. Several words are used in a unique way in the Rhind and Moscow Mathematical Papyri, and then they may only be used a few times. For example, if we examine the words used for geometric dimensions in the Rhind Mathematical Papyrus (Table 8.2), it is clear that their meaning has to be derived from the context in which they

186   Cultures of silence

Table 8.2 Words used for geometric dimensions in the Rhind Mathematical Papyrus

Word	Translation	Rhind Mathematical Papyrus problem numbers
3w	length	44
wsḫ	width	43, 44
wr r wr	base (of a pyramid)	56, 57, 58
wḥ3 ṯbt	height (of a pyramid)	56, 57, 58
pr-m -wš	height	43, 44
k3	height	43, 44
k3 = f-n-ḥrw	height (of a cone?)	60
rḫt	amount, dimension	46
ḥr mryt	on the long side (of a triangle)	51, 52
h3k	the truncation (of a trapezoid)	52
sntt	base width (of a cone)	60
skd	gradient (of a pyramid)	56, 57, 58
tp -r	the short side (of a triangle)	51, 52
dbn	round (diameter of a circle)	41, 42, 50

appear in the problem. The mathematical workings of the text can be used to derive the meaning of some of the words as they refer to dimensions of a geometrical shape.

With such a small corpus of extant literature it is impossible to give a completely accurate translation of any word. For instance, Problem 10 of the Moscow Mathematical Papyrus has two examples of the names for dimensions which can only be deciphered through their mathematical context. The translation hinges around the meanings of two words for the dimensions of a basket – the *tp-r* and the *ꜥd*. Because of possible mistakes by the scribe and damage to the papyrus in a particularly crucial part of the problem, the problem could refer to the area of a semicircle, the curved surface of half a cylinder or the surface of a hemisphere. If this problem does refer to the hemisphere then it would be a remarkable achievement for a text of this date, yet controversy surrounds its translation and interpretation (Cooper, 2010). The extent to which the mathematical workings should influence our opinion is part of this controversy, as well as the extent to which factors such as our knowledge of Egyptian basketry and units of measurement should colour that opinion.

The Egyptian texts also use everyday words and phrases for mathematical computations. For example, in Problem 17 of the Moscow Mathematical Papyrus translated above, the phrase 'that which you put on the length, you put a third and a fifteenth of it on the breadth', tells us that the breadth is one and two fifths the length. Translating it in this way makes the problem appear cumbersome when we are so used to modern algebraic notation. However, increasing acquaintance with the texts mean that they become easier to read and phrases that at first appear awkward become familiar. The type of language used should not become a reason for presuming that the ancient Egyptians lacked a precise, mathematical mind.

Indeed, the verb forms used show that, contrary to being written in the everyday language implied by the English translations, they use a specialised language that cannot be translated. The particular verb form they use is known as the *sḏm-ḥr = f* form. In mathematical texts it lends a feeling of a continuing process as it has a very strong narrative sense because the particle *ḥr* is used to show what comes next in order. It is a specialised verb form, not particularly common in other types of text, yet it is common in Egyptian mathematical texts. The feeling that this verb form would have imparted to the ancient Egyptians is uncertain, but suggests that they were specialist texts and would have been read as such.

An example of a word that is frequently used in Egyptian mathematical texts where the translation of that word changes the entire feeling of the text is the word . It is transliterated as *nfr* and pronounced nefer. It is a relatively common word in ancient Egyptian texts. Faulkner's *Dictionary of Middle Egyptian* defines the word as 'beautiful', 'fair', 'good', 'fine' and 'happy', depending on the context of the word (Faulkner, 1962). Gardiner's *Egyptian Grammar*, a standard textbook for learning to read hieroglyphics (Gardiner, 1957), gives the meaning as 'good, beautiful, happy, well'. The word forms part of the name of Nefertiti, subject of one of the most striking busts that survive from ancient Egypt, now in the Neues Museum in Berlin. The name is translated as 'a beautiful woman has come'. In the narrative format of ancient Egyptian mathematical problems it is common to finish off the solution of the problem with the phrase *gm = k nfr*, pronounced gem-ek nefer. The word *gm* is a verb usually translated as found, and the *= k* indicates the second person singular, you. In the context of Egyptian mathematics the word *nfr* is often translated as 'correct' or 'rightly' (e.g. Cooper, 2010, Gillings, 1972). In the first translation of the Moscow Mathematical Papyrus it is translated as 'richtig' (Struve, 1930); in French it is translated as 'juste' (Couchoud, 1993). The translations in this chapter are by the author and use the word 'good'. If the word was translated as beautiful, rather than the functional 'correct', then it changes the meaning of the entire phrase. Rather than merely following a method to the correct endpoint, the mathematician uncovers something pleasing, suggesting an appreciation of mathematics as an aesthetic, rather than a purely practical, pursuit.

Modern maths as a yardstick

In works written about the mathematical achievements of the ancient Egyptians, modern mathematical notation, algorithms and representations have been used as a yardstick against which to measure the achievements of others. It is assumed that modern methods have evolved, honed and developed over time, and thus represent an ideal, any deviation from which reveals a deficiency in

188 *Cultures of silence*

method and consequently a deficiency of thought. This form of historical writing is quick to identify and label differences. Without considering the possibility of other forms of silence, their only option is to interpret difference as absence.

Even recent interpretations which have attempted to take a more inclusive approach in recognising sensitivities around non-European sources can become mired in an attempt to shoehorn mathematical artefacts into modern terminology. For example, debates still continue about whether Egyptian *aha* problems can be considered algebraic or otherwise (Immhausen, 2009). These problems are named after the Egyptian word ꜥḥꜥ meaning heap. It stands for an unknown quantity in a series of problems, including Moscow Mathematical Papyrus Problem 19:

> Example of working out a heap. Make one and a half times together with four, it has come as ten. What heap says this?
> You shall work out the excess of the ten over the four; it becomes six.
> You shall work out one and a half in order to find one, it becomes two thirds.
> You shall work out two-thirds of the six, it becomes four. See, it is four, says it. You will find it good.
>
> (Moscow Mathematical Papyrus Problem 19)

Rendered into modern notation this problem would read:

$$\frac{3}{2}x + 4 = 10$$

$$\frac{3}{2}x = 6$$

$$x = 6\left(\frac{2}{3}\right)$$

$$x = 4$$

In this particular ꜥḥꜥ problem, the scribe uses a method practically identical to the way that is taught today. While the problem is not written in a symbolic way, using instead the narrative format that is common in Egyptian mathematical texts, it still contains the concept of an unknown quantity – a heap – after which the set of problems is named. Although this example follows a method that is closer to the modern method of solving the problem, there are several other methods of solving problems like these in the extant mathematical texts (Immhausen, 2002). Despite showing the range of methods that an Egyptian scribe could employ to solve problems, one point of debate is whether they can be considered equations in the modern sense, and whether it is appropriate to render the problems into modern algebraic notation as equations (Immhausen, 2009). Such debate focuses far more on modern ideas of mathematics than on the achievements contained in a text that was written around 2,000 years before Muhammad ibn Musa al-Khwarizmi wrote the *Kitab al-Jabrw'al-Muqabala* from which we derive the word algebra.

Silences in the reconstruction of ancient mathematical practices are a property of the way in which we talk about mathematics. We use a vocabulary that is highly specialised because this is necessary when dealing with highly abstract and precise mathematical objects. Without a way of talking about texts such as Moscow Mathematical Papyrus Problem 19 in a historically neutral way, modern ideals and practices will be used as the yardstick by which we articulate the achievements of a mathematical culture that is far removed from us in time and values.

The discussion about whether or not the ꜥḥꜥ problems of the Egyptians can be called algebraic equations is not merely a question of mathematical philosophy, it is a question of social and cultural meaning. To label the ꜥḥꜥ problems as algebraic equations affords the Egyptian civilisation a particular respect, because with this label comes a set of mathematical aesthetics.

Greek opinion

While there are many silences embedded in the interpretation of ancient Egyptian mathematical texts, there are also voices that need to be listened to and considered. Several of these voices come from ancient Greece where thinkers such as Herodotus wrote of their debt to the ancient Egyptians and traced the origins of geometry back to them. For instance, Herodotus wrote:

> The king moreover (so they say) divided the country among all the Egyptians by giving each an equal square parcel of land, and made this his source of revenue, appointing the payment of a yearly tax. And any man who was robbed by the river of a part of his land would come to Sesostris and declare what had befallen him; then the king would send men to look into it and measure the space by which the land was diminished, so that thereafter it should pay in proportion to the tax originally imposed. From this, to my thinking, the Greeks learned the art of geometry.
>
> (Herodotus, 1954, *The Histories*, 2.109)

Similarly, Aristotle wrote:

> Hence it was after all such inventions [the practical arts] were already established that those of the sciences which are not directed to the attainment of pleasure of the necessities of life were discovered; and this happened in the places where men had leisure. This is why the mathematical arts were first set up in Egypt; for there the priestly caste were allowed to enjoy leisure.
>
> (Aristotle, 1920, *Metaphysics*, 981^{b}20–25)

The Greeks acknowledged the debt they had to the Egyptians. It has been claimed that the Greeks were exaggerating when they recognised geometry as an invention of the Egyptians. They certainly cannot be considered to be contemporary sources. Herodotus was writing in the 5th century BCE and Aristotle in the 4th, over 1,000 years after the surviving texts were written. Yet, when we are so keen to trace the history of our mathematics back to ancient Greece because we believe that they were the first to treat mathematics as an abstract subject in its own right, their own opinions on the origins of the subject are significant.

Conclusion

There are many silences apparent in the reconstruction of the mathematical culture of ancient Egypt. Some are unavoidable because of issues around survival of evidence and the decipherment of the Egyptian language. Other silences are products of the way in which mathematics has been studied and the assertions about what is desirable in a mathematical text. To fully appreciate the achievements of the ancient Egyptians, and to place their mathematics within its proper context, we need an understanding of the entirety of Egyptian culture and its world view. With this understanding original silences can be teased apart from those that result from our historiography. For example, the desirability, or otherwise, of unit fractions can only be understood within the cultural necessities that forged them; it is not a mathematical question. The precepts for the 2/n table show a developed and considered system, yet the origins of that system are presently matters for pure conjecture. The silences are generated by not linking mathematical objects to the more tangible remains that can provide commentary on the civil service, taxation and life in a workers' town such as Kahun from which we have surviving mathematical fragments.

Where Egyptian mathematical texts are silent is on the reasoning behind the techniques that they use in their calculations. Equations such as the sum of a geometrical progression are pursued with no further explanation. Again, it is a matter of conjecture where this silence arises; is it the result of never having been written down or of the evidence not having survived? Some argue that this particular silence is due to ignorance on the part of the Egyptian scribe, who they say would have arrived by their methods through trial and error, rather than abstract reasoning. This accusation leads to further charges that the Egyptians did not recognise mathematics as a separate subject. One silence, which if labelled as a conceptual lack, has the effect of placing all other silences into the same category: a hole in a papyrus at a crucial point in a complex problem is translated in the most pessimistic way because of the overlying assumption of ignorance.

At the heart of understanding and interpreting these silences is the question of the extent to which we are indebted to different cultures for the discovery of mathematics. This question has many different philosophical and political contexts. As long as neutrality in the language of description is impossible, silences will remain. Egyptian temples are not criticised for their lack of stained glass, yet the mathematics needed to build them is criticised for not containing modern elements of mathematics. Mathematics is prone to shifts in aesthetics and desirability. While pure neutrality may be impossible to achieve, recognising that different cultures have different driving factors when producing mathematical artefacts would assist in producing an interpretation where silences can be treated as something other than a lack. Such an approach requires a new collaboration across research schisms.

Notes

1 There is a discrepancy between the diagram and the narrative explanation in the original papyrus, which probably stems from an unstated change in the unit of measurement.
2 Sic. This should read 2401 and may be a copying error.
3 The Hekat was an Egyptian measure for a quantity of grain. Its use here is not meant as a measurement, but a convenient label for the final term of the progression.

References

Aristotle, 1920. *Metaphysics*. Trans. Godley, A.D. London: Heinemann.
Clagett, M., 1989. *Ancient Egyptian Science Volume 1: Knowledge and Order*. Philadelphia: American Philosophical Society.
Claggett, M., 1999. *Ancient Egypian Science A Source Book Volume 3: Ancient Egyptian Mathematics*. Philidephia: American Pilosophical Society.
Cooper, L., 2010. A new interpretation of Problem 10 of the Moscow Mathematical Papyrus. *Historia Mathematica*, 37, pp. 11–27.
Couchoud, S., 1993. *Mathématiques Egyptiennes: Recherches sur les Connaissances Mathématiques de l'Egypte Pharaonique*. Paris: Editions Le Leopard d'Or.
Faulkner, R.O., 1962. *A Concise Dictionary of Middle Egyptian*. Oxford: Griffith Institute.
Gardiner, A., 1957. *Egyptian Grammar*. 3rd edn. Oxford: Griffith Institute.
Gillings, R.J., 1972. *Mathematics in the Time of the Pharaohs*. New York: Dover Publications.
Hawking, S. (ed.), 2006. *God Created the Integers: The Mathematical Breakthroughs that Changed History*. London: Penguin Books.
Herodotus, 1954. *The Histories*. Trans. De Selincourt, A. Harmondsworth: Penguin.
Hornung, E., Krauss, R. and Warburton, A.D., 2006. *Ancient Egyptian Chronolgy*. Leiden: Brill.
Immhausen, A., 2002. The algorithmic structure of the Egyptian mathematical problem texts. In: Steele, J.M. and Immhausen, A. (eds), *Under One Sky: Astronomy and Mathematics in the Ancient Near East*. Münster: Ugarit-Verlag, pp. 147–166.
Immhausen, A., 2009. Traditions and myths in the historiography of Egyptian mathematics. In: Robson, E. and Stedall, J. (eds), *The Oxford Handbook of The History of Mathematics*. Oxford: Oxford University Press, pp. 781–800.
Joseph, G., 2000. *The Crest of the Peacock: Non-European Roots of Mathematics*. London: Penguin Books.

Kemp, B., 1989. *Ancient Egypt: Anatomy of a Civilization*. London: Routledge.
Leach, B. and Tait, J., 2000. Papyrus. In: Nicholson, P.T. and Shaw, I. (eds), *Ancient Egyptian Materials and Technology*. Cambridge: Cambridge University Press, pp. 227–253.
Neugebauer, O., 1952. *The Exact Sciences in Antiquity*. Princeton, NJ: Princeton University Press.
Robins, G. and Shute, C., 1987. *The Rhind Mathematical Papyrus*. London: British Museum Publications.
Struve, W.W., 1930. *Mathematischer Papyrus des Staatlichen Museums der schönen Künste in Moskau*. Berlin: Springer.

9 Meditations on silence
The (non-)conveying of the experiential in scientific accounts of Buddhist meditation

Brian Rappert, Catelijne Coopmans and Giovanna Colombetti

This volume attends to an under-appreciated aspect of science communication: its silences.[1] As discussed in the Introduction, one way of opening up silence as a phenomenon is to note that what counts as silence can vary across situations. Just when the absence of sound becomes treated as 'silence' varies depending on whether one is listening to the radio, working in an open plan office, at home enjoying a cup of tea before bed, or talking with a friend over the phone. Thus, the question 'What is silence?' needs to give way to questions like 'Silent for whom?' 'When?' and 'In what manner?' Silences are *enacted* in specific contexts and also help to define those contexts. Being produced through contingent and situated practices, silences come in many forms: what goes without saying, what is unspeakable, what has been censored, what is known but not sayable, and so on. Accordingly, to be silent can be recognized as a way of demonstrating deference or defiance, understanding or lack of understanding, giving or avoiding offence, as well as displaying or suppressing emotion. In this respect, silences are pregnant with possibilities.

In this chapter we develop these points by exploring the struggle to render present what is considered silent in a specific context: the recent (renewed) attempts to scientifically demonstrate the effects of Buddhist meditation on the human brain and behaviour. Silence and meditation seem natural companions: through stilling the body and quieting the mind, people create the conditions for experiencing what is happening in the present moment. Yet silence, in various guises, also mediates accounts *about* meditation, which must grapple with how best to do it justice as a kind of practice. Efforts to convey what meditation 'is' face a number of tensions deriving from the idea that the best, and ultimately only, way of knowing it is through first-hand experience. By shifting the focus from 'telling' to 'showing', scientific approaches using brain imaging and other physiological methods have expanded the possibilities for conveying meditation experiences. In the process, these approaches invite, necessitate and shape particular forms of silence.

Overall, we are interested in how silences work across and in the collaboration between the varied traditions that speak about meditation. Unlike mere gaps that hinder the development of knowledge, we argue that the interlacing of what is said and what is not said about the experiential helps cohere different traditions. In doing so, silences are consequential in helping define notions of expertise and producing forms of scientific accounting.

The next section begins by introducing some of the general complications associated with describing lived experience. The third section considers how such complications play out within Buddhist contemplative traditions, indicating how and why accounting (in words) for experiences of meditation is seen as problematic. The fourth section examines scientific and popular-scientific accounts of neuroimaging and other experimental studies on the effects of meditation, in particular by attending to the case of a Tibetan Buddhist meditator who has assumed a high prominence. In this examination, we focus on what is *not* present in write-ups of experiments and on the manner in which the lived experience of meditators is rendered present and absent in such accounts. The final section surveys how accounts of lived experience of meditation have been positioned within the fields of neuroscience and so-called 'neurophenomenology' in the last few years and how this might develop in future.

Putting experience into words

We can begin by noting some of the general curiosities associated with offering accounts of lived experience, or what we will refer to as 'the experiential'. Noting these curiosities sets an overall intellectual backdrop for the discussion in subsequent sections about what resists being said about meditation and how, despite these issues, meditation is spoken about all the same.

From bodily sensations to desires, memories, emotions and thoughts, experiences would seem to be unquestionable and always present. Famously, Descartes reasoned that one thing that cannot be doubted is the reality of our conscious thinking activity. And yet, as subsequent reflections on the nature of experience (developed for example in philosophical phenomenology and psychotherapy) have highlighted, experiences are neither always clearly or fully present, nor always readily available for self-reports. To begin appreciating this point, consider bodily sensations (such as an itch or a headache). At times, these can be overwhelming and take up all our attention; but this is not always so (see Petitmengin, 2009). To see this, take a moment now proprioceptively to 'scan' your body. You might notice stiffness and even pain in your lower back, or tautness associated with your position, slumped shoulders, and so on. Now close your eye lids and direct attention to the muscles at the back of your eyes. Can you locate them? How do they feel? Is this feeling easy to characterize? Try now relaxing those muscles. Could they be relaxed? Was there built up tension that you had not previously noticed?

The point of this exercise is to show that, by directing attention to different parts of one's body, it is possible to become aware of ongoing happenings that are otherwise not noticed. In the philosophical phenomenological tradition, experiences that are generally unheeded, but that can nevertheless be brought to attention, reflected on and reported into words, are said to be *pre-reflective*.

Pre-reflective experiences are generally 'tacit' or implicit, but can become reflective or explicit (see Zahavi, 2005).

In addition, consider that even when one becomes reflectively aware of certain experiences, staying with their unfolding can prove challenging. This is perhaps most apparent in the case of 'difficult experiences', such as intense negative emotions (anxiety, grief, etc.) or acute physical pain. It is challenging to pay close and sustained attention to these experiences over time and our natural tendency is to 'block them out', think of something else or construct narratives about them. Keeping sustained attention even on ordinary and not unpleasant sensations is highly demanding, as anyone who has tried to focus on, for example, the proprioceptive and tactile sensations of the breath (such as the perception of the belly rising and falling or the passage of air through the nostrils) will testify. Very soon judgments (e.g. 'my breath is so shallow') or other thoughts (planning one's dinner, remembering a previous quarrel, etc.) take one's attention away from the bodily sensations.

On top of these complications, even when one can attend closely to one's experiences as they unfold, rendering those experiences into words often amounts to a gross reduction. Commonly used descriptions – say, 'I'm bored' – stand as coarse and generic labels for what are often varying and complex experiences. Arguably, producing different labels to capture as exhaustively as possible different varieties of boredom, fear, concentration, etc. is likely to be inadequate to speak to the diversity, richness and uniqueness of individual experiences. As another exercise, for instance, you could try to list how many different forms of boredom you can name and consider whether this list adequately captures the quality of your experience of boredom.

As we will see in the course of this chapter, there are different ways of responding to and dealing with these complexities. Generally, however, it is acknowledged that becoming aware of pre-reflective experiences and offering verbal accounts of them is a skill that needs to be honed. In psychotherapy, one method for doing this is known as 'focusing' (Gendlin, 1996). This method recognizes that people are often not aware of (at least some of) their emotional experiences and tend to describe their behaviours rather than their feelings, or what 'they are supposed to feel' in a certain situation according to social conventions or personal schemas. To put clients in touch with their lived experience, the focusing psychotherapist explicitly invites them to attend to their bodily sensations and to how a certain situation feels 'in' their body.[2] This psychotherapeutic technique thus recognizes words as limiting: people can get 'stuck' in self-descriptions that prevent them from tapping their lived experience. To be (re)directed toward the latter, people need to move away from a detached descriptive view of their condition. Yet at the same time, the therapist has to use words to facilitate this redirection, and clients need words to communicate how their body feels and how they interpret these bodily feelings.

Reference to an experience that cannot be captured through words is pervasive in works of literature as well. As Timothy Walsh (1998: p. 111) contends, a passing phrase – such as William Faulkner's reference to the 'smell of rotting cucumbers and something else which had no name' – invites readers to move beyond words through words, but in a manner that has no determinate ending for meaning. In the hands of writers such as Virginia Woolf or D. H. Lawrence, such simple allusions mesh with layers of partial descriptions, absent referents and incomplete storylines to produce sometimes subtle and sometimes overt evocations, through language, of what is and is not expressible about experience.

Perhaps nowhere are the tensions associated with the desire to express while acknowledging inexpressibility more evident than in the case of existential, contemplative and spiritual matters. Here, while much is regarded as too profound for words to capture, considerable effort is directed towards doing just that – indeed volumes and volumes have been written on such limits of expression (Burke, 1961).

The obstacles to expressing experiential understandings, comprehensions, awareness, etc. (what we lump together as 'insight' for short) are many. Foundational religious texts that might provide the basis for deep understanding – such as the Bible or Qur'ān – have been subject to much debate regarding their literalness, who can interpret their meaning, what remains outside of them and the extent to which divine reasoning can be accessed by human thought (e.g. Saeed, 2006). The limits of human concepts impinge upon both what could be written about and what can be read from sacred texts. In other words, how can we know the limitless nature of God through mundane human concepts? To refer to an experience as 'transcendent' is to gesture towards a sense of how it necessarily 'lies beyond'. If experiential insights resist comprehension by those undergoing them, then their expression to others is even more contestable.

Refraining from attempts to express is one way to acknowledge such predicaments, but dynamic questioning and debate has often followed instead. In the tradition of speaking of the ineffable, Franke (2007: p. 2) commented that:

> The irrepressible impulse to 'speak' essential silence is a constant (or close to a constant) of human experience confronted ever anew with what surpasses saying. While what is experienced remains inaccessible to speech, there is no limit to what can be said about – or rather from within and as a testimonial to – this experience which, nevertheless, in itself cannot be described except as experience of ... what cannot be said.

The recognition of such tensions has inspired many attempts to exceed the limitations of language by working with language.

Having highlighted some general complications associated with describing lived experience, the remainder of the chapter turns to how experiential insight is accounted for in descriptions of Buddhist meditation. Both first-person accounts of those undergoing experiences and secondary analyses of them by academics, writers, religious authorities and others, are patterned by what is and is not said about experience, and the interrelations between the two.

Moon pointing: 'can nots' and 'should nots' of recounting meditation

Today the label 'Buddhism' is often used to lump together systems of thought and local practices that have varied significantly over time and space. In different periods and places, practitioners and scholars understood and promoted sutras[3] and other discourses central to understanding experience within Buddhism in one of four ways: as literal, as containing hidden messages, as subject to multiple interpretations depending on the proficiency of readers or as simply beyond comprehension (Mair, 2013; Teeuwen, 2006). Moreover, both at the level of philosophical doctrine and cultural practice, such contests over interpretative authority have been shaped by prevailing political and social structures of the day (Rambelli, 2006).

In this section, we invite readers to consider how the experiential has figured as a source of insight – though a complicated one – in many instantiations of Buddhism. The points in this section, and the spirit of 'probing the possible' that accompanies it, will serve as a prefatory basis for examining in the next section recent scientific efforts to document the effects of meditation on the brain.

We begin with the observation that in providing detailed self-accounts of what it is like to meditate, practising meditators themselves often refer to challenges and tensions in speaking about their experiences. The paradoxical contention that words are incapable of obtaining and conveying insight is routine in contemporary written first-person accounts of meditative journeys, notably those written in English for general audiences not necessarily steeped in Buddhist traditions.[4] A common refrain is that, as a practical undertaking, the experience of meditation cannot be reduced to language. This applies particularly to what can be characterized as 'higher' states of awareness that are non-conceptual (Thompson, 2015: p. 94). For instance, whether the experience of 'nirvana' can be described is a matter of debate, not least because of the different ways it is understood across and within Buddhist traditions (Albahari, 2011).

What can be described also depends on who is speaking to whom – some readers may not be able to understand because of where they are in their own meditation progression (Herrigel, 1999: p. 23). Indeed, whatever the possibility of description, the advisability of offering it can be doubted because of the risk of glamorizing meditation practice, setting up unrealistic expectations, fostering misunderstanding and doubt, and divorcing readers from attending to their own experiences in favour of striving to achieve externally inspired states (like someone else's sense of compassion, bliss, or one-ness; see Kornfield, 2001).

By their own acknowledgement, first-person accounts are often said to capture something of importance (an 'experience') and at the same time not to do so (because of the inadequacy of language to describe or the inability of the reader to understand). While such *disavowals* may undermine confidence in authors' descriptions of their lived experience, they do signal an awareness of being aware of experiences – and thus shore up a particular kind of claim to proficiency.

We can try to make sense of the above-mentioned cautions about first-person accounts by aligning them with foundational doctrine. Consider for example the following passage from one of the best known sutras. In volume 2 of the Shurangama Sutra, the Buddha likened the habitual orientation to his teachings of those working within a non-enlightened mind as one of fixating on the finger that points to the moon, rather than on the moon itself. So rather than following the words of the Buddha toward enlightenment (represented by the moon), the common tendency is to look at the finger pointing at it (the teachings of the Buddha), thus mistaking the finger for the moon.

In the same vein, many traditions of Buddhism express caution about seeking to put experiences into words. Words are potentially dangerous because they are never experiences – either of the dissatisfaction of the world or of the awakenings of enlightenment. Wrapping words and concepts around experience is likely to reinforce habitual intellectual tendencies toward grasping at the world through greed, aversion and delusion. Labels given to experiences – such as 'pain', 'pleasure', 'old', 'ugly', 'beautiful' – become mistaken for the experiences themselves and condition subsequent thought and action; labels are used to produce theories about experience, and these theories become unquestioned doctrines. Speaking to the dangers of doctrine, the Buddha offered another simile in the Alagaddupama Sutra, here likening his teachings to that of a raft. A raft is necessary for crossing a vast expanse of water (i.e. a journey of awakening). Once that journey has been undertaken, though, it would be foolish to continue to carry the raft. Similarly, even the words of the Buddha should be seen as temporary aides, rather than abiding truths.

One recurring danger with the elaboration, explication and evaluation of experience is that the reading of words creates expectations and engenders striving for certain sensations. Another danger is that knowledge of what is written gets mistaken for the experiential understanding obtained through moment-to-moment engagement in the world. Because of the potential for words to reinforce commonplace delusional thinking, the Buddha recommends sticking to experience itself. To use modern terminology, it is not simply that experience is 'ineffable', but also that it is *better* left undescribed.

The points in the previous two paragraphs play out in different ways across different traditions in Buddhism. Consider two extremes. In an exceptionally in-depth analysis of what he labelled 'the phenomenology of meditative experience', Brown (1977: p. 238) argues that 'meditative and mystical states are rarely ineffable. Ineffability is largely a function of inadequate data sources, inadequate methods of analyzing the texts, and lack of verbal skills.' Examining the Tibetan *Muh Gmudru* tradition, he works through extensive language developed for describing meditative experiences. The rendering of this textual language into one understandable to Western researchers is presented as unproblematically accomplished through drawing on the 'semantic field method' in cultural anthropology. And yet, against this literal rendering, elsewhere Brown also speaks of the difficulty of comprehending, let alone expressing, the notion of 'subtle energy' discussed in the *Muh Gmudru*. Certain higher levels of meditation are regarded as too complex and subtle for this extensive analysis, even if they themselves are merely a 'starting point of meditative insight' (ibid.: p. 261). In addition, Brown's analysis only presents a description of general states meant to be reached in meditative practice rather than trying to recount the first-person lived experience of those who have reached them.

Elsewhere, routine rejections of the possibility of describing insight give way to attempts to do just that. Claims about the inability of words to capture insight are perhaps most pronounced in Zen Buddhist traditions, one of the most written-about forms of Buddhism in the West. Here denials that insights can be communicated typically accompany accounts of Zen. As D. T. Suzuki (1949: p. 267) writes:

> Zen refuses even tentatively to be defined or described in any manner. The best way to understand it will be, of course, to study and practice it at least some years in the meditation Hall. Therefore, even after the reader has carefully gone over the Essay, he will still be at sea as to the real significance of Zen.

In book after book, as part of his influential work in bringing Zen to Western audiences, Suzuki strives to undermine the belief that he is conveying anything substantial (ibid.: p. 235) – or at least that is the way we read his text as relative novices to the meditation hall. Stated disavowals place his accounts in a precarious status, as both pointing to and pointing away from something significant. This is so because these accounts reinforce the established intellectual conditioning instead of developing awareness of one's ongoing lived experiences (Dass, 1988: p. x). In presenting intellectualizing as impeding the cultivation of awareness (Kasulis, 1981), the writing of Suzuki and others could be interpreted as *distracting* readers from obtaining awareness.[5]

In short – as has been said more than once – the more that is said about Zen, the further one is from it. To be open to the contradictions of such a statement is to open to the contradictions of offering accounts of the experiential.

How to see and describe the moon then?

Meditation under trial: what is not being rendered in words

Experimental approaches have provided a way to reorient the complications of speaking what cannot or should not be spoken. To indicate the basic logic that undergirds these experiments: if it can be empirically demonstrated that experienced meditators regulate their cognitive, physical and emotional responses better than 'normal' subjects do (see, e.g., Goleman, 2003; Davidson and Harrington, 2002), this facilitates understanding how meditation slowly but surely *imprints* on behaviour, brain function and even brain structure. In this section and the next two, we contrast some of the varying ways in which the experiential is expressed, or left in silence, in the encounters between Buddhism (particularly its Tibetan forms) and scientific traditions. Our goal is to consider how the interlacing of what is said and what is not said facilitates the collaboration between varied traditions concerned with meditation. Silences, we will argue, are consequential in helping define notions of expertise and producing forms of scientific accounting.

That science can help elucidate the nature of meditative experience has been a contentious matter in the past.[6] On the one hand, some Buddhist practitioners have welcomed efforts to map and measure bodily and brain activity, for these measures might render meditation intelligible as a distinct form of practice with real consequences. On the other, there have been concerns that the translation of meditation into a phenomenon amenable to scientific study might lead to impoverished understandings or misguided conclusions.[7]

For example, in his autobiography published in 1987, Tibetan monk Lobsang Tenzin[8] wrote with bemusement about the efforts of Western doctors and scientists to measure the *tummo* or 'psychic heat' he generated as part of meditation practice. Reflecting on a visit to the US, he commented:

> With their instruments the scientists could see such things as heat, movement of wind and blood, and my brain waves. There are certain things, however, which cannot be seen by their instruments, such as the non-dual wisdom of great bliss and emptiness, which is the root of all realisation. Therefore the tantric texts say that it is the basic principle of Tantra to generate a subjective blissful wisdom realising emptiness through the withdrawal, abiding and dissolution of the winds in the central psychic channel [...] [W]ithout the subjective wisdom of great bliss realizing emptiness, it would be impossible to gain control over the winds and channels.
>
> (Tenzin, 1987: p. 111)

Tenzin's article appeared in the periodical *Chö-Yang: The Voice of Tibetan Religion and Culture*, so he could presume his audience to be at least somewhat familiar with the canonical texts, the hermit life and the spiritual journey in the context of which his meditation practice took place. In the piece, he acknowledges the limits of words and 'conceptual thoughts' in conveying actual experience (Tenzin 1987: p. 110), but he also offers rich detail about the trials and tribulations of solitary meditation, using a Buddhist vocabulary of 'bliss', 'winds', 'channels' and 'emptiness'. Western science appears in his account as both alien and alienating. The scientists got what they wanted out of the encounter ('I could tell from the scientists' faces that they were satisfied, and they said so too' [Tenzin, 1987: p. 111]), but, as the earlier quote indicates, their methods and measurements missed what Tenzin considered to be most salient about his practice.

Collaborative research

In the early 1990s, under the auspices of the fourteenth Dalai Lama, collaboration was initiated that sought to place Buddhist monks and Western scientists on more equal terms. A mixed group of scientists and scholars set out to study Tibetan monks who lived in retreat on Bhagsu Mountain in the Himalayas (Houshmand et al., 2002). Senior monks were invited to 'provide insights both from the formal teachings of their own tradition and from their own direct experience of meditative practice to help shape the ultimate design of the experiments' (p. 4). These monks, by virtue of their intensive and long-term meditation practice, were seen as highly advanced in the investigation and training of their own minds. Scientists such as Richard Davidson wanted to test and measure their proficiency in relation to theories of neuroplasticity (the brain's ability to change in structure and functionality in response to training). Such testing and measuring required the monks to share details about what they were doing when meditating, as well as the changes they had experienced with the advancement of their practice. This proposal, however, did not reckon with the various forms of silence that accompany the practice of meditation in the context of the Tibetan monastery.

As Sara Houshmand and colleagues recount in a book chapter on the early days of this cross-cultural collaboration, the designated monks were not readily prepared to provide insight into their meditation practices: 'This is a sensitive area, because many of the practices are traditionally held as secret, to be discussed only with one's teacher or perhaps with others who are similarly initiated' (Houshmand et al., 2002: p. 11). Even more fundamentally, the monks who were approached by the research team 'all denied having achieved any special spiritual progress' (ibid.), despite having had many years of training in Buddhist thought and having devoted much of their lives to intensive meditation. In Tibetan Buddhism, those who are recognized as having achieved spiritual attainment are not at all interested in letting this be known or in convincing others of it; also pretending to have high spiritual attainment is a major breach of monastic rule (for instance, see Revel and Ricard, 1998: p. 51). As a result, 'it is almost unheard of for a monk to speak of his own accomplishment or progress in the practice' (Houshmand et al., 2002: p. 11).

Scientists, however, did not give up on the idea that advanced meditators make especially valuable experimental subjects. Further and later efforts to observe the effects of long-term meditation on the brain went ahead, but compromises were made. For example, the trend in scientific publications became to qualify experimental subjects on the basis of 'total number of hours' they had spent meditating.[9] In this way, the touchy issue of proclaiming advancement could be sidestepped; indeed, beyond the impression conveyed by the large numbers (10,000 hours and upward), the question of who qualifies as an advanced meditator (and why) is routinely not unpacked in scientific reports today.

In 2000, again under the Dalai Lama's influence, another round of experiments was initiated, involving a Buddhist monk who has since become an iconic experimental subject in accounts of neuroscientific meditation research in popular writings and the international media. The earliest of these accounts, Daniel Goleman's book *Destructive Emotions and How We Can Overcome Them* (2003; hereafter *Destructive Emotions*), presents 'Lama Öser' as a 'European-born convert to Buddhism [who] trained as a Tibetan monk in the Himalayas for more than three decades, including many years at the side of one of Tibet's greatest spiritual masters' (Goleman, 2003: p. 3). He agreed to a series of laboratory tests designed to study his brain activity, emotional response, physiological arousal in situations of disagreement and startle response reflex while variously engaged in different forms of meditation. Goleman provides a lively description of the rationale behind and the setup of these experiments, of Lama Öser's demeanour throughout and of the scientists' rush to analyse their data in time for a laboratory visit by the Dalai Lama. While describing the results as preliminary and not conclusive, the author conveys that Lama Öser performed exceptionally well on all the tests.

Since the publication of *Destructive Emotions*, the story of this particular set of experiments has circulated widely in popular accounts about science and Buddhism.[10] In the process, it became known that 'Lama Öser' was, in fact, Matthieu Ricard, a French monk with a Ph.D. in molecular genetics. Ricard started working with the Dalai Lama as his French interpreter in 1989. For several decades now he has been involved in raising awareness and understanding of Buddhist thought and of meditation among Western audiences. This he does through his photography, translation work and popular writing on subjects such as Buddhism and philosophy (with his father Jean-Francois Revel, 1998), Buddhism and science (with Trinh Xuan Thuan, 2001), Buddhism, neuroscience and economics (co-edited with Tania Singer, 2015), how to meditate (2010) and happiness (2007).

Ricard's complex identity as a Buddhist monk (and part of the Dalai Lama's entourage), a scientist and, increasingly, a celebrity who can connect with Western audiences has made his role as an experimental subject in meditation research especially significant. Here is someone who can skilfully operate within Tibetan Buddhism as well as Western science, someone with the integrity and know-how to make Buddhist insights and experiences available to a wide audience. Unlike the reluctant monks of Bhagsu Mountain, Ricard understands the project and the stakes 'from both sides'. This allows him to provide helpful suggestions to scientists for how to make meditation amenable to experimental study – suggestions that maintain the integrity of Tibetan Buddhist practices while also understanding the parameters of scientific research.

Commentators indeed make much of Ricard's distinctive positioning. Goleman, for example, in his foreword to Ricard's book *Happiness* (2007) credits the latter with 'unparalleled authority' because of his combined Buddhist and scientific attainments (Goleman, 2007: pp. xi–xii). The same book includes Ricard's own account of those early experiments in which he participated as an experimental subject and collaborator. Introducing a reprint of that chapter in a later volume, editors Barry Boyce and colleagues note that:

'[s]ince he [Ricard] himself was a laboratory subject in one of these major scientific studies, he has a unique vantage point from which to describe the historic encounter between science and meditation' (Boyce et al., 2011: p. 127). To what extent are Ricard's experiences of meditation conveyed as part of this unique vantage point?

Blanking

In relation to the themes of this chapter, of particular interest is the interlacing between what is said and not said in these and other descriptions of Ricard as an expert experimental subject. Despite making much of Ricard's unique positioning, the accounts by Goleman in *Destructive Emotions*, Ricard himself in *Happiness*, and others elsewhere do not make mention of Ricard's first-person experience of meditation during the experiments (nor of such experiences on the part of any other monk). While it is reported that Ricard brought about certain mental states – such as concentration and compassion – there is no attempt to describe his lived experience during these states of awareness.[11] Instead, what we find in these texts is disengagement from experience even when it is held up as relevant. We discuss this dynamic as various forms of *indirection*.

The first form of indirection relates to the claim that experienced meditators have greater insight into the nature of the mind, and hence greater discipline over thoughts and emotions, than the average person (e.g. Lutz and Thompson, 2003: p. 37; Lutz et al., 2007: p. 521). Such insight and discipline are expected to allow a person to provide more precise descriptions of what he or she is experiencing at any moment in time. In the literature on meditation research, a subject's ability to furnish such descriptive first-person accounts is highlighted as essential for the design of meaningful experiments (Lutz et al., 2007) and also as a mark of expertise (Fox et al., 2012). Yet for scientists to acknowledge, let alone make space for, such experiential accounts in their writing is not straightforward, as these do not mesh with the wider conventions of scientific publication. On rare occasions, the importance of first-person accounts has been flagged in scientific writing, as an aside. For example, in a recent article on the experiments that focused on Ricard's startle response reflex,[12] authors Robert Levenson and Paul Ekman state:

> We showed Ricard a number of emotionally arousing films, of the sort we typically use in our emotion research […] Our general experience is that, when allowed to give a free response describing their subjective experience, almost all subjects respond quite briefly. Ricard's descriptions, in contrast, were *much more elaborated, rich in detail and recounting the moment-to-moment changes that occurred in his emotional state.*
> (Levenson et al., 2012; p. 656; italics added)

And yet, even in this exceptional instance of commenting on first-person accounts in a scientific article, not one example is provided that might illustrate the richness of Ricard's descriptions. Indeed, a negotiated sense of presence and absence is especially vivid here. The authors make a point of highlighting, as something of interest, both the richness of Ricard's experience and his ability to recount such richness, but do not go on to share this richness with the reader. They may well have considered this out of place in a scientific publication,[13] and indeed it would appear that popular writings like *Destructive Emotions* and *Happiness* can do this more easily (though see below). Our point is that indirection occurs in the way the richness of Ricard's moment-to-moment experiential descriptions is rendered present and not.

A second form of indirection relates to how, especially in *Destructive Emotions*, Ricard's experience as a meditator is made salient to the design of the experiments. The scientists rely on Ricard to distinguish in theory, and perform in practice, six different varieties of meditation: visualization, one-pointedness, open state, compassion, devotion and fearlessness. Descriptions of what these mental states entail and how they differ from one another are essential to making sense of the experimental data generated via functional magnetic resonance imaging and other means. However, as detailed in *Destructive Emotions* and also in *Happiness*, this happens in a particular and limited way. The scientists and their experimental subject give general labels to achieved mental states (e.g. compassion), but they provide no account of the lived mental and bodily experiences of those states from the perspective of the meditator. General labels are made to stand in, as a gloss, for lived experience.

A third form of indirection relates to the dislodging of any religious connotations from meditation training. The absence of marks of lived experience from a first-person point of view in *Destructive Emotions* and *Happiness* makes it easy to leave religion out of the account. This also resonates with the expressed intention of Ricard and others to present meditation as a mental technique rather than a Buddhist religious practice. As Ricard states: 'If such meditation techniques are valid and address the deepest mechanisms of the human mind, their value is universal and they don't have to be labelled Buddhist even though they are the fruit of more than twenty centuries of Buddhist contemplatives' investigations of the mind' (Ricard, 2007: p. 201). In the context of experimental testing, the dislodging of religious connotations also makes less relevant concerns about validity that might be raised if accounts of meditation were framed through – indeed as squarely derived from – a Buddhist vocabulary and training (as in the case of Lobsang Tenzin's account above).[14] More generally, the lack of experiential accounting sidesteps questions about whose understanding of religious experiences should count – no explanation needs to be given for *which* introspective accounts of the experiences of meditation by *which* meditators from *which* traditions on *which* occasions need to be taken with what measure of significance.

A fourth and final form of indirection relates to how Ricard as a subject is presented in various written accounts of the experiments. As mentioned above, the particularities of his background and profile have been heralded as making him the ideal subject-collaborator. It is interesting, then, that even and especially in Ricard's own writing, he comes across as hardly present at the scene of the experiment. In *Happiness*, Ricard acknowledges: 'I happened to be the first "guinea pig"' (Ricard, 2007: p. 190). However, he then proceeds to describe these experiments in highly generalized terms, never once referring to the tests as personally experienced. Passive, scientific language prevails ('A protocol was developed'; 'The meditator alternates thirty-second neutral periods with ninety-second periods in which he generates one of the meditative states' [ibid.: pp. 190, 191]). There is a strong emphasis on results ('Using fMRI, Lutz, Davidson, and their colleagues also found that the brain activity of the practitioners meditating on compassion was especially high in the left prefrontal cortex' [ibid.: p. 195]). Notwithstanding his 'unique vantage point' as an experimental subject and scientific collaborator, Ricard talks about himself mostly in the third person, for example when recounting the responses of 'the first meditator' in the experiment on the startle reflex (ibid.: pp. 197–198).[15] All these examples point away from the tradition and struggles of writing about meditation in the first person; they foreground science as a way to represent what meditation really does. Ricard's subjectivity thereby slips from view – it is as though there is a blank figure at the centre of his account.

Kevin Hetherington and Nick Lee (2000) have defined the 'blank figure' as an entity that helps broker new possibilities because of the way it, itself, remains undefined.[16] They mention the joker in a game of cards and the double-blank domino as examples. The key characteristic of a blank figure is that, as an 'underdetermined element', it helps 'stitch incommensurabilities together' (p. 175). Ricard's 'blankness' as an experiencing subject, similarly, may be a remedy for some of the perceived incommensurabilities or tensions between the experience of meditation within Buddhism and scientific or other forms of secular and rationalist discourse that appeal to Western audiences (see Andresen, 2000).

Taken together, these four forms of indirection keep engagement with the lived experience of meditation at bay by hinting at it without describing it, by glossing it, by stripping away religious connotations, and by both emphasizing and effacing the importance of Ricard as a person. In no instance are we told what Matthieu Ricard feels at any particular moment throughout his meditative practice. This is particularly noteworthy because, based largely on the neuroimaging experiments done on him, Ricard has been repeatedly dubbed by others as 'the happiest man alive' (Barnes, 2007; Simons, 2010). Given the various forms of indirection when it comes to conveying the experiential, it is perhaps no surprise that these accounts do not engage the 'can nots' and 'should nots' of recounting meditation discussed in an earlier section of this paper. Whether Ricard's or other monks' experiences are beyond description ('ineffable') or better left undescribed is beside the point as long as these experiences are not featured in the accounts.[17]

Recent experiences in contemplative neuroscience and neurophenomenology

Since *Destructive Emotions*, the effects of meditation on the brain and body have been the topic of a burgeoning interest and have led to the establishment of what is now called 'contemplative neuroscience'. Much of the initial research in this latest period of scientific interest in mediation was cultivated through the Dalai Lama and the Mind and Life Institute, which he co-founded with entrepreneur Adam Engle and neuroscientist Francisco Varela.[18] In the last several years, scientific studies on meditation also derive from a wider 'semi-secular' attention to the benefits of so-called 'mindfulness' (see, in particular, the Mindfulness-Based Stress Reduction and Mindfulness-Based Cognitive Therapy programmes, e.g. Kabat-Zinn, 2005).

Within this field, we can distinguish different strands and attitudes to experience and self-reports, in which silence takes different forms and accordingly matters in different ways. In this section we first discuss a strand that is in line with the mainstream approach in cognitive neuroscience and that, as such, pays little (if any) attention to the subjective experience of meditating and to related self-reports. We then turn to 'neurophenomenology' – an emerging approach that explicitly calls for the integration of first-person reports into mainstream cognitive neuroscience. Here we point to a different form of silence; in neurophenomenology there is little consideration for what can or should *not* be described about the experience of meditating. Both types of silences are noteworthy for how they are bound up with the production of claims to scientific knowledge. We conclude the section by delineating what we think may be a possible further development of contemplative neuroscience that takes into account the difficulties and caveats discussed earlier in this chapter.

The cognitive neuroscience of meditation

Much of contemplative neuroscience reflects the mainstream approach in cognitive neuroscience, which typically looks at what the brain does in response to various stimuli and at how brain activity correlates with performance on cognitive-behavioural tasks. In this approach, participants' self-reports about their experiences are usually not included, because they are considered 'subjective' and thus uncertain, questionable and unreliable. Likewise, cognitive-neuroscientific studies of meditation side-line participants' experiences of meditating and do not sample any description of these experiences (see, for example, Ahani et al., 2013). Some researchers specifically identify complications with self-reports due to mindfulness topic-specific considerations, such as the potential variability of questionnaire responses due to prior familiarity with mindfulness-based practices (e.g. Keng et al., 2011).

This situation is similar to what one of us has highlighted elsewhere for the field of 'affective neuroscience'. Here, even when self-reports are collected as part of research, reliance on them is 'extremely cautious and minimized'

(Colombetti, 2014: p. 144). We can point to several parallels between the neuroscience of emotion and that of meditation with regard to how self-reports are sidelined even if they are noted:[19]

1. *Lack of incorporation of first-person data into the experimental set-up*: In affective neuroscience, when first-person accounts are obtained, this is often primarily as a form of control for 'more objective' behavioural and neurological measures. In addition, self-reports often do not play a meaningful role in the design of experiments. The same goes for first-person accounts of meditation, which are often collected to validate other measures (e.g. Marzetti et al., 2014; Xu et al., 2014), and whose scientific standing tends to be downplayed (e.g. Levenson et al., 2012). As noted above, in terms of recruitment and categorization of participants, individuals' proficiency is routinely gauged through the quantitative measure of total meditation hours and not informed by first-person evidence.

2. *First-person data collected by questionnaires*: Within the study of emotion, experiences are typically assessed through questionnaires that both transform them into numerical measures, and understand them through assumptions informed by theoretical frameworks. As a result, fine-grained descriptions of what it is like to undergo an experience are removed. The use of generic questionnaires for the purpose of gauging first-person experience is also prevalent in meditation research (e.g. MacCoon et al., 2014). At times, the standard experimental ways of gauging experience appear to neglect basic Buddhist understandings regarding the aims and effects of meditation. For instance, in one experiment participants (experienced meditators and non-meditators) were asked retrospectively to rate their emotional responses to a set of stimuli on a Likert scale, from 'Not at all' to 'Very' angry, sad, happy or disgusted (McCall et al., 2014). Here, emotions are treated simply as categorical experiences one has in degrees. Yet a central consideration in Buddhism is that emotions are complex experiences that arise under certain conditions and can be met with different levels of self-identification and attachment. One aim of meditation is to recognize the appearance of the 'ego' in emotion and the craving or suffering that this involves. From our position as authors with some knowledge of, but not a specialized expertise in, Buddhist philosophy, a more nuanced approach stemming from these considerations would be to consider whether meditators and non-meditators experience the arising of emotion differently and whether emotions come with different levels of self-awareness and attachment across these groups.

3. *The use of standard clinical scales*: In affective neuroscience, the experience of emotion is typically measured via standard scales (such as the Beck Depression Inventory or the PANAS scales). Participants are generally not allowed to describe their experiences in their own words. This also happens in the study of meditation (e.g. Desbordes et al., 2012).

4. *Experience as a static phenomenon*: Because of the considerations raised by points 2 and 3 as well as other simplifications during measurement, experience is often treated as a static phenomenon rather than one that unfolds and changes over time. Likewise, studies of meditation pay little attention to how participants' experiences unfold from moment to moment (e.g. Lutz et al., 2008; Xu et al., 2014) and rather treat them as fixed and uniform. Neuroscientists use specific scales (particularly the Mindful Attention Awareness Scale and the Five Facet Mindfulness Questionnaire) to measure the effects of meditation, but these scales are designed to capture general overall levels of mindfulness ('traits') rather than the qualitative character of experience as it unfolds.

Neurophenomenology

Although first-person accounts have rarely played a significant role within the design, analysis or write-up of contemplative neuroscience studies to date, they are not wholly absent. Recent years have witnessed the emergence of 'neurophenomenology', a research programme aiming to incorporate into the scientific study of brain activity rigorous methods for the generation of rich first-person data. First heralded by Francisco Varela (1996) – one of the cofounders of the Mind and Life Institute – this approach has gradually been making its way into more mainstream cognitive-neuroscientific research (see Lutz and Thompson, 2003). The label 'neurophenomenology' explicitly refers to a combination of methods: standard methods of brain-imaging during the performance of some mental activity (fMRI, EEG, etc.), as well as 'phenomenological' methods aimed at sampling descriptions of experience.[20] Supporters of this approach typically seek ways of producing self-accounts that meet standards of scientific validity, such as being reproducible through established methods and being verifiable or falsifiable (e.g. Varela and Shear, 1999; Hendricks, 2009; Maurel, 2009; Petitmengin, 2009; Vermersch, 2009). Much effort has been directed not only at establishing and refining methodologies to collect first-person accounts, but also at refuting critiques from those sceptical about the scientific validity and reliability of such accounts (e.g. Petitmengin and Bitbol, 2009).

It is possible to distinguish two types of neurophenomenologically inspired contemplative neuroscience. The first one introduces an explicit reliance on 'experiential categories' in cognitive-behavioural neuroscience and uses them to interpret brain activity. For example, an early study by Antoine Lutz and colleagues (2004) compared brain activity in long-term Buddhist meditators and people with only a few hours training while they generated an experience of 'pure compassion'.[21] The experimenters found that in long-term meditators, this experience was associated with unusually high degrees of synchrony in gamma activity (thought to reflect attention and emotional processes). This study can be considered already 'neurophenomenological' because brain activity here is measured not as participants are exposed to specific stimuli, but as they generate specific experiences (as Ricard did during the experiments described in *Destructive Emotions*). The experimenters relied on participants' self-reports to know when these experiences occurred and their article also mentions (albeit briefly) that self-reports were used to make sense of the temporal profile of brain activity. To borrow Gallagher's (2003) useful term, this study *front-loads* phenomenology in the experimental design – that is, it uses an experiential category identified in long-standing Buddhist meditation practices (e.g. 'pure compassion') to guide interpretation of brain data.

It can be noted that in this approach the phenomenological component is still very limited and cautious, in the sense that subjects are not asked to describe their experience in much detail (if at all). We can compare this 'thin' neurophenomenological approach with a more recent and 'thicker' one, which involves asking meditators freely to report on their experiences in their own words, then using these self-reports to generate specific categories of experience to aid the interpretation of brain activity. A recent study that takes this 'thick' approach is by Yair Dor-Ziderman and colleagues (2013) on the neural activity associated with different states of self-awareness in long-term Vipassana meditators. Here, the experimenters asked meditators to enter three different 'self-related' states, including a 'selfless condition' in which participants had to try to be aware of the present moment without, however, experiencing themselves as subjects of experience. Importantly, in addition to adopting standard measures of experience involving Likert scales, the experimenters also interviewed the participants, asking them to describe their experience while meditating 'freely and in their own words, without reflection or judgment' (Dor-Ziderman et al., 2013: p. 4). For the 'selfless condition', participants gave descriptions that experimenters grouped into three categories – the one labelled 'lack of ownership' turned out to be associated with a distinct neural signature. This study is thus notable in that it does not only use pre-established categories of self-awareness to interpret brain activity, but it also relies on participants' self-reports, produced in the course of the study, to generate new categories that provide further neuroscientifically relevant information.

Another 'thick' neurophenomenological study is by Kathleen Garrison and colleagues (2013a, b), who combined 'real-time fMRI' (an fMRI technique that provides feedback in real time about brain activity via some form of visual representation) with grounded theory (a widespread inductive qualitative approach in the social sciences, that entails starting from data about experience to build conceptual categories and theoretical schema). Long-term meditators, including some non-Buddhists, were asked to describe their experience in their own words via open-ended questions (presumably to keep the data more manageable, they were asked to provide concise descriptions and to focus on the 'highlights' of the meditation experience). They were also asked to pay attention to the real-time feedback – provided in the form of a graph – from their brain activity and to consider whether this feedback corresponded with their experience of meditation. If they found that the graph did not correspond to their experience, they were asked to explain how they knew this. By analysing these self-reports with grounded theory, the experimenters were able to extrapolate several categories of experience, which they used to identify and interpret specific patterns of brain activation. The researchers then went on to outline hypotheses derived from this neurophenomenological design.

These two studies thus overcome the silences about experience and self-reports we identified earlier in mainstream cognitive-neuroscientific approaches to meditation, and even in the earlier 'thinner' neurophenomenological ones, by making room for moment-to-moment experience to be recounted in meditators' own terms. And yet, on the basis of our earlier considerations about the difficulties of putting experience into words, we can note here a different form of silence that (still) characterizes the thick neurophenomenological approach. The latter is particularly concerned with elucidating, establishing and 'explicitating' lived experience, and with providing means of verbalizing it. This emphasis is accompanied by a lack of attention to, and discussion of, what is, or even needs to be, *absent* from first-person accounts. In other words, this approach generally does not recognize the need for caution about the possibility or advisability of trying to capture through words the experience of meditation.[22] Nor does it treat what participants choose *not* to describe about the experience of meditation as indicating an awareness of their awareness. Instead, the overall orientation of the thick neurophenomenological approach is one of developing scientific means for making pre-reflective aspects of experience explicit and for analysing them in a rigorous way.

A noteworthy feature of the studies just outlined (and of the cognitive-neuroscientific approach more generally) is what they do *not* make reference to. They do not give any consideration to what the meditators might think should be left undescribed or even not be subject to descriptive attempts. They do not mention either whether the meditators examined felt any reluctance in being identified as accomplished meditators. In other ways, too, these studies give primacy to what can be verbalized over what cannot. The use of qualitative methods (such as grounded theory) to extrapolate categories of experience tends to rely on what gets said, rather than on what is not or cannot be said. There is also a point to be made about vocabularies and their indebtedness, or lack thereof, to Buddhist thinking. Dor-Ziderman et al. (2013) used first-person descriptions by experienced meditators to identify three different forms of experience. These experiences are recounted through concepts and descriptors that (largely Western and non-Buddhist) readers of this volume are likely to regard as commonplace. Arguably, this would have been a harder and more question-begging task if it had involved meditators who had interpreted their experience in a language steeped within a Buddhist imaginary (as in the case of Lobsang Tenzin).

This is not to say that the neurophenomenological approach does not show any awareness at all of the difficulties associated with putting experience into words. Echoing some of the Buddhist themes we noted above, Petitmengin and Bitbol (2009: p. 389), for instance, note that 'words [...] don't display experience, they only point at it'. Like Gendlin (1996), they warn that, in becoming absorbed in verbal descriptions, individuals may lose sight of their experiences. And yet, this consideration does not lead Petitmengin and Bitbol to recognize the ineffable or contemplate that practitioners might see a limit to what should be described. Rather, it is for them a reason to strive for the development of ever more refined techniques enabling people to become aware of experiences not ordinarily noted and to describe them in detail whilst remaining 'close' to them. Arguably, language is taken here as a tool for representing and reporting experiences in a direct, mostly non-problematic way.

A similar tendency can be found in Thompson's (2015) recent book *Waking, Dreaming, Being*, which closely examines a variety of Indian yogic and Tibetan Buddhist accounts of several states of consciousness, including 'non-dual' ones such as the 'clear light' said to be experienced by trained meditators in the dying process. These states of consciousness are by definition 'object-less' (that is, they are not about anything in particular) and they are non-conceptual. This makes them very difficult to report, even retrospectively – because as they occur, they cannot, arguably, be described without losing their non-dual and non-conceptual status. Thompson is particularly interested in whether and how neuroscience can find the neural signatures of these states and reviews several studies that have attempted this for various experiential states (as well as giving suggestions for how to do so). Although his account is unquestionably sensitive to the complexities and subtleties of experiential states achievable by meditators, it is primarily oriented to clarifying these experiential states for the purpose of identifying features of experience that lend themselves to neuroscientific investigation. His account does not talk about what can or should *not* be talked about in meditation. As a result, it is not clear whether Thompson thinks that there are experiential states, or aspects thereof, that do not lend themselves to neurophenomenological investigation precisely because of their non-dual and non-conceptual nature.

A possibility for a third-stage neurophenomenology would be to integrate recognition of these difficulties and struggles explicitly into the experimental approach. This approach would neither sideline experiential reports, nor be silent about the risks and paradoxes inherent in the project of describing experience during meditation. Such a neurophenomenology would engage with participants in a yet more thorough way, by assessing their attitude to the tasks of reporting on their experience, by enquiring into their background and conceptual frameworks, by asking explicitly to report on features of experience that are harder to put into words than others, and to explain why this is the case. This project of explicitation and reflection would aim at making more room for the constructive and generative role of silences and absences in the meditation experience – one that, as we saw, has been long pointed out in non-scientific traditions of writing about Buddhist meditation.

Concluding remarks

This chapter has sought to understand how silences contribute to the production of knowledge claims in attempts to subject meditation practices within Buddhism to scientific scrutiny. In this intersection between scientific and Buddhist traditions much is up for negotiation. We have been struck by a particular recurring dynamic that plays out in different ways in a variety of settings: namely, that examinations of the experiential practice of meditation rarely seek to offer an account of the lived experience of meditation, including the silences and ineffable aspects that characterize this experience.

As has been developed elsewhere in this volume, silences can both frustrate and facilitate inter-subjective agreement, act as a mark of expertise and of ignorance, stifle the exchange of ideas and stimulate the production of knowledge. In this chapter, silence has not been conceived as a state or space, a gap in what has been said, or a kind of social glue that holds science and Buddhism together.

To better characterize silence in the spotted and ambivalent engagements with the lived experience of meditation in the accounts examined in this chapter, we can perhaps turn to music as an analogy. Just as the denoting of sounds in the composition of music brings with it what is in-between the notes, so too the telling or showing of experiences of meditation entails the production of a sense of what is outside. Both what is 'there' and 'not' are constitutive of music scores and meditation accounts. But neither are simply fixed properties. Silences might be structured by composers and musicians in such a way that they take centre-stage, or are pervasive but not intended to be noticed. Then again, '[l]isteners may engage with silences to whatever degree they like, depending on how attentively they feel like listening, how actively they are engaging with the material they are hearing, or not at all' (Losseff and Doctor, 2007: p. 11). What is there and what is not are interrelated in the manner that the interpretation given to either depends on the other. It is, then, through silence as well as explication that experience (and music) is performed and made sense of. As argued in the case of accounts of meditation, within these relations knowledge of the experiential is variously denied, defended, deferred or made subject to deference.

Notes

1 Our thanks to the editors for their helpful corrections and comments on an earlier draft of this chapter.
2 Gendlin (1996, pp. 29–30), for example, illustrates the case of a client initially complaining that she often misses school, does not hand in papers and does not have a romantic relationship even if she would like one. When asked to pay attention to how her body feels when thinking of these situations, she reports feeling 'jittery' and also 'pulling back', and subsequently comes to the realization that she is scared.
3 The term 'sutras' (or 'suttas' in Pali) refers to Buddhist canonical scriptures, many of which are considered records of the Buddha's teachings.
4 For instance, see Lerner (1977: pp. 60, 72, 100, 127), Austin (1998: ch. 13) and Kasulis (1981: p. 41).
5 This theme occupies a particularly prominent place within some traditions of Buddhism, see Burton (2000).

6 In focusing on the different ways in which silence is negotiated at the intersection of Buddhist meditation and neuropsychology, this section cannot do justice to the variegated history of the scientific study of meditation. For more extensive accounts, see Harrington (2008: ch. 6) and Andresen (2000).
7 For a scientist's perspective on this work, see Benson (1991).
8 A different Lobsang Tenzin from the one who was prime minister of the Tibetan government-in-exile in 2001–2011.
9 For example: 'Buddhist practitioners underwent mental training in the same Tibetan Nyingmapa and Kagyupa traditions for 10,000 to 50,000 hours over time periods ranging from 15 to 40 years. The length of the training was estimated based on their daily practice and the time they spent in meditative retreats. Eight hours of sitting meditation was counted per day of retreat' (Lutz et al., 2004: p. 16369).
10 Examples include Geirland (2006), Davidson (2009), and Ricard (2007; 2011). Sometimes accounts of the experiment serve to signpost the beginning of rigorous studies of meditation; at other times they are used to illustrate the notion of neuroplasticity (as in Ricard, 2007, 2011).
11 For a contrasting orientation that attempts to make sense of first-person experience through a scientific language, see Austin (1998: pp. 353, 479).
12 These were described more informally in Goleman (2003).
13 The quote is included in a section of the paper reserved for observations on Ricard's performance in tasks and tests that 'lacked the experimental rigor necessary for scientific publication, [yet] we still feel they may be of interest to others and may help place the startle findings in their larger context' (Levenson et al., 2012: p. 656).
14 For a discussion of these validity concerns, see Lutz et al. (2007). For attempts to mix science with Buddhist philosophy, see Hanson and Mendius (2009) and Austin (1998).
15 Especially curious is that Ricard's writing about the experiments relies heavily on quotes from other published sources. For instance, Ricard brings in the voice of Richard Davidson, the scientist in charge of the experiments, by quoting newspaper articles for which Davidson was interviewed.
16 See also Hetherington (1997), who in turn draws on the work of Michel Serres.
17 The accounts we have read also certainly do not attempt to represent what might be dubbed as the 'higher (non-dualist) states of awareness' that Ricard or other monks participating in these experiments might achieve (such as 'sublime consciousness' or 'pure awareness').
18 As recounted in, for example, Harrington (2008: ch. 6) and Thompson (2015: ch. 3).
19 This section is based on a review of fifty-one articles returned after a PubMed search of the terms 'neuroscience' and 'meditation' in January 2015. Not all returned studies are strictly neuroscientific ones (some are behavioural-physiological). However, to the best of our knowledge all the features we identify below do apply to the neuroscience of meditation (as well as to many behavioural-physiological and other scientific studies).
20 Phenomenology is a philosophical tradition usually traced back to Husserl and associated with works by Heidegger, Merleau-Ponty, Sartre and many others. For present purposes, we can characterize phenomenology as the study of phenomena, i.e. of what 'appears' or 'is given' in first-person experience.
21 Characterized here as 'unrestricted readiness and availability to help living beings' (Lutz et al., 2004: p. 16369).
22 This is particularly noteworthy, we think, given that many supporters of the neurophenomenological approach are themselves experienced meditators.

References

Ahani, A., Wahbeh, H., Miller, M., Nezamfar, H., Erdogmus, D. and Oken, B., 2013. Change in physiological signals during mindfulness meditation. *International IEEE EMBS Conference on Neural Engineering*, pp. 1738–1781, doi: 10.1109/NER.2013.6696199.

Albahari, M., 2011. Nirvana and ownerless consciousness. In: Siderits, M., Thompson, E. and Zahavi, D. (eds), *Self, No Self? Perspectives from Analytical, Phenomenological, and Indian Traditions*. Oxford: Oxford University Press, pp. 79–113.

Andresen, J., 2000. Meditation meets behavioural medicine. *Journal of Consciousness Studies*, 7(11–12), pp. 17–73.

Austin, J., 1998. *Zen and the Brain: Toward an Understanding of Meditation and Consciousness*. Cambridge, MA: MIT Press.

Barnes, A., 2007. The happiest man in the world? *Independent*, 21 Jan. Available at: www.independent.co.uk/news/uk/this-britain/the-happiest-man-in-the-world-433063.html. Accessed August 2015.

Benson, H., 1991. Mind/body interactions including Tibetan studies. In: Goleman, D. and Thurman, R. (eds), *MindScience: An East-West Dialogue*. Boston: Wisdom, pp. 37–48.

Boyce, B. and the editors of the Shambhala Sun (eds), 2011. *The Mindfulness Revolution: Leading Psychologists, Scientists, Artists, and Meditation Teachers on the Power of Mindfulness in Daily Life*. Boston and London: Shambhala.

Brown, D.P., 1977. A model for the levels of concentrative meditation. *International Journal of Clinical and Experimental Hypnosis*, 25(4), pp. 236–273.

Burke, K., 1961. *The Rhetoric of Religion: Studies in Logology*. Berkeley: University of California Press.

Burton, D., 2000. Wisdom beyond words? Ineffability in yogācāra and madhyamaka Buddhism. *Contemporary Buddhism*, 1(1), pp. 53–76.

Colombetti, G., 2014. *The Feeling Body: Affective Science Meets the Enactive Mind*. London: MIT Press.

Dass, R., 1988. Foreword. In: Goleman, D., *The Meditative Mind: The Varieties of Meditative Experience*. New York: St. Martin's Press.

Davidson, R.J., 2009. Transform your mind, change your brain: neuroplasticity and personal transformation. Google TechTalk, 23 Sep. Available at: www.youtube.com/watch?v=7tRdDqXgsJ0. Accessed 14 May 2011.

Davidson, R. and Harrington, A. (eds), 2002. *Visions of Compassion: Western Scientists and Tibetan Buddhists Examine Human Nature*. Oxford: Oxford University Press.

Desbordes, G., Negi, L.T., Pace, T.W., Wallace, B.A., Raison, C.L. and Schwartz, E.L., 2012. Effects of mindful-attention and compassion meditation training on amygdala response to emotional stimuli in an ordinary, non-meditative state. *Frontiers in Human Neuroscience*, 6, doi: 10.3389/fnhum.2012.00292.

Dor-Ziderman, Y., Berkovich-Ohana, A., Glicksohn, J. and Goldstein, A., 2013. Mindfulness-induced selflessness: a MEG neurophenomenological study. *Frontiers in Human Neuroscience*, 7, doi: 10.3389/fnhum.2013.00582.

Fox, K.C.R., Zakarauskas, P., Dixon, M., Ellamil, M., Thompson, E. and Christoff, K., 2012. Meditation experience predicts introspective accuracy. *PLoS One*, 7(9), e45370, pp. 1–9.

Franke, W. 2007. *On What Cannot Be Said Apophatic Discourses in Philosophy, Religion, Literature, and the Arts*. Volume 1: Classic Formulations. Notre Dame, IL: University of Notre Dame Press.

Gallagher, S., 2003. Phenomenology and experimental design: toward a phenomenologically enlightened experimental science. *Journal of Consciousness Studies*, 10, pp. 85–99.

Garrison, K.A., Santoyo, J.F., Davis, J.H., Thornhill, T.A., Kerr, C.E. and Brewer, J.A., 2013b. Effortless awareness: using real time neurofeedback to investigate correlates of posterior cingulate cortex activity in meditators' self-report. *Frontiers in Human Neuroscience*, 7, doi: 10.3389/fnhum.2013.00440.

Garrison, K.A., Scheinost, D., Worhunsky, P.D., Elwafi, H.M., Thornhill, T.A., Thompson, E. et al., 2013a. Real-time fMRI links subjective experience with brain activity during focused attention. *NeuroImage*, 81, pp. 110–118, doi: 10.1016/j.neuroimage.2013.05.030.

Geirland, J., 2006. Buddha on the brain: the hot new frontier of neuroscience: meditation! (Just ask the Dalai Lama.). *Wired*, 14(2). Available at www.wired.com/wired/archive/14.02/dalai.html?pg=11&topic=dalai&topic_set=. Accessed 10 March 2010.

Gendlin, E., 1996. *Focusing-Oriented Psychotherapy*. New York: Guilford Press.

Goleman, D., 2003. *Destructive Emotions and How We Can Overcome Them: A Dialogue with the Dalai Lama*. London: Bloomsbury.

Goleman, D., 2007. Foreword. In: Ricard, M., *Happiness*. London: Atlantic, pp. xi–xiv.

Hanson, R. and Mendius, R., 2009. *Buddha's Brain: The Practical Neuroscience of Happiness, Love, and Wisdom*. Oakland, CA: New Harbinger.

Harrington, A., 2008. *The Cure Within: A History of Mind–Body Medicine*. New York: W.W. Norton.

Hendricks, M., 2009. Experiencing level: an instance of developing a variable from a first person process so it can be reliably measured and taught. *Journal of Consciousness Studies*, 16(10–12), pp. 129–155.

Herrigel, E., 1999 [1953]. *Zen in the Art of Archery*. London: Penguin.

Hetherington, K., 1997. Museum topology and the will to connect. *Journal of Material Culture*, 2(2), pp. 199–218.

Hetherington, K. and Lee, N., 2000. Social order and the blank figure. *Environment and Planning D: Society and Space*, 18(2), pp. 169–184.

Houshmand, Z., Harrington, A., Saron, C. and Davidson, R.J., 2002. Training the mind: first steps in a cross-cultural collaboration in neuroscientific research. In: Davidson, R.J. and Harrington, A. (eds), *Visions of Compassion*. New York: Oxford University Press, pp. 3–17.

Kabat-Zinn, J., 2005 [1990]. *Full Catastrophe Living: Using the Wisdom of Your Body and Mind to Face Stress, Pain, and Illness*. New York: Delta Trade Paperbacks.

Kasulis, T.P., 1981. *Zen Action/Zen Person*. Honolulu: University of Hawaii Press.

Keng, S.L., Smoski, M.J. and Robins, C.J., 2011. Effects of mindfulness on psychological health: a review of empirical studies. *Clinical Psychology Review*, 31(6), pp. 1041–1056, doi: 10.1016/j.cpr.2011.04.006.

Kornfield, J., 2001. *After the Ecstasy, the Laundry: How the Heart Grows Wise on the Spiritual Path*. London: Bantam.

Lerner, E., 1977. *Journey of Insight Meditation: A Personal Experience of the Buddha's Way*. New York: Schocken Books.

Levenson, R.W., Ekman, P. and Ricard, M. 2012. Meditation and the startle response: a case study. *Emotion*, 12(3), pp. 650–658, doi: 10.1037/a0027472.

Losseff, N. and Doctor, J., 2007. Introduction. *Silence, Music, Silent Music*. Aldershot and Burlington, VT: Ashgate, pp. 1–13.

Lutz, A., Brefczynski-Lewis, J., Johnstone, T. and Davidson, R.J., 2008. Regulation of the neural circuitry of emotion by compassion meditation: effects of meditative expertise. *PLoS One*, 3(3), doi: 10.1371/journal.pone.0001897.

Lutz, A., Dunne, J.D. and Davidson, R.J., 2007. Meditation and the neuroscience of consciousness: an introduction. In: Zelazo, P.D., Moscovitch, M. and Thompson, E. (eds), *The Cambridge Handbook of Consciousness*. Cambridge and New York: Cambridge University Press, pp. 499–551.

Lutz, A., Greischar, L.L., Rawlings, N.B., Ricard, M. and Davidson, R.J., 2004. Long-term meditators self-induce high-amplitude gamma synchrony during mental practice. *Proceedings of the National Academy of Sciences of the United States of America*, 101(46), pp. 16369–16373.

Lutz, A. and Thompson, E., 2003. Neurophenomenology: integrating subjective experience and brain dynamics in the neuroscience of consciousness. *Journal of Consciousness Studies*, 10, pp. 31–52.

MacCoon, D.G., MacLean, K.A., Davidson, R.J., Saron, C.D. and Lutz, A., 2014. No sustained attention differences in a longitudinal randomized trial comparing mindfulness based stress reduction versus active control. *PLoS One*, 9(6), e97551, doi: 10.1371/journal.pone.0097551.

Mair, J., 2013. *On Not Being Buddha*. London: Palgrave.

Marzetti, L., DiLanzo, C., Zappasodi, F., Chella, F., Raffone, A. and Pizzella, V., 2014. Magnetoencephalographic alpha band connectivity reveals differential default mode network interactions during focused attention and open monitoring meditation. *Frontiers in Human Neuroscience*, 8, doi: 10.3389/fnhum.2014.00832.

Maurel, M., 2009. The explicitation interview: examples and applications. *Journal of Consciousness Studies*, 16(10–12), pp. 58–89.

McCall, C., Steinbeis, N., Ricard, M. and Singer, T., 2014. Compassion meditators show less anger, less punishment, and more compensation of victims in response to fairness violations. *Frontiers in Behavioral Neuroscience*, 8, doi: 10.3389/fnbeh.2014.00424.

Petitmengin, C., 2009. Editorial introduction. *Journal of Consciousness Studies*, 16(10–12), pp. 7–19.

Petitmengin, C. and Bitbol, M., 2009. Listening from within. *Journal of Consciousness Studies*, 16(10–12), pp. 363–404.

Rambelli, F., 2006. Secrecy in Japanese esoteric Buddhism. In: Scheid, B. and Teeuwen, M. (eds), *The Culture of Secrecy in Japanese Religion*. London: Routledge, pp. 107–129.

Revel, J.F. and Ricard, M., 1998. *The Monk and the Philosopher: A Father and Son Discuss the Meaning of Life*. New York: Schocken Books.

Ricard, M., 2007. *Happiness: A Guide to Developing Life's Most Important Skill*. London: Atlantic.

Ricard, M., 2010. *Why Meditate? Working with Thoughts and Emotions*. Carlsbad, CA: Hay Inc.

Ricard, M., 2011. This is your brain on mindfulness. In: Boyce, B. and the editors of the Shambhala Sun (eds), *The Mindfulness Revolution*. Boston and London: Shambhala, pp. 127–135.

Ricard, M. and Trinh, X.T., 2001. *The Quantum and the Lotus: A Journey to the Frontiers Where Science and Buddhism Meet*. New York: Crown Publishers.

Saeed, A., 2006. *Interpreting the Qurʾān: Towards a Contemporary Approach*. London: Routledge.

Simons, J.W., 2010. The happiest men in the world. *The Times*, 8 Feb., p. 9.
Singer, T. and Ricard, M. (eds), 2015. *Caring Economics: Conversations on Altruism and Compassion, Between Scientists, Economists, and the Dalai Lama*. New York: Picador.
Suzuki, D.T., 1949. *Essays in Zen Buddhism, Volume 1*. New York: Grove Press.
Teeuwen, M., 2006. Introduction: Japan's culture of secrecy from a comparative perspective. In: Scheid, B. and Teeuwen, M. (eds), *The Culture of Secrecy in Japanese Religion*. London: Routledge, pp. 172–203.
Tenzin, L., 1987. Biography of a contemporary yogi. *Chö-Yang: The Voice of Tibetan Religion and Culture*, 3, pp. 102–111.
Thompson, E., 2015. *Waking, Dreaming, Being: Self and Consciousness in Neuroscience, Meditation, and Philosophy*. New York: Columbia University Press.
Varela, F., 1996. Neurophenomenology: a methodological remedy for the hard problem. *Journal of Consciousness Studies*, 3(4), pp. 330–335.
Varela, F.J. and Shear, J., 1999. First-person methodologies: what, why, how? In: Varela, F. and Shear, J. (eds), *The View from Within: First-person Approaches to the Study of Consciousness*. Exeter: Imprint Academic, pp. 1–14.
Vermersch, P., 2009. Describing the practice of introspection. *Journal of Consciousness Studies*, 16(10–12), pp. 20–25.
Walsh, T., 1998. *The Dark Matter of Words: Absence, Unknowing, and Emptiness in Literature*. Carbondale: Southern Illinois University Press.
Xu, J., Vik, A., Groote, I.R., Lagopoulos, J., Holen, A., Ellingsen, O., Håberg, A.K. and Davanger, S., 2014. Nondirective meditation activates default mode network and areas associated with memory retrieval and emotional processing. *Frontiers in Human Neuroscience*, 8, doi: 10.3389/fnhum.2014.00086.
Zahavi, D., 2005. *Subjectivity and Selfhood: Investigating the First-Person Perspective*. Cambridge, MA: MIT Press.

Part III
Silences in the public sphere

10 The silent introduction of synthetic dyestuffs into nineteenth-century food

Carolyn Cobbold

Unintended uses

Sometimes what we don't know, and what we choose not to know more about, tells us as much about our society or past societies as the study of knowledge itself. So argued historian of science Robert Proctor in his book *Agnotology: The Making and Unmaking of Ignorance*. Proctor and his colleagues described incidents in history where doubt is deliberately used to create confusion and scepticism to refute or discredit scientific evidence or avoid regulation, such as witnessed in the debates surrounding climate change, genetically modified food or the link between tobacco and lung cancer. Proctor also highlighted areas where knowledge is not sought for cultural reasons or is subsequently forgotten or overlooked. In this chapter, I will argue that lacunae in knowledge can arise for many reasons, both deliberately and inadvertently. I believe the latter can happen when objects, substances, processes or inventions already assimilated into a culture take on a use or agency for which they were not intended and so escape recognition and monitoring.

Synthetic dyes and their use in industrial food production are a good example of this. Most western countries now tightly regulate the chemical colourings that food and drink manufacturers use, either through a list of permitted dyes or a list of prohibited ones. However, many countries have only introduced regulations in recent decades, more than a century after their first appearance in our food, and countries continue to disagree as to which synthetic dyes should be permitted and which prohibited.[1] Why did it take so long for society to investigate and regulate the use of new and, in many cases, dangerous chemicals used in food production?

During the second half of the nineteenth century, a period when food scares were reported almost daily in the press, hundreds of aniline and other coal-tar-based dyes began to be used in food and drink products with no oversight and little widespread concern, particularly from the scientific community. Synthetic dyes were one of a kaleidoscope of new substances, including drugs, flavourings, sweeteners and perfumes, synthesised by European chemists from coal-tar waste from the mid-1850s. The Victorian media greeted these new chemical substances as wonders of science and their discovery led to the formation of chemical and pharmaceutical companies such as BASF, Bayer and Ciba-Geigy. Historians have identified synthetic dyes as a valuable site for debating the rising power of industrial chemistry and the transition from pure scientific research to industrial practice (Garfield, 2001; Haber, 1969; Beer, 1958; 1959; Homburg et al., 1998; Travis, 1983; 1993). However, few historians have examined an area of dye consumption that impacted everyone during this period and whose repercussions are still felt today – the use of chemical dyes in food.

Manufacturers successfully – and silently – introduced synthetic colourings into food at a time when social reformers, politicians and the media were wringing their hands about the dangers of food adulteration. On the surface, it seems surprising that the introduction of completely new chemical substances, including aniline dyes which later proved to be highly poisonous, was ignored during a period when food legislation was being continually updated and when public analysts were being paid to identify harmful and fraudulently applied food additives. Examining the introduction of these new chemicals into Victorian food sheds light on the public perception of science at the time, the struggle for status and authority among those arbitrating the food supply, and the ways in which food processes and practices come to be seen as legitimate or fraudulent.

A discovery of wonder

William Perkin stumbled across the first aniline dye, mauveine, in 1856 in his quest to manufacture a synthetic version of quinine, a well known treatment for malaria, a disease that was the scourge of the British Empire. He probably had little idea that his new purple dye would herald the start of a technicolour scientific revolution, but he did realise the commercial potential of the new vibrant dye. He applied for a patent, left the Royal College of Chemistry where he had been working under the tutelage of the German chemist August Wilhelm von Hofmann, and began to manufacture the dye on a commercial basis (Garfield, 2001: p. 19). Although it took Perkin about two years to turn his discovery into a commercial product, as soon as it hit the market it was hailed by the press as evidence of Britain's supremacy in industry and science.

The news of Perkin's breakthrough was reported extensively in newspapers and periodicals, and within a few years Perkin had become the embodiment of British scientific and technological achievement. This is evident in articles throughout the British media at the time, such as that in praise of 'Perkins's [sic] purple' published in Charles Dickens' edited journal *All the Year Round* (Anon., 1859: p. 468). During this period of 'scientific wonder' chemistry was producing visible and practicable products, such as the new dyes, that made eye-catching exhibits at the exhibitions and trade fairs that were becoming increasingly popular. Describing the 'exquisite dyes produced from coal' in an article about the 1862 International Exhibition, the *Ladies' Treasury* waxed lyrical about the array of colours released from 'that imprisoned life which thousands of years long past was encased in decaying vegetable substance – [that] has at last sprung into light and beauty, at the magic touch of science' (Anon., 1862: p. 342).

The image of trapped sunbeams and imprisoned life released by scientists after millions of years to delight and provide service for an industrial world was one that was painted across the press. In 1865, an article in the *Ladies' Treasury* shows that the sense of wonder had not diminished – the article described the 'brilliant hues which our modern chemists have eliminated from the "black stones" [as] the imprisoned sunbeams of ages ago'. But there was also recognition of just how widespread the use of coal-tar-derived products had become and of the democratising potential of chemistry in bringing products to the masses. The article noted that 'the brightest colours that deck the form of beauty, or that lend a charm to the peasant maiden, are obtained from coal. The delicate flavouring of many confections, of custards, and of corn flour puddings come from this ancient mineral' (Anon., 1865: p. 242). Articles in the press during the 1860s suggest that the 'scientific' coal-tar-derived additives were, at this point in time, perceived as harmless, an example of science's ability to improve food products and their availability and accessibility.

Within a few years, entrepreneurs and chemists across Europe, particularly in France, Germany and Switzerland, began producing dozens of new chemical dyes on an industrial scale. The vibrant new colours continued to be a source of wonder and amazement, but by the 1870s concern about adverse reactions to the dyes, both from workers in dye factories and wearers of the new brightly coloured socks, hats and dresses, began to appear in newspapers and journals. Interestingly, and perhaps surprisingly, concern about the use of the dyes in food began to appear sporadically in the general media only at a much later stage, despite evidence that their use in food had begun almost as soon as the dyes were sold commercially for textiles and paints.

Even in the case of textiles and paints, the transition of synthetic dyes from substances of wonder to problematic and complex concoctions was not straightforward. Many articles in the popular press blamed the danger of these new synthesised substances on rogue contaminants and not the scientifically discovered substance of aniline itself. The *Ladies' Treasury*, which until now had still been praising science for releasing the 'trapped sunbeams of ages past', was by 1875 noting that 'many of the colours derived from coal-tar are known to possess poisonous qualities, and all of them are looked upon with suspicion by ultra-careful housewives'. However, while it noted that one type of aniline green, prepared with picric acid and arsenic, was poisonous and the dyes of rosaline and coralline also often contained poisons, it reassured its readers that aniline dyes 'are all quite harmless when pure'. It suggested that dyed goods should only be sold with certificates stating their freedom from arsenic (Anon., 1875: p. 42). Arsenic, it seemed, had become the universal scapegoat, blamed by analysts and observers for any adverse affects associated with the aniline dyes.

Artificial food colouring before aniline dyes

Until the late 1870s there is little mention of aniline dyes being used in food, even though by this time the new dyes were being used extensively for this purpose. This seems surprising considering the debate that had been raging in the media for most of the nineteenth century surrounding artificial colourings in food. This earlier debate is instructive for its contrast with the silence that greeted the introduction of the aniline dyes.

The use of substances such as arsenic, red lead and the mercury-based vermillion to colour food and drink had been highlighted in pamphlets such as Friedrich Accum's 1820 *Treatise on Adulterations of Foods and Culinary Poisons*. In 1830, an article in the medical journal the *Lancet* highlighted the use of minerals, such as the arsenic-based Schweinfurt green, in French confectionery. Subsequent investigations indicated similar use of potentially toxic minerals to colour British sweets (Whorton, 2011). By the 1850s, the *Lancet* was running a series of articles and investigations on food adulteration, led by Arthur Hill Hassall, a chemist, physician and microscopist who would spend his career working on public health and food safety issues (Hassall, 1855).

In an 1854 report into the use of minerals to colour foods, Hassall identified many toxic materials such as lead chromate, lead oxide, copper arsenite or Scheele's green and other arsenical compounds. Hassall described food colouring as: 'a very prevalent adulteration, and it is the most objectionable and reprehensible of all, because substances are frequently employed, for the purpose of imparting colour, possessing highly deleterious and even in some cases, poisonous properties, as various preparations of lead, copper, mercury and arsenic' (Hassall, 1855: p. 4). Hassall's concerns also embraced both loss of income to the state due to lower food costs and deception of the public with a subsequent loss of faith in the commercial system:

> It is clear that the sellers of adulterated articles of consumption, be they manufacturers or retail dealers, are in a position to enhance their profits by the practice of adulteration, and are able to undersell, and too often to ruin, their more scrupulous and honest competitors.
>
> (ibid.: p. xxx)

The *Lancet* articles and investigations were widely reported in the popular press. Throughout the 1850s newspapers and journals also recounted many instances where children and adults were poisoned, often by toxic mineral colourings used to decorate cakes (Whorton, 2011). In 1858 *Punch* published a cartoon showing Death as the 'Great Lozenge-maker', making sweets beside a barrel of arsenic. Adulteration and the deception of the public was also tackled by novelists, including Charles Dickens in *David Copperfield* and Charles Kingsley in *The Water Babies*, as well as the poet Christina Rossetti in her allegorical poem *Goblin Market*, written in 1859 and published in 1862 (see Stern, 2003). These works were part of a large Victorian literature, including ballads, books, articles, pamphlets and cartoons, that commented on poisoning, market corruption and food adulteration.

The public and press uproar surrounding food adulteration in the aftermath of the *Lancet*'s investigation led to a parliamentary inquiry in 1855.[2] The subsequent report of the Select Committee was covered extensively in the British, Continental European and American press. The report spoke of Venetian red (ferric oxide) being used to colour the liquid surrounding Dutch, French and Sicilian fish in order for them to be passed off as anchovies; cocoa and chocolate coloured with Venetian red, red ochre and other 'ferruginous earths'; custard and egg powders coloured with chrome yellow or chromate of lead; curry powder coloured with red lead; snuff dyed with chromate of lead, umber, yellow ochre; and tea coloured with most things from black lead to Prussian blue, or ferro-cynanide of iron. It is clear from this list that a flourishing practice of food colouring was commonplace in food manufacturing and commerce by the middle of the nineteenth century before the introduction of synthetic aniline dyes, and that it was a practice that was creating considerable public anxiety.

Aniline dyes silently replace metals

Given the public concern over the use of mineral colours in food, it is interesting that there is little mention in the press before the 1870s of the substitution of these colours by the new aniline dyes. I suspect this is because of a lack of awareness rather than a lack of presence. Certainly, as a result of the uproar surrounding Hassall's investigation, there seems to have been less use of minerals to colour food. However, there is no mention of commercially prepared food losing its colour, which suggests that some form of food colouring was still being employed. Meanwhile, cookery books also promoted 'spectacle', highly-coloured food where presentation was deemed more important than taste (Humble, 2006). Advertisements from wholesalers of food colours and essences talk about new 'harmless' substances that could be used in minuscule amounts to colour food safely, although there was seldom any mention of what these new dyes were made from.

For the first few decades of their existence, aniline and other coal-tar-derived dyes, manufactured specifically for the textile industry, were being incorporated into food with few questions being asked of their suitability, whether by food producers, retailers, ingredient wholesalers, public analysts, physicians, the public, politicians or the press, despite the controversy surrounding poisonous metallic food colourings that raged for most of the nineteenth century. It seems, therefore, that these new substances were being introduced into the food supply with no publicity, no concern, and no monitoring, an example of the recurring pattern identified by historian James C. Whorton (2011) whereby new chemicals are freely and unsuspectingly 'welcomed' into the domestic environment before their dangers are recognised.

That aniline dyes were being used in food is evidenced by references in newspapers and periodicals such as the *Bradford Observer*, which in 1869 listed the wide range of uses for aniline dyes including inks, painting, photographs, 'the soaking of tissues of objects for microscopical and anatomical purposes ... the tinting of ... white vinegar and syrup of raspberry, the blueing of linen and the colouring of confectionery'. The article also pointed out the use of aniline dyes in cosmetics, which it claimed 'has been of undoubted service. They have superseded the metallic substances – the preparations of mercury, of bismuth and of lead – which were almost all injurious to health' (Anon., 1869). It seems that the aniline dyes became a commonplace and unchallenged substitute for the tarnished metallic colours.

An early mention of concern about the possible toxicity of aniline dyes was in an 1871 article in the *Health Reformer*, published by the American Health Reform Institute. Entitled *Poisoned Candies*, the article warned that 'the candy makers' of New York are spreading death among children since 'various cheap devices are employed as substitutes for cochineal and saffron'. According to the journal, the 'red colour is usually produced by amboline, which is obtained in a crystallized form from coal tar during its process of refining'. Sold for $2 an ounce, amboline 'will equal in colouring twenty times its weight in cochineal'. The article noted that other red dyes used included another aniline colour, fuchsine (Anon., 1871).

Fears centred on foreign food

This article again shows that aniline dyes had made their way into food by the early 1870s, but there was little critical mention of their use in consumables in the British press at this time. It is only from the late 1870s that one can find any evidence of concern in the British media with the identification of the synthetic food colourants as a new form of adulteration and not necessarily a safe replacement for the poisonous substances previously used to colour food such as copper, lead and arsenic. It is hard to pinpoint exactly why the use of synthetic dyes in food began to be questioned. It may well be linked to the rise of the French, German and Swiss chemical industries at this time and the increase in imported food, such as the newly patented oleomargarine, a cheap substitute for butter, which was often dyed yellow to make it appear more like butter. Certainly, one of the noticeable trends in reporting from this time was the association made between these new food adulterants and foreign imported food. An 1878 edition of *Funny Folks*, a weekly magazine of humorous stories and jokes, stated:

> A German chemist is now experimenting with the view of giving us our mutton, beef and pork of as many different colours. The eye, he thinks, is weary of the monotonous tints of fat and lean and why, he would seem to be anxious to know, should not a mutton-chop of azure, with magenta fat, let us say, be a possibility? Why should not a purple haunch of mutton be kept in countenance by an emerald-green round of beef, flanked by cutlets of a vivid orange, and sweetbread of a deep blood-red? Judging from the results of experiments he has made with pig, fed partly

on aniline dye, it will be practicable to cut a slice of beef or pork, vying in variety of colour with the disc of a kaleidoscope.

(Anon., 1878)

By the 1880s, tinted foreign produce was under sustained attack with the *Country Gentleman* warning that cheap French red wine, full of aniline dyes, was the cause of ill health among the British population. Meanwhile, the whimsically titled *Moonshine* pointed out in 1886 that in Paris, 'they make quite a nice *genuine* raspberry jelly' made from 'oil of vitriol, impure glucose, algine, aniline and raspberry ether'. In 1890 the *Sheffield and Rotherham Independent* mentioned a report on Bologna sausages that are coloured with 'garnet red and Bismarck red'. It is stated, the newspaper noted, 'that purchasers "like a bit of colour" but one cannot but suspect that these aniline dyes are used to prevent them knowing what they are eating.' According to the *Friendly Companion: A Magazine for Youth*, orange growers in Florida 'manufacture' blood oranges by 'piercing the rind of ordinary oranges with a fine syringe and injecting an aniline dye which quickly permeates the pulp' (Anon., 1882, 1886, 1890a, 1890b). The widespread criticism of foreign food, and claims that it was adulterated and possibly poisoned, are tied up with prevailing concerns over the absence of provenance of food for an urbanising population. Fears over Britain's increasing dependency on imported foods raised the spectre of the nation's inability to supply the wants of its population, with dependency upon foreign foods and increased overseas competition leading to a perception that Britain's position in the world was declining and a prevailing fear of national, individual and social degeneration (Olson, 2008).

By the 1890s, the Victorian media were routinely presenting aniline dyes as dangerous adulterants. Within four decades, synthetic dyes had thus been transformed by the popular press from a wonderful new substance of science into a toxic tool for commercial deception; an out-of-control danger in need of managing. Interestingly, during this period of growing awareness and doubt expressed in the press, it is still very unclear from articles or advertisements in the popular press which producers were using aniline dyes in food. While manufacturers were increasingly likely to advertise their food as 'pure' and free from adulterants during the nineteenth century, evidence from company archives and parliamentary investigations suggests that aniline and other coal-tar-derived dyes were in fact being extensively used in food preparation, despite the fact that nobody publicly laid claim to such practices. Indeed, evidence strongly suggests that many of the food manufacturers were unaware that they were using aniline dyes, or, at least, failed to ask their suppliers, or chose not to ask them, what the 'harmless dyes' they were buying actually were composed of (Committee on Food Preservatives, 1901).

Scientists ignore public concern

While it is surprising that food manufacturers managed to add chemical dyes into food and drink for several decades without attracting much comment or criticism, it seems even more remarkable that the public analysts charged with overseeing the nation's safe and proper food supply remained unconcerned about the practice even when doubts began to be expressed in the press.

An examination of the *Analyst,* the monthly journal published by the Society of Public Analysts, indicates that the use of aniline and azo dyes by food and drink manufactures was not a high priority for Britain's food chemists or regulators. Most of the articles mentioning their use, detection or possible harm were extracts from overseas chemical or food analysis journals, typically French, German or American. French chemists were worried about the use of fuchsine and other aniline dyes in wine, the Germans worried about magenta and other red coal-tar dyes being used in sausages, and the Americans, influenced by the powerful US dairy lobby, worried about 'butter yellow' and other synthetic dyes being used to disguise margarine as butter. While the British press was complaining about the use of aniline dyes in imported foreign food, British food analysts were far more reluctant to address the issue than their overseas peers, despite evidence that the coal-tar colours were being used by British food manufacturers as liberally as by foreign food producers.

Instead, British public analysts at the time were busy focussing on the watering-down of milk, a deceit allegedly practised widely to boost producer and retailer's profits at a time when vast quantities of milk were having to travel by train or cart to the rapidly growing cities. Although public analysts knew that aniline dyes were being used to mask the thin blue appearance of diluted milk, the analysts' main priority was in determining the fat content of the milk in order to prove that it was watered down. The desire to have a predetermined set percentage of fat content, laid down by regulation, in order to define natural versus adulterated milk became a matter of contentious debate between the public analysts employed by local authorities and the central government excise chemists based in Somerset House. The practice of *colouring* milk, or indeed other drink and food, with synthetic dyes became completely hidden in the midst of a noisy public battle over fat content between two groups of chemists, each seeking to establish their credentials and status during a period when professional scientists were struggling to find their position in society (Atkins, 2010).

Indeed, in 1900, Otto Hehner, one of Britain's leading public analysts and a sharp critic of adulterated food including the use of new chemical ingredients, informed a parliamentary committee investigating the use of chemical preservatives and colourings in food and drink that the use of aniline dye was 'continually present – almost invariably in the London milk'. While critical of the practice, he did not advocate banning it:

> I think it is a deceptive practice, and undoubtedly was due originally to the milkman being anxious to hide the blueness of his milk. At the same time, it is such a universal practice now that the consumer would probably refuse the natural milk if he got it. … I do not like the practice. At the same time it is not a matter which I think calls for urgent attention.
>
> (Committee on Food Preservatives, 1901: pp. 191–194)

It seems anomalous that analytical chemists, who from the outset were highly critical of the use of toxic metals used to colour foods, ignored the concern by now being expressed in the press about the toxicity of aniline and other coal-tar-derived dyes. Certainly, the publicity surrounding the poisonous metallic colouring of confectionery and other foodstuffs had helped promote food legislation to protect the public and to establish the public analyst's role as a gatekeeper of food safety. But while concern about artificial food colouring played a crucial part in early attempts to regulate food production, the new aniline and azo coal-tar colours seemed to pass comparatively unnoticed for several decades. Even when concern began to be raised in the press about the health effects and unmonitored usage of coal-tar dyes, many public analysts remained oddly indifferent to their use.

I believe that the analysts' reluctance to tackle the increasing use of chemical dyes in food is tied up with the insecurity of the analysts themselves and their desire to boost the authority and credibility of science and scientific expertise in public life. Synthetic dyes were a public proof of the expertise of chemists and their ability to transform society; to criticise the dyes might have cast doubt on one of the century's wonders of chemical achievement and, in turn, on chemists and chemistry. Public analysts also worried about their lack of status among other chemists in an era when professional chemists were jostling for position and credibility.

In recent years historians have shown that public analysts were very much at the centre of disputes surrounding authority, standards and methodology in public health in the second half of nineteenth-century Britain (Atkins, 2010; Hamlin, 1986; 1990; Steere-Williams, 2014). Acknowledging the depth of these disputes and the contested position of public analysts, together with an appreciation that chemicals synthesised from coal-tar waste were emblematic of the progress of science and were being used in food in such small amounts as to be almost impossible to detect, helps us understand why the new dyes were not an area of concern for public analysts.

Public analysts fight for recognition

The Society of Public Analysts (SPA) was formed in 1874, in response to political and public criticism about the inexperience of public analysts and their inconsistent decisions, as well as a lack of consensus over what constituted adulteration. As the geographer and historian Peter Atkins and the society's early biographers have pointed out, the SPA was formed from a primarily defensive and aggrieved position. Dr Theophilus Redwood, Professor of Chemistry at the Pharmaceutical Society and Public Analyst for the County of Middlesex, who took the chair for the SPA's first meeting on 7 August 1874, described the objectives of the meeting as:

> first, the refutation of unjust imputations; secondly, the repudiation of proposed measures of interference with our professional position and

independence; and thirdly, the formation of an association having for its objects the promotion of mutual assistance and co-operation among public analysts.

(Atkins, 2010; Chirnside and Hamence, 1974)

Such words, as well as correspondence and reports of the SPA's meetings published in the *Analyst*, reveal the degree to which these new public appointees felt the need to justify their position and demonstrate their expertise. So-called practising chemists during the nineteenth century – that is those chemists not chiefly employed in academia, in industry or as dispensing chemists – generally earned their living in private practice as consulting and analytical chemists. This often meant working for many different clients in diverse sectors of commerce. As William Brock has shown in his work on William Crookes and other practising chemists of the period, success came from taking jobs whenever and wherever available, alongside a vigorous self-promotion and public endorsement of the chemical profession and of chemistry in general (Brock, 2008). Crookes himself set up and edited *Chemical News*, one of many chemists to use the burgeoning Victorian media to promote chemistry and the work of chemists.

For decades, members of the Society of Public Analysts felt under attack from all directions, including from members of the public who complained that taxes were being paid to public analysts who did little to prevent adulteration (Anon., 1876). However, disputes between the public analysts themselves and other men of science, notably fellow chemists, attracted much more comment in the pages of the *Analyst*. The role of the chemist in adulteration was both ambiguous and contested. Over time, many food and drink ingredients came to be viewed as adulterants, including liquorice, ginger, turmeric, alum and sulphuric acid. Such products, including the 'harmless' chemical dyes, were supplied to the Victorian food and drink trade by chemists and druggists (Burns, 2011). This was a time when professional chemists were attempting to define who they were and to differentiate themselves from the retail druggists and other practical chemists. The profession of the chemist, and the scientist in general, was in the midst of a rapid and broad transformation. The experience and training of many members of the scientific profession spanned across many areas that only later would be classified as separate disciplines. Not only did these burgeoning professionals have their toes in many disciplines, operating in both academia and commerce and bridging both public and private practice, but they invariably knew each other and indeed had often been trained by the same mentors. This, in many ways, made their disputes more vitriolic, as they strove to assert their own opinions about the best way to secure their professional interests.

The public analysts kept their sternest rebukes for the government scientists employed by the government's Customs and Excise department, who were able to overrule them in disputed adulteration cases. Being challenged by the government chemists, whose primary duty was to protect the government's customs coffers, was only one part of the battle public analysts faced in their early years. From 1872 public analysts were appointed and funded by local authorities in order to ensure safe and honest food supplies. However, these posts were often underfunded and shared jointly with district medical officers of health. Many public analysts in the early years were doctors with little chemical experience.

Chemical dyes prove elusive

Meanwhile, during the latter decades of the nineteenth century, the understanding of, and practical developments in, organic chemistry were rapidly changing. Academic and industrial chemists were synthesising new compounds from coal-tar waste by the dozen, but identifying and understanding individual substances was not an easy task, even for their discoverers. The difficulty for the analytical chemists, using time-consuming and detailed elimination tests to detect the new substances, was that, by the 1880s, there were hundreds of dyes in the European and American marketplace – some known, many unknown – and food and drink manufacturers often added several colouring additives to one product in order to make the detection of discrete dyes harder. The new dyes were used to colour a wide range of food products, but because of the tiny quantities of dye used, even expert chemists were often unable to detect their use, still less identify the individual dyes.

The dyes, meanwhile, passed through a complex international supply chain from chemical factory to wholesale and retail chemists to the food industry, with their physical and chemical origins becoming more and more obscured. The situation in assessing chemical dyes was particularly complex, as these were completely novel substances, created by chemists, who themselves were unable to agree on their nomenclature or chemical composition. The dyes were rapidly permeating, invisibly and in tiny amounts, many different products being consumed throughout society (Hepler-Smith, 2015; Green, 1908).

The sheer number, complexity and confusion of names and classification of synthetic dyes in the nineteenth century is immediately apparent when reading analysts' reports from this period. Following a trip to Paris in 1885, the British public analyst John Muter described tests used by French chemists to detect the use of rosaniline, safranine, aniline violets, mauvaniline, chrysotoluidine, amidonitrobenzol, roccelline, foundation red, Bordeaux red and blue, Ponceau red and blue, Biebrich reds, Tropeoline 000, 1, 2 and O, chrysoine, helianthine, eosine B, eosine JJ, safrosine, ethyleosine and red coralline (Muter, 1885). This extensive list of some of the new synthetic substances being used to colour food and drink makes clear the difficulties analysts faced. The analytical tests employed by chemists to detect different dyes in food and drink relied on elimination techniques or tests for the presence of specific individual dyes. The ability of food producers and retailers to use multiple dyes from a choice of literally hundreds of new substances made both these testing strategies pragmatic impossibilities.

By 1904, a compendium of dyestuffs listed nearly 700 dyes, with little consistency in how they were named and classified (Green, 1908). Some names were given to dyes because of their chemical construction, others described their colour such as 'butter-yellow', while some reflected the dye's inventor or place of manufacture such as 'Martius brown' and 'Manchester brown'. Two – magenta and sulpherino – even celebrated Napoleonic victories. Many of the earliest aniline dyes were named using classical terms for colours (*flavin* – yellow), flowers (*safranin* – saffron, *fuchsine* – fuchsia, *mauvein* – mallow, *rhodamin* – rose), minerals (*auramin*), or animal dyes (*purpurin*). Later names tended to indicate their chemical composition, applications or properties, such as *Methylviolett, Benzingelb* and, as the number of dyes continued to expand, letters referring to colouration were added such as R (for rot) or G (for gelb or grün), and so on. In other cases, dyes were named in groups, sometimes linked to the manufacturer or a brand name (Jones, 2013; Schaeffer, 1951).

Moreover, chemically identical dyes had different names in different countries or when sold by different wholesalers. Manufacturers would often keep the chemical formulae of their new dyes secret, leading to uncertainty and confusion, while producers and retailers frequently re-used established names for commercial reasons, resulting in the same name being applied to several different types of dye. So, for example, chrome yellow, Turner's yellow, yellow lead oxide and some azo dyes, among others, were all called *Neugelb*, while *Kaisergelb* was used for yellow ochre, cadmium and chrome yellow, and aurantia, a nitro dye (Jones, 2013: p. 132).

According to the German chemist Theodore Weyl, 'the trade names of the coal-tar colors are mostly fanciful, since the scientific titles are cumbersome and difficult to remember.' For example, aurantia's technical name was hexanitrodiphenylamine. Meanwhile, Weyl complained that colours were given different names by different suppliers, noting that crocein orange, ponceaux 4 GB and brilliant orange were all the same colour, while Bismarck and Manchester brown, phenylene brown and canella were also identical dyes. Weyl had no doubt that this practice led to confusion in the market which traders used to their advantage:

> Different colors are often designated by the same name, especially with a view of substituting a cheap for a costly product. In this way ... the low priced Martius' yellow is called naphthol yellow S, although the latter name belongs to a more expensive preparation. Finally mixtures of familiar colors necessary to produce peculiar tints are frequently sold under new names, with deceptive intent. Cardinal, for example, is a mixture of chrysoidin and fuchsin.
>
> (Weyl and Leffman, 1892: pp. 89–90)

This was a deception that could harm health as well as quality and commerce, since Weyl's experiments indicated that Martius yellow, for example, was toxic, whereas Napthol yellow S could be safely consumed in small doses.

The German bacteriologist Ferdinand Hueppe, who, like many German scientists, increasingly used the new dyes to stain tissue samples, also highlighted this situation:

> In attempts to group the aniline colours a kind of uncertainty appears even among color chemists. The same trade name does not always correspond to the same preparation. Many preparations are not chemical individuals, but mixtures of related colors. Many preparations are 'standardised' for the trade; for example, with dextrin. On account of the patent laws, factory secrets surround the production of many coloring matters, and frequently statements are met with which are directly intended for the purpose of misleading competition.
>
> (Hueppe, 1891, cited in Hesse, 1912)

For the under-funded and poorly resourced public analysts, trying to detect and prove adulteration and then defend their position in court against producers, retailers and expert witnesses, including government scientists or other chemists, was not an easy task, even when dealing with tried and tested adulterants. The identification of tiny amounts of the new chemical additives in food, most of which were unknown or described under a variety of different names or chemical formulae, was impossible to accomplish.

Meanwhile, there was a lack of experimentation and research into the health effects of the hundreds of dyes entering the food market. This situation had arisen for several reasons, including a lack of initial concern and awareness of the extent to which these new substances were being used in food and a lack of standardised tests to identify the dyes or test their toxicity. This had left the analysts playing continual catch-up with the food and drink manufacturers.

For a public analyst to try to identify and prove in court the presence of individual, specified, synthetic dyes would be to risk his credibility, and that of science, at a time when such chemists were desperately trying to assert themselves as reputable scientists. At this time, analysts did not have a sufficiently secure public credibility to question the use of synthetic dyes, which had become a beacon of the significance of chemistry in Victorian society. The absence of agreed-upon tests for the dyes meant that to stake one's reputation in court upon the presence of particular dyes in food was risky. The proliferation of the colours meant that chemists who claimed that a particular dye was present in a particular food were liable to find themselves countered by claims that a different colour had been used.

In a debate in London in 1890 on artificial colourings used in sugar, leading analysts agreed that it would be counterproductive to go beyond the duty of the analyst simply to prove adulteration and to attempt to identify particular additives. August Dupré argued that a public analyst was not bound to state the exact composition and character of the adulterant, and emphasised that doing so increased the chance of being proved wrong. 'It would be a most dangerous thing for the public analyst to bind himself down to a particular composition', he advised (Cassal, 1890). Dupré, like many of his contemporaries in Britain, had completed part of his training in Germany, a country that led the field in organic chemistry and where chemists were already highly respected and where most of the dyes were being discovered and manufactured (Anon., 1907).

The ability to determine the exact synthetic dye used in any food product would have been very difficult for most public analysts. The historian Catherine Jackson has shown in her studies of mid-nineteenth-century organic chemistry that assessing the exact molecular structure of any organic substance was not an accurate science during this period (Jackson, 2014). Historians have also described how numerous international conferences were held from the 1860 Karlsruhe Congress of Chemistry through to at least eight international congresses between 1894 and 1914 in attempts to create some uniformity of nomenclature and standardisation in chemistry (Crosland, 1998; Hepler-Smith, 2015). While organic analytical chemistry had developed significantly during the nineteenth century, the detection of individual, new synthetic substances in food was thus extremely problematic. Identifying and condemning synthetic dyes in food and drink potentially risked not only the reputation of analysts but the credibility of chemistry itself. Rather than criticising chemicals for adulterating food, chemists were anxious to promote science and chemistry as a means of improving the nation's health, sanitation and food supply.

Food is a political issue

Feeding the public was a contentious and political issue. Concerns surrounding adulteration and lack of provenance were great and so too was anxiety over the nation's ability to feed its increasing population. As the government's Somerset House chemist Richard Bannister remarked in a turn-of-the-century essay on the state of civilisation, the 'food of the people is related to its civilisation' and 'the food industries of different nations have a more direct bearing upon the moral and material condition.' While Bannister recognised the importance of anti-adulteration legislation, most important to him was the ability to feed the population of Britain, 'where the greatest concentration is to be found of the food products of the world, and where the need for a large, wholesome supply is indispensable to its advance in civilisation.' To Bannister, free trade, the opening up of foreign markets, and improvements in science, technology, food preservation techniques and transport had allowed the provision of cheaper food for the masses, as well as more choice and greater and more uniform quality. While the dairy industry lobbied against the colouring of margarine, Bannister argued that, as long as it was not passed off fraudulently as butter, coloured margarine should not be discriminated against but welcomed as an important addition to the cheap food supply (Samuelson, 1896).

The introduction of newly synthesised chemicals into food production in the late nineteenth century demonstrates the difficulty of determining the boundaries between legitimate and illicit interventions in food. The question as to whether certain ingredients, particularly the new chemical substances, should be regarded as food improvers or adulterants divided analysts, as well as food manufacturers, the public, politicians and the growing sanitarian and health and food reform movements. For instance, oleomargarine was a novel food product, invented to provide a cheap fat substitute during a period when consumption of fat was considered important to provide sufficient energy to feed the human working body. As a new food type, it did not have a history of norms and practices associated with it. This conundrum was evident in the conclusions reached by the government committee tasked at the turn of the twentieth century to examine the use of colourings and preservatives used in food. While the committee recommended banning the use of colouring in milk, because of the large quantities consumed and expectation among consumers that it is a 'natural' product, it concluded that colouring should remain permitted in butter and cheese:

> In the butter trade and still more so in the cheese trade artificial colouring has long been established. Highly coloured goods find favour in some markets, uncoloured or lightly coloured goods in others. We have not found that in the interest of the consumer any interference is necessary with the customs of the trade in this respect.
> (Committee on Food Preservatives, 1901: p. xxix)

However, the Committee's response to the use of colouring in the new product of margarine was more ambivalent.

> In regard to margarine, we have to deal with a cheap and relatively inferior article coloured to resemble a more costly and superior article, and probably the only means of protecting the public from imposition would be to prohibit the introduction of any colouring into margarine which shall cause it to resemble butter. ... But as the margarine may be assumed to be a perfectly wholesome article of diet, it does not fall within the terms of our reference to make any recommendation upon a practice which is not attended with risk to the public health.
> (p. xxix)

For most analysts, the main concern was the misuse of colourings to disguise a food product or to deceive the consumer, rather than the possible harmful effects of the colourings themselves. Dupré claimed that he often had artificially coloured sugars submitted to him, noting, however, that he would be doubtful about stating that such sugars were adulterated unless they were being passed off as something they were not, such as beet sugar sold as demerara. As he explained:

> Butter was often coloured, so was milk, but many analysts do not consider the colouration of butter to be an adulteration. It had been a long-continued custom, and the public liked to see their butter yellow. If an artificial

butter was coloured with a view to its being passed off as a genuine butter that was a different matter, but the mere fact of its being coloured was not an adulteration.

(Cassal, 1890)

Such comments demonstrate the problematical function of additives like dyes and the difficulties attached to assimilating them into the category of 'adulterants'.

Passionate about their role as guardians of food safety, public analysts acted as crusaders in a battle against both fraudulent traders and bureaucrats within the Inland Revenue Laboratory. At the same time, they also regarded themselves as pioneers of the new science of organic chemistry, forging a new path towards a scientifically controlled and enhanced food supply for the nation at a time when food scarcity and concerns about the reliance of the nation's food supply on foreign imports were omnipresent. As SPA president Dr Alfred Hill proclaimed in 1885: 'few societies can lay a stronger claim to the *raison d'être* than ours, for it is concerned with a department, or an application of chemico-physical science which, until thirty years ago, had received comparatively little attention.' Britain, Hill added, was the 'leader in the anti-adulteration crusade' and

> was not only the first country to move in the direction of repressing adulteration, but her chemists, who at the same time are or have been members of this Society, have had the distinction of being foremost in the invention and perfection of processes for the analysis of articles of food and drink.
>
> (Hill, 1885: p. 42)

Looking behind their veneer of crusading zeal and claims of scientific prowess and precision, it is possible to glimpse a group of men struggling to forge careers and reputations as a new breed of chemist. These were men who had different educational and working backgrounds to both the prominent academic chemists of the early to mid-nineteenth century and the government chemists of the second half of the century. The public analysts were a small group of chemists, some self-taught, others who had attended the new technical colleges in Britain before travelling to Germany to extend their knowledge. They knew each other and had been taught or had worked under the same people. Unlike the government chemists, they were unable to rely on a life-long state salary and found themselves touting their skills to different companies, local authorities and members of the public. The public analysts sought stricter standards and uniform methods to strengthen their authority and credibility and saw their role as protectors of the safety and morality of the nation's food supply.

Conclusion

Synthetic dyes, created and manufactured by a burgeoning chemical industry, entered the food supply system in Britain during a period when public analysts were clearly fighting to secure their place as creditable and trustworthy analytical chemists, as well as seeking to persuade the public that chemistry was the foremost sanitary and socially useful science. As such their professional status was aligned with that of the new chemical substances and it is perhaps not surprising that they would choose to remain silent about the new chemical colours being used in food, while they struggled to reach agreement between themselves and other professional chemists and scientists about existing and long-standing food additives and disputed adulterations. It would not, perhaps, have been in the public analysts' interest or that of the reputation of chemistry itself to have questioned the safety of what the media was claiming to be one of nineteenth-century chemistry's greatest contributions to science, industry and society – the 'miracle' wonder dyes extracted from coal-tar waste.

It is also to be noted that these dyes were being produced and discovered by organic chemists, many of whom the public analysts may have known or at least shared mentors with. Moreover, the ability to determine the exact synthetic dye used in any food product would have been very difficult for most public analysts. While organic analytical chemistry had developed significantly during the nineteenth century, the detection of individual, new synthetic substances in food was extremely problematic. At a time when public analysts were in constant dispute with other scientists, including the government chemists, over the accuracy of their analyses, tackling the issue of synthetic dyes, which were being created by specialist organic chemists in well-equipped industrial laboratories, may have been a step too far. Reticence on the matter, if not silence, was the more prudent strategy.

It should also be recognised that synthetic dyes produced by organic chemists were seen by many public analysts as a safer alternative than the mineral-based dyes formerly used to colour food. Meanwhile, the inclusion of chemical dyes in food was not widely advertised or acknowledged by food manufacturers. All these contextual factors need to be considered in order to assess objectively why public analysts in Britain did not regard the use of coal-tar-derived dyes to be of serious concern and why chemical dyes continued to be silently added to food during a period of increasing food legislation and concern about food adulteration.

Exploring the introduction of new chemical dyes into food in the late nineteenth century demonstrates many issues experienced by the introduction of new scientific products or processes. These new products were hailed as miracles of science and the solution to existing problems. For many years they were added to food and drink, without many people being aware of their presence. By the time concern was raised their use had become widespread and normalised. Consumers expected certain, uniform colouring in their food, manufacturers had adopted specific techniques and ingredients to meet consumer expectations and market conditions, politicians recognised the need to adequately feed the nation and chemists saw science as the means to do this. Vested interest in the continued use of the chemical dyes was widespread encompassing consumers, producers, politicians, regulators and scientists.

As I have shown, synthetic dyes, many of which were subsequently found to be highly toxic and never intended as food additives, were introduced into food unannounced. Their continued use for so long was the result of a constellation of many factors, including a lack of consensus and status among scientists, political anxiety about food security, consumer expectations and economic competition in the marketplace. I believe that, like many scientific risks we face today, the silence surrounding this application of science centres on a complex social and cultural situation in which regulation and control becomes more and more complicated as the practice is normalised and vested interest increases and broadens.

Notes

1 The European Food Safety Authority regulates and lists all the food additives permitted in the European Union and assigns to them an E number. Permitted E numbers were introduced by the EFSA in 1962 specifically to regulate the use of dyes in food. E numbers have since been applied to many other types of food additives. Other countries, such as the USA, have similar regulations permitting or banning specific substances. While there is a growing consensus as to which additives should be permitted, there are still differences in national regulation. For example, carmoisine (or azorubine E122) is approved in the EU, but is being phased out in the UK and is banned in Canada, Japan, Norway and the US. Conversely, fast green (E143) is permitted in the US but banned in the EU. Meanwhile, the EFSA announced in 2014 a reassessment of all 45 food colours currently permitted for use in food in the European Union, giving priority to the reassessment of synthetic dyes. In 2013, the Authority recommended new tests for six sulphonated mono azo dyes, allura red AC (E129), amaranth (E123), ponceau 4R (E124), sunset yellow FCF (E110), tartrazine (E102) and azorubine/carmoisine (E122). For more information and current regulation see the websites of national and international food agencies; for example: www.fda.gov/Forindustry/ColorAdditives/default.htm, www.food.gov.uk/science/additives/foodcolours/#.U4LvXy_gLZs, www.efsa.europa.eu/en/topics/topic/foodcolours.htm, and www.efsa.europa.eu/en/faqs/faqfoodcolours.htm (All accessed 26 May 2014.)
2 The inquiry led to an Act for the Adulteration of Articles of Food and Drink in 1860, later superseded by the 1872 Adulteration Act, which led to the appointment of public analysts.

References

Accum, F., 1820. *Treatise on Adulterations of Foods and Culinary Poisons*. London: Longman, Hurst, Rees, Orme and Brown.
Anon. 1859. Perkins's purple. *All Year Round*, September issue, p. 222.
Anon., 1862. A ramble into the eastern annexe of the International Exhibition. *The Ladies' Treasury*, 1 Nov., p. 342.
Anon., 1865. Some very ancient things. *The Ladies' Treasury*, 1 Aug., p. 242.
Anon., 1869. Aniline colours. *Bradford Observer*, Issue 1892, 14 Jan., p. 3.
Anon., 1871. Poisoned candies. *The Health Reformer*, 6–7, p. 131.
Anon., 1875. The useful book. *The Treasury of Literature and the Ladies Treasury*, 1 Jan., p. 42.

Anon., 1876. The public and 'public analysts'. *Analyst*, 1(9), pp. 155–156.
Anon., 1878. Meat-tints for the millions. *Funny Folks*, Issue 174, 30 Mar., p. 101.
Anon., 1882. Not very many years ago. *The Country Gentleman: Sporting Gazette and Agricultural Journal*, Issue 1076, 23 Dec., p. 1322.
Anon., 1886. Talk about depression in trade. *Moonshine*, 7 Aug., p. 72.
Anon., 1890a. Chit-chat. *Sheffield and Rotherham Independent*, Issue 11278, 9 Oct., p. 5.
Anon., 1890b. Blood oranges. *The Friendly Companion: A Magazine for Youth and the Home Circle*. 1 Nov., p. 309.
Anon., 1907. Obituary. *Nature*, Issue 76, 1 Aug., p. 318.
Atkins, P.W., 2010. *Liquid Materialities: A History of Milk, Science and the Law*. Farnham: Ashgate Publishing.
Beer, J.J., 1958. Coal tar dye manufacture and the origins of the modern industrial research laboratory. *Isis: Journal of the History of Science in Society*, 49(2), pp. 123–131.
Beer, J.J., 1959. *The Emergence of the German Dye Industry*. Urbana: University of Illinois Press.
Brock, W.H., 2008. *William Crookes (1832–1919) and the Commercialization of Science*. Farnham: Ashgate Publishing.
Burns, E., 2011. *Bad Whisky: The Scandal That Created the World's Most Successful Spirit*, 3rd revised edition. Glasgow: Neil Wilson.
Cassal, J., 1890. On dyed sugar. *Analyst*, 15, pp. 141–149.
Chirnside, R.C. and Hamence, J.H., 1974. *'Practising Chemists': History of the Society for Analytical Chemistry, 1874–1974*. 1st edn. London: Society for Analytical Chemistry.
Committee on Food Preservatives, 1901. *Report of the Departmental Committee appointed to inquire into the use of preservatives and colouring matters in the preservation and colouring of food: together with minutes of evidence, appendices and index*. London: HMSO.
Crosland, M., 1998, The organisation of chemistry in nineteenth-century France. In: Knight, D. and Kragh, H. (eds), *The Making of the Chemist: The Social History of Chemistry in Europe, 1789–1914*. Cambridge: Cambridge University Press.
Garfield, S., 2001. *Mauve: How One Man Invented a Colour That Changed the World*. London: Faber.
Green, A.G., 1908. *A Systematic Survey of the Organic Colouring Matters*. London: Macmillan. Available at: http://archive.org/details/asystematicsurv00juligoog. Accessed 17 July, 2012.
Haber, L.F., 1969. *The Chemical Industry During the Nineteenth Century: A Study of the Economic Aspect of Applied Chemistry in Europe and North America*. Oxford: Clarendon Press.
Hamlin, C., 1986. Scientific method and expert witnessing: Victorian perspectives on a modern problem. *Social Studies of Science*, 16(3), pp. 485–513.
Hamlin, C., 1990. *A Science of Impurity: Water Analysis in 19th Century Britain*. Los Angeles: University of California Press.
Hassall, A.H., 1855. *Food and Its Adulterations: comprising the reports of the Analytical Sanitary Commission of 'The Lancet' for the years 1851 to 1854 inclusive, revised and extended: being records of the results of some thousands of original microscopical and chemical analyses of the solids and fluids consumed by all classes of the public*. London: Longman, Brown, Green and Hurst.
Hepler-Smith, E., 2015. 'Just as the structural formula does': names, diagrams, and the structure of organic chemistry at the 1892 Geneva Nomenclature Congress. *Ambix*, 62(1), pp. 1–28.

Hesse, B.C., 1912. *Coal-Tar Colors Used in Food Products.* Washington, DC: Government Printing Office.
Hill, A., 1885. Inaugural address by Dr Hill, president (delivered at the Annual Meeting, 28th January 1885). *Analyst*, 10, March, p. 42.
Homburg, E., Travis, A.S. and Schröter, H.G. (eds), 1998. *The Chemical Industry in Europe, 1850–1914: Industrial Growth, Pollution, and Professionalization.* London: Kluwer Academic.
Humble, N., 2006. *Culinary Pleasures: Cook Books and the Transformation of British Food.* London: Faber.
Jackson, C., 2014. The curious case of coniine. Constructive synthesis and aromatic structure theory. In: Klein, U. and Reinhardt, C. (eds), *Objects of Chemical Inquiry.* Sagamore Beach, MA: Science History Publications, pp. 61–103.
Jones, W.J., 2013. *German Colour Terms: A Study in their Historical Evolution from Earliest Times to the Present.* Amsterdam: John Benjamins.
Muter, J., 1885. On the processes and standards in use at the Municipal Laboratory at the City of Paris. *The Analyst*, 10, pp. 195–199.
Olson, R., 2008. *Science and Scientism in Nineteenth-Century Europe.* Urbana: University of Illinois Press.
Proctor, R.N. and Schiebinger, L. (eds), 2008. *Agnotology: The Making and Unmaking of Ignorance.* Stanford, CA: Stanford University Press.
Schaeffer, A., 1951. *Die Entwicklung der künstlichen organischen Farbstoffe.* Horheim, Germany: self-published.
Samuelson, J., 1896. *The Civilisation of Our Day: Essays.* London: Sampson, Low, Martson & Co.
Steere-Williams, J., 2014. A conflict of analysis: analytical chemistry and milk adulteration in Victorian Britain. *Ambix*, 61(3), pp. 279–298.
Stern, R.F., 2003. 'Adulterations detected': food and fraud in Christina Rossetti's 'Goblin Market'. *Nineteenth-Century Literature*, 57(4), pp. 477–511.
Travis, A.S., 1983. *The Colour Chemists.* London: Brent Schools and Industry Project.
Travis, A.S., 1993. *The Rainbow Makers: Origins of the Synthetic Dyestuffs Industry in Western Europe.* Bethlehem, PA: Lehigh University Press.
Weyl, T. and Leffman, H., 1892. *The Coal-Tar Colors: With Especial Reference to their Injurious Qualities and the Restriction of their Use.* Philadelphia, PA: P. Blakiston.
Whorton, J.C., 2011. *The Arsenic Century: How Victorian Britain Was Poisoned at Home, Work, and Play.* Oxford: Oxford University Press.

11 Having it all

Ownership in open science

Ann Grand

If, as the communications scholar Miriam Metzger (2007) suggested, in the digital environment nearly anyone can be an author, in the open science environment nearly anyone can be a producer, consumer or analyser of data. At its most complete, the philosophy of 'open science' commits researchers to revealing and sharing the entirety of their practice: questions, data, methodologies, results, models, speculations, wrong turns and all (Nielsen, 2009). Open practice has the power to reveal formerly closed and silent parts of the scientific process, so that anyone, from professional colleagues to interested members of the public (at least, those with internet access), can have direct, unmediated access to research: can indeed 'have it all'. This chapter argues that breaking the silences of scientific practice raises a number of concerns for both professional and non-professional participants in open science.

The ideal of open science

The first practitioners of what we now call science seemed to prefer to work in secrecy; Hooke famously encoded what we now call Hooke's law in an anagram and it took Newton almost 40 years to publish a full description of his calculus. However, since the first learned societies and scientific journals came into being in the seventeenth century, one could argue that sharing the results of work has been a first principle for scientists, offering a ladder that enables each – to go back to Newton – to climb onto the shoulders of earlier giants. The etiquette of modern research, of priority, publication and citation, is based in the concepts of trust and civility current in seventeenth- and eighteenth-century England (Shapin, 1994). From these principles, scientists developed a convention, a culture of sharing through journals, conferences, symposia and workshops. However, despite the *politesse* and shared values, full disclosure mostly remained a privilege of the few – and those few were almost certainly other scientists. Those outside the academy, unable to afford or unable to get journal subscriptions, had to rely on reading about the results of research via books and, more recently, newspapers, television and radio. Moreover, these mass media are largely channels for communication after the fact; it is the results of work that are published, leaving details of how the science was done as something of a mystery.

With the development of the internet from about 1972 and of the World Wide Web from about 1990, new and more rapid ways of communication became possible, creating tools that enable scientists, if they choose to, to share the whole of their research and members of the public, if they choose to, to have access to the constantly evolving, dynamic, tentativeness that is the hallmark of all research. Open practice not only opens up the journal pages, nor even just the laboratory; as Burton (2009) suggests, it opens up the scientist's thought processes: the ultimate silent space.

And people are choosing to enter those previously closed spaces. One of the dominant lexical trends of the early twenty-first century is the rise of 'open' as a prefix for a range of activities and institutions: government, culture, archives, knowledge, source code, data, research, democracy, science and more. Digitally mediated emergent communities and interest groups demand open and instant access to information – and to complete information – which if they can't achieve legally, they may get to via less legitimate means. One notorious example is provided by the climate sceptics who demanded access to climate data collected by scientists at the University of East Anglia, UK. When the sceptics believed the scientists were failing to respond to Freedom of Information requests in a helpful way, their emails were later leaked to several websites, in what became known as 'Climategate'. In his report on the affair, the chair of the official inquiry, Muir Russell, described calls for open data as indicative of 'a transformation in the way science has to be conducted in the twenty-first century' (Russell, 2010).

In other post-Climategate examinations of research culture, British funders, government and learned societies have laid out a common ground where demands for openness and transparency are accepted as unarguable and legitimate. This goes beyond the 'Climategate' sceptics' demands for open access to results. The House of Commons Science and Technology Committee (2010) asked publicly-funded researchers to 'consider whether they are being as open as they can be, and ought to be, with the details of their methodologies', while the Royal Society's 2012 report *Science as an Open Enterprise* made a powerful argument that open enquiry lies at the heart of the scientific enterprise and, indeed, is absolutely necessary if we are to be able, in a rough translation of the Society's own motto, to 'take nobody's word for it' (*nullis in verba*):

> Publication of scientific theories – and of the experimental and observational data on which they are based – permits others to identify errors, to support, reject or refine theories and to reuse data for further understanding and knowledge. Science's powerful capacity for self-correction comes from this openness to scrutiny and challenge.
>
> (Royal Society, 2012: p. 7)

The thrust of the argument from these three reports is that open inquiry is crucial to the success of the scientific process, both as an academic activity and a social enterprise. Such open inquiry goes beyond open access to publications to include access to research outputs (open data) and the process by which knowledge is made (open science) (Beck, 2014). Researchers who commit to a philosophy of 'open science' must become, in Burton's (2009) words, 'open scholars':

> The 'Open Scholar', as I'm defining this person, is not simply someone who agrees to allow free access and reuse of his or her traditional scholarly articles and books; no, the Open Scholar is someone who makes their intellectual projects and processes digitally visible and who invites and encourages ongoing criticism of their work and secondary uses of any or all parts of it – at any stage of its development.
>
> (Burton, 2009)

Open practice not only makes new ways of working possible, but also new collaborations. Scientists – indeed researchers in general – are not necessarily only to be found in universities nowadays; they may be part of a civil organisation, community group or special interest community, belong to no organisation or work in their private home (Hess, 2010). Instead of the workroom of 'popular mythology [in which] science is a lonely activity, undertaken by eccentric boffins in dark laboratories late at night' (Leadbeater, 2009: p. 154), the research space has become a 'collaboratory' (Wulf, 1993), where – potentially – professional researchers can work alongside non-professionals, using software that enables them to access any library, use instruments in laboratories around the world (or even in space), to co-create questions and collaboratively analyse data, to share new thoughts and setbacks, detours and disagreements, new discoveries and new directions for exploration.

This is an ideal, golden recognition of the utility and value of open practice. This is not to say it is unachievable or even undesirable. It may be inevitable (Russell, 2010) but researchers also acknowledge its ethical value, its importance in holding science accountable, its potential for enhancing the value, repeatability and scrutiny of research, its support for increased collaboration and, finally, that it is their duty as scientists (Grand, 2012).

Despite all these valuable qualities, there are concerns. One area of concern relates to how producers and consumers can be supported to cope with the vast quantities of data that can emerge from research projects, both in terms of quantity, completeness and credibility, and also in terms of how data can be curated to be accessible to public participants. A second concern is that the quality of data produced by non-professional participants might not be good enough to be used alongside professionally produced data. This is linked to concerns about the role of experts and the position of expertise. Given the increasingly fluid boundaries between producers and consumers of science, and the consequent opening of formerly closed professional spaces, might scientists lose their role? Finally, there are concerns about ownership, both of the process and the results of research. There is clear potential for conflict between the adoption of open science and companies' and institutions' tendency to see protection of intellectual property as vital for its successful exploitation. As Helen Gurley Brown (1982) was perhaps the first to suggest, 'having it all' requires hard graft.

Addressing these concerns means professional researchers must invite and encourage criticism of their work, even as it is in progress, and devote time and skill to contextualising and framing their outputs. Non-professional participants must develop new skills in gathering, interpreting and understanding data and creating high-quality analyses. Finally, all participants will have to resolve the tensions over ownership of research, both its processes and its outputs.

These data are mine

The first fear for some researchers is that if anyone who wants it can get their data, the researchers who produced it in the first place might be 'scooped' as other researchers take advantage of the openly accessible data to beat the originators to publication. As discussed above, there are time-honoured communal mores safeguarding priority, precedence and reputation; as a result, the scientific community shares a strong sense of the way things should be done. This ethical structure is bequeathed by teachers and inherited by each successive generation of scientists (Research Information Network, 2010); overturning long-standing ways of working can take both courage and time.

Communally accepted practices do not exist in a vacuum or without reason. The desire to establish and maintain ownership of work is unsurprising: among academics, professional reputations depend on establishing precedence and priority; in private industry, successful development requires control of ideas, knowledge and data and the asserting of potentially valuable intellectual property rights. The classic pattern of 'work, finish, publish' – attributed to Michael Faraday in J.H. Gladstone's 1874 biography – is a long-standing archetype for the scientific process; a neat, linear triad from hypothesis to theory that keeps information under the researchers' control until it is let out into the public domain. Sharing that information might be to the advantage of science in general but not necessarily to a researcher's personal status.

Against this, it can be argued (Beck, 2014; Grand, 2012) that no two researchers will dip into a well of data in the same way and come up with the same results. Even if a researcher is scooped, it is unlikely that the scooper will publish exactly what the originator would have written. There is a balance to be struck between protection and access; should the majority of data be kept in silence and researchers be denied access to rich, potentially reuseable datasets because of a minority who might abuse the privilege? More practically, advocates contend that openness offers ways to protect priority and precedence (Wald, 2009). Online documents, for example, can be automatically time-stamped, making it easy to show precedence over potential 'scoopers'; open information establishes its priority from the moment it is published (Waldrop, 2008). However, this could require a change of culture on the part of researchers. Hand-written notes in a personal research diary lend themselves well to protective privacy; open science calls for the use of tools such as wikis, blogs, open notebooks and automatic data collectors and loggers. Not only are these tools electronic, by their nature, they are communal and collaborative, something that not all researchers will be comfortable with, whether because of a lack of perceived skills or because of those traditional ways of doing science passed down from preceding generations and down the hierarchy (Research Information Network, 2010).

Science has always been a social practice. Science functions because of the social mores by which scientists share models, discuss ideas and re-analyse others' data (De Roure et al., 2008). As noted above, scientists expect to share; since the seventeenth century when, within the space of a few months, the Royal Society's *Philosophical Transactions* was first published in the UK and the *Journal des Sçavans* appeared in France, members of the research community have co-operated to validate their work by peer review, a process which discourages 'secrecy with respect to the content of science, at least on the part of other scientists' (Hull, 1985: p. 10). What has changed is that they are now expected – and largely are willing – to share with communities beyond their scientific colleagues.

It's not real work …

Open practice cracks the boundaries of the closed and silent parts of the scientific process, allowing 'people with varied backgrounds and scientific expertise [the means to] articulate and contribute their perspectives, ideas, knowledge, and values in response to scientific questions or science-related controversies' (McCallie et al., 2009: p. 12). However, the engagement nourished by open science could also be the means of exposing scientists to a white noise of continuous commentary, which rather than enhancing and enriching the dialogue of science, blocks progress by taking their time and attention from their 'real work'. As David Crotty (a former research scientist who is now a senior editor with Oxford University Press Journals) wrote in a post on his personal blog, *The Scholarly Kitchen*: 'every second spent blogging, chatting on FriendFeed or leaving comments on a PLoS paper is a second taken away from other activities [that] have direct rewards towards advancement' (Crotty, 2010).

Open practice demands that scientists develop a wide and eclectic array of skills and the ability to act competently across a range of media, activities and dialogues (Burns et al., 2003). Developing the skills of listening, communication, public relations, management and delegation, in addition to the tools of their scientific trade, may not be something all scientists want to do (Russo, 2010).

Why did I bother getting this Ph.D?

There are also concerns that opening the scientific process to a greater range of actors could sacrifice scientists' (often painfully acquired) expertise; fears that too much dialogue could mean scientists lose their role or that decisions about science somehow become subject to popularity polls. Certainly, the traditional markers of academic respectability and expertise will be affected by open publication and access of research outputs. Under present models of comment and peer review, researchers can generally assume that their work, whether that be methods, results, papers or presentations, will be commented on by their peers, whose level of expertise can be judged against existing criteria and who have common levels of attitude, work and professionalism. By contrast, comment from non-professional commenters lacks those useful markers and must each be judged – at the cost of time and effort – on their own merit. There have been experiments to extend and modify the publishing model, particularly the classic model of blinded peer review (in which the identities of the writer and reviewer remain unknown to each other), but the outcomes of such experiments have been inconsistent. For instance, journals such as *Nature, Cell, PLoS One* and the *British Medical Journal* (*BMJ*) have tested both open pre-publication peer review (where authors and reviewers are openly named) and open review (publishing papers on a website and making them available for free comment). Responses were mixed. In one experiment, in the journal *Atmospheric Chemistry and Physics*, two papers made available for comment received just five comments between them (Gura, 2002). Similarly, when the journal *Nature* conducted an experiment with open review, they noted that while researchers expressed a significant level of interest in the new practice, they displayed a marked reluctance actually to write anything (*Nature*, 2006a; 2006b). In contrast, the *BMJ*'s 'rapid response' model, which allowed readers to reply to and comment on articles published online, was welcomed by its readers, who posted 30,000 rapid responses between 2002 and 2005. The editors concluded the system was a success but it eventually became a victim of that success, the massive response forcing the editors to implement criteria to constrain comments, such as timeliness and word counts (Davies and Delamothe, 2005).

Of course, these experiments took place before the enormous increase in the use of social media that has made online commenting and virtual dialogue a commonplace. The *Nature* (2006a) experiment would probably have a very different outcome ten years on. Social media are not an unmixed blessing; while they can encourage timely and well-directed commentary in an atmosphere of mutual support, their rapid pace and often necessarily brusque and clipped tone can feel intimidating and personally attacking (Mandavilli, 2011). Nevertheless, the peer review and comment model could undoubtedly be reconstructed. Lessons might be learned from 'citizen journalism', the movement that since the late 1990s has radically changed how we gather, produce and disseminate news. Cheap and ubiquitous technologies have made every citizen a potential news producer, press photographer or film-maker, instantly on the spot in a way that professional reporters can never be. Earthquakes, wars, civil uprisings and protests have all reached us via citizens' phones and computers. While openness here has increased the amount of unsubstantiated information in the public realm, it has also brought increased vigour and critical debate to the reporting of news, giving professionals and non-professionals the power to contribute to political discourse and potentially to affect political events (Niblock, 2010). Open practice could do the same for science, either by reflecting alternative perspectives in analysis or contributing to the accessibility of research outputs for particular user groups (Staley, 2009).

The quality of data

Opening up science to the participation of a wider range of actors raises particular concerns about the quality of their input. Science has become a mature professional field, yet, as both a practice and a body of knowledge, it owes its very beginnings to skilled amateurs. Men such as Robert Hooke, Charles Darwin and Benjamin Franklin, all gentleman scholars, pioneered the creation of the foundations of modern enquiry. This amateur tradition has continued; in the early twentieth century, the Audubon Society (2011) conducted its first annual bird count, gathering the results by post. Since then – particularly in domains where skilled and devoted observation matters more than the quality of equipment, fields such as astronomy, archaeology and natural history (Silvertown, 2009) – devoted legions of interested volunteers have contributed their time and energy to the accumulation of data. In the twenty-first century this is particularly noticeable in the growth of 'Citizen Science', 'scientist-driven public research projects' (Bonney et al., 2009: p. 15) in which volunteers collect or process data in projects specifically designed, normally by scientists, to support amateur input (Silvertown, 2009). Citizen Science projects – including famous examples such as Galaxy Zoo, which has evolved into the Zooniverse (Zooniverse, 2013), FoldIt (FoldIt, n.d.) and various garden bird, insect, butterfly and other 'watches', such as the Royal Society for the Protection of Birds' annual 'Big Garden Birdwatch' (RSPB, 2015) – have drawn on the time and energies of hundreds of thousands of participants. While Citizen Science is not necessarily open science (Wiggins and Crowston, 2011), there is a considerable element of public participation and the difficulties in quality control encountered in Citizen Science projects means these are likely to be reflected in any projects that encourage open input.

To enable their results to be considered for classic journal publication, Citizen Science projects have had to address the issue of data quality. There are, broadly, three routes for ensuring that the quality of community-generated data is acceptable to scientific peer reviewers: pre-submission training, post-submission cleaning and shared methods. Many Citizen Science projects, such as those in the Zooniverse, ask participants to undergo training to test and improve their identification skills before their data can be accepted (Worthington et al., 2011; Zooniverse, 2013). Other projects, for example iSpot (Clow and Makriyannis, 2011) have systems for identifying possible errors and cleaning up the data. In iSpot, 'expert' users confirm identifications made by amateur natural historians. At the beginning of the iSpot project, the experts were largely people with formal academic qualifications, but the programming allowed for especially gifted amateur identifiers to 'rise through the ranks' and themselves become badged experts (literally 'badged' – iSpot rewards its participants with digital badges of various shades). Yet other projects, for example the DART open archaeology project (DART Project, 2010), share their methodologies and structures, to allow participants to replicate methods independently. However, shared practice goes beyond following a prescribed method; methods can go on to be enhanced by modifications arising from situational or experiential expertise. As both Trumbull et al. (2000) and Brossard et al. (2005) noted, where participants consider a prescribed scientist-designed methodology inappropriate – for example, where local environmental conditions differ radically from those existing where the research was designed – they may adapt the methodology, potentially rendering their data unusable. Openness has to be two-way: open to promulgating methods but also open to these being adapted in the light of local or personal expertise.

Through such means, publicly sourced data unquestionably meets rigorous professional standards. However, although adherence to prescribed methods, training or cleaning means the quality of the data is less in dispute, the 'mechanical Turk' nature of the work, in which participants may be limited to data-gathering and data-organising, means they are not necessarily deeply intellectually involved in the full process of science, in all its inventiveness and creativity. In Citizen Science, at present, researchers are still largely in control; the move to open practice, and the extending of the shared practices and social mores of science to professional and non-professional participants, will require a certain loss of control.

Navigating the data flow

As well as quality, there are questions about how participants will cope with the sheer quantity of information likely to emerge from open projects. Open science allows participants to be more than data collectors and organisers; to analyse, synthesise and contribute their thoughts; to create the questions as well as supply the information from which the answers can be derived. However, science is, in its formative stages, messy, dynamic, tentative and uncertain. Raw, unpolished information is difficult to handle; to be digestible, it has to be collated, filtered and, to a certain extent, interpreted. How far can (or should) professional scientists take public needs into account? One option would be to do nothing, to present the raw data with no feedback or context, but this risks overwhelming participants with a flood of information. Moreover, as Alexander Pope put it, 'a little learning is a dangerous thing'; without context (*Nature*, 2011) and narrative, information is an unformed and unusable stream of 1s and 0s.

I have already suggested that 'having it all' is no easy option. Contextualisation – making the outputs of open science useful – requires 'simple analytical structures, more common vocabulary and user interfaces that demand minimal domain knowledge' (Borgman, 2003: p. 165). Setting up a website, writing a blog or keeping an open electronic laboratory notebook can provide that context and structure. However, setting up such tools may require scientists to learn new techniques (although it should be said that increasingly, they are activities that many researchers, like many people, accommodate in their daily lives). A more subtle skill that scientists may need to acquire is the ability to mediate information for a variety of users. To be useful, information has to be accessible, and accessible means more than simply available; for complete accessibility, information has to be mapped so that users can navigate their way through it.

Scientists are not the only participants who will be called upon to develop new skills; without the skills to access it and the skills to sift and use it in interesting ways, information is redundant. To navigate the data flood effectively, non-professional participants may need to develop new skills of gaining access to, and interpreting and understanding the structures of, the digital 'collaboratory' (Wulf, 1993). Fortunately, the greater availability of data may itself offer them both the opportunity to develop those new skills and the material on which to practise them. However developing these new skills and creating new contexts takes time; time that may be difficult for hard-pressed participants to find.

Conclusion

The principle that publicly-funded research should be freely accessible and publicly available has been described as 'compelling and fundamentally unanswerable' (Finch Report, 2012: p. 5). In addition to this philosophical carrot, researchers are also called on to respond to the stick of increasing demand from citizens, civic groups and non-governmental organisations for the evidence that will enable them to scrutinise conclusions and participate effectively in research (Royal Society, 2012). Together, this push and pull have the potential to blur the professional/amateur divide and shift the social dynamics of science.

The practice of open science has the potential to unseal the previously silent spaces of science. It offers new participants, both professional and non-professional, novel routes for unmediated access to active research projects. Open science supports honesty and transparency, both among colleagues and in the relationship between scientists and their publics. By adopting an open science philosophy, scientists are supported to demonstrate the public value (Moore, 1995) of their work, not only its economic value but also its less tangible value for citizens and communities (Talbot, 2008). However, open science will require a change of culture among professional researchers; practically, because it will entail devoting time and skill to the contextualisation and framing of information; philosophically, because it requires researchers to invite and encourage criticism of their work even as it is in progress (Burton, 2009).

Open science requires no less of a culture change among its non-professional participants. For members of the public, open science offers access to a greater range of evidence and can provide the context against which published information can be judged; when the raw dataset and the methodologies by which the research was conducted are made available, the published and polished conclusion can be compared and contrasted against it. However, a public used to carefully considered, normalised and well-ordered conclusions may be confused, or even bored, by the mass of raw data that is the first output of many research projects. Making effective use of the information made available through open science and moving beyond a reliance on polished and normalised research outputs may require the development of new skills, either when using new tools, such as social networks and other web-based media, or practising new processes, such as gathering data or creating analyses that are of acceptable quality.

The social mood of the moment (at least in Europe and North America) – a mood supported by governments, funders and policy-makers – strongly favours greater transparency and greater public influence in decisions about the issues in science and technology that affect people's lives. Open science is well suited to be a medium to support public engagement and public dialogue, to bring new voices to the conversation about science and to break the silence that has previously surrounded the scientific laboratory.

References

Audubon Society, 2011. *History of the Christmas Bird Count*. Available at: http://birds.audubon.org/history-christmas-bird-count. Accessed July 2011.

Beck, A., 2014. Technology and data driven collaboration: archaeological practice in the 21st Century. *Engaging Research*. Blog. Available at: www.open.ac.uk/blogs/per/?p=3080. Accessed August 2015.

Bonney, R., Shirk, J.L., Ballard, H.L., Wilderman, C.C., Phillips, T., Wiggins, A. and Jordan, R., 2009. *Public Participation in Scientific Research: Defining the Field and Assessing Its Potential for Informal Science Education*. Washington, DC: Center for Advancement of Informal Science Education.

Borgman, C., 2003. *From Gutenberg to the Global Information Infrastructure: Access to Information in the Networked World*. 2nd edn. Cambridge, MA: MIT Press.

Brossard, D., Lewenstein, B. and Bonney, R., 2005. Scientific knowledge and attitude change: the impact of a citizen science project. *International Journal of Science Education*, 27(9), pp. 1099–1121.

Burns, T., O'Connor, D. and Stocklmayer, S., 2003. Science communication: a contemporary definition. *Public Understanding of Science*, 12(2), pp. 183–202.

Burton, G., 2009. *The Open Scholar*. Available at: www.academicevolution.com/2009/08/the-open-scholar.html. Accessed March 2015.

Clow, D. and Makriyannis, E., 2011. iSpot analysed: participatory learning and reputation. In: Proceedings of the 1st International Conference on Learning Analytics and Knowledge, 28 Feb.–1 Mar., Banff, pp. 34–43.

Crotty, D., 2010. Science and web 2.0: talking about science vs. doing science. *The Scholarly Kitchen*. Blog. Available at: http://scholarlykitchen.sspnet.org/2010/02/08/science-and-web-2-0-talking-about-science-versus-doing-science/. Accessed March 2015.

DART Project, 2010. Methodology. *The DART Project*. Available at: http://dartproject.info/WPBlog/?page_id=40. Accessed March 2015.

Davies, S. and Delamothe, T., 2005. Revitalising rapid responses. *British Medical Journal*, 2 Jun., 330, doi: http://dx.doi.org/10.1136/bmj.330.7503.1284.

De Roure, D., Hooper, C., Meredith-Lobay, M., Page, K., Tarte, S., Cruickshank, D. and De Roure, C., 2008. myExperiment: defining the social virtual research environment. *Proceedings, Fourth IEEE Conference on eScience*. Indianapolis, IN: IEEE, pp. 82–189.

Finch Report, 2012. Working Group on Expanding Access to Published Research Findings, chaired by Professor Dame Janet Finch. Available at www.researchinfonet.org/wp-content/uploads/2012/06/Finch-Group-report-FINAL-VERSION

FoldIt, n.d. The science behind FoldIt. Available at: http://fold.it/portal/info/science. Accessed March 2015.

Grand, A., 2012. Open science and public engagement: exploring the potential of the open paradigm to support public engagement with science. Ph.D. thesis. University of the West of England.

Gura, T., 2002. Scientific publishing: peer review, unmasked. *Nature*, 416(6878), p. 258.

Gurley Brown, H., 1982. *Having It All*. New York: Simon & Schuster.

Hess, D., 2010. To tell the truth: on scientific counterpublics. *Public Understanding of Science*, 20(5), p. 627.

House of Commons Science and Technology Committee, 2010. *The Disclosure of Climate Data from the Climatic Research Unit at the University of East Anglia*. London: The Stationery Office.

Hull, D., 1985. Openness and secrecy in science: their origins and limitations. *Science, Technology, and Human Values*, 10(2), p. 4.

Leadbeater, C., 2009. *We-Think*. 2nd edn. London: Profile Books.

Mandavilli, A., 2011. Trial by Twitter. *Nature*, 469(7330), p. 286.

McCallie, E., Bell, L., Lohwater, T., Falk, J.H., Lehr, J.L., Lewenstein, B.V., Needham, C., and Wiehe, B., 2009. *Many Experts, Many Audiences: Public Engagement with Science and Informal Science Education. A CAISE Inquiry Group Report*. Washington, DC: Center for Advancement of Informal Science Education.

Metzger, M., 2007. Making sense of credibility on the web: models for evaluating online information and recommendations for future research. *Journal of the American Society for Information Science and Technology*, 13(58), p. 2078.

Moore, M., 1995. *Creating Public Value: Strategic Management in Government*. Cambridge, MA: Harvard University Press.

Nature, 2006a. Nature's trial of open peer review. *Nature*. Available at: go.nature.com/N67mFk. Accessed March 2015.

Nature, 2006b. Editorial: peer review and fraud. *Nature*, 444, pp. 971–972.

Nature, 2011. Editorial: a little knowledge. *Nature*, 472, p. 135.

Niblock, S., 2010. *Journalism: A Beginner's Guide*. Oxford: Oneworld Publications.

Nielsen, M., 2009. Doing science in the open. *Physics World*, 22(5), p. 30.

Nielsen, M., 2012. *Reinventing Discovery: The New Era of Networked Science*. Princeton, NJ: Princeton University Press.

Research Information Network, 2010. If you build it, will they come? How researchers perceive and use web 2.0. Available at: www.rin.ac.uk/our-work/communicating-and-disseminating-research/use-and-relevance-web-20-researchers. Accessed March 2015.

Royal Society, 2012. *Science as an Open Enterprise*. Royal Society Science Policy Centre Report 02/12. Available at: royalsociety.org/policy/projects/science-public-enterprise/report/. Accessed March 2015.

RSPB, 2015. *Big Garden Birdwatch*. Available at: https://www.rspb.org.uk/discoverandenjoynature/discoverandlearn/birdwatch/. Accessed March 2015.

Russell, M., 2010. *The Independent Climate Change Emails Review*. Available at: www.cce-review.org/pdf/final%20report.pdf. Accessed March 2015.

Russo, G., 2010. Backlash against multitasking. *Nature*, 29 Jul., 466(7306), p. 655.

Shapin, S., 1994. *A Social History of Truth: Civility and Science in Seventeenth-century England*. Chicago: University of Chicago Press.

Silvertown, J., 2009. A new dawn for citizen science. *Trends in Ecology and Evolution*, 24(9), pp. 467–471.

Staley, K., 2009. *Exploring Impact: Public Involvement in NHS, Public Health and Social Care Research*. Eastleigh: INVOLVE.

Talbot, C., 2008. *Measuring Public Value: A Competing Values Approach*. London: The Work Foundation.

Trumbull, D., Bonney, R., Bascom, D. and Cabral, A., 2000. Thinking scientifically during participation in a citizen-science project. *Science Education*, 84(2), pp. 265–275.

Wald, C., 2009. Scientists embrace openness. *Science Careers*, 9 Apr. Available at: http://sciencecareers.sciencemag.org/career_magazine/previous_issues/articles/2010_04_09/caredit.a1000036. Accessed March 2015.

Waldrop, M., 2008. Science 2.0 – is open access science the future? *Scientific American*, May. Available at: www.scientificamerican.com/article/science-2-point-0/. Accessed March 2015.

Wiggins, A. and Crowston, K., 2011. From conservation to crowdsourcing: a typology of citizen science. In: *Proceedings of the Forty-fourth Hawai'i International Conference on System Science*. Kauai, HI: IEEE, pp. 1–10.

Worthington, J., Silvertown, J., Cook, L.M., Cameron, R.A.D. and Dodd, M.E., 2011. Evolution MegaLab: a case study in citizen science methods. *Methods in Ecology and Evolution*, 3 Nov., doi: 10.1111/j.2041–2210X.2011.00164.x.

Wulf, W., 1993. The collaboratory opportunity. *Science*, 261(5123), pp. 854–855.

Zooniverse, 2013. The Zooniverse. Available at: https://www.zooniverse.org/. Accessed March 2015.

12 Shocking silences

The management and distribution of silences around TASER™

Abi Dymond[1]

This chapter develops a case study of the silences around the use of the TASER™ electric-shock weapon by the police in England and Wales.[2] Of particular interest are the silences regarding TASER's relative safety compared to other so-called 'less lethal weapons' that the police have at their disposal. I argue that an understanding of these silences is not only valuable in its own right but can also point to broader insights into how silences can be analysed within science and technology studies (STS) more generally. Building on the work of Rappert (2001) and Lee (1999) on the performance and distribution of ambiguities, I suggest that the regulatory particularities of TASER and other less lethal weapons in policing highlight the need to consider how silences are resolved, by whom, and with what effects – issues which have been under-examined in the STS literature to date. This approach, in turn, helps point to new ways of conceptualising the winners and losers within a given socio-technical network – with neither the former nor the latter being easily identifiable *a priori*.

My argument develops as follows. First of all I provide a brief outline of the STS literature and how it currently deals with silences. I then introduce and discuss the merits of an alternative approach concerned with the management and distribution of ambiguities and suggest that this approach may also be useful in analysing silences and their effects. I then provide some conceptual comments on silence, before applying this approach to discussions of the relative safety of TASER and contrasting it with more conventional accounts of the weapon. I start, however, by giving a brief introduction to TASER technology and exploring why the silences around the weapon's relative safety might prove a valuable empirical study for those interested in silences in science.

Characterising TASER

The term TASER refers to electric-shock weapons manufactured by the American company TASER International. Whilst other projectile electric-shock weapons are available, TASER models are in use in over a hundred countries (TASER International, 2012) and are the only projectile electric-shock weapon currently authorised for use by law enforcement agencies in England and Wales. The TASER, often described as a 'less lethal' weapon, delivers a potentially incapacitating electric-shock via tethered wires and probes to subjects up to 6.4 metres away. The current TASER model in use in England and Wales, the X26™, is programmed to automatically deliver a 5-second shock of electricity when the trigger is depressed, although police officers can override this and deliver a longer shock by keeping the trigger held down for longer. Other modes of use exist. The weapon can also be used in 'drive-stun' mode where the weapon is pressed directly up against the skin to deliver a non-incapacitating electrical shock, and in 'red-dotting' mode where the red-dot laser sight is aimed at an individual (College of Policing, 2014a).

This overview, whilst designed to orientate readers, nevertheless risks slightly misleading by defining the weapon and the debate around it solely in technical terms through reference to a core of non-social elements (Grint and Woolgar, 1992). So before proceeding further we must note that many of these features – such as the 5-second shock, or the use of the red-dot laser as a warning – have developed out of complex socio-technical interactions, or 'dances of agency' (Pickering, 2005), between human and non-human actors. Significant debate also exists around the efficacy of the weapon, its ability to incapacitate and the term 'less lethal'.

Several technologies could have been chosen as the focus of this chapter. However, a study of the silences around the relative safety of TASER seems appropriate for three reasons. First, the focus on TASER is timely in the light of calls for a more explicit acknowledgment of the silences around the weapon (Sussman, 2012) and as one response to the Defence Scientific Advisory Council Sub-Committee on the Medical Implications of Less-Lethal Weapons in the UK (DOMILL), which has acknowledged the gaps in our knowledge around TASER technology (DOMILL, 2012). (DOMILL has now been replaced by the Scientific Advisory Committee on the Medical Implications of Less-Lethal Weapons, or SACMILL, which has endorsed its predecessor's statements on TASER. Both committees are referred to throughout the chapter.)

Second, it is important that we understand the relative effects and safety of the weapon for those on whom it is used. This issue, often raised by proponents of less lethal weapons (Rappert, 2001), has been highlighted as particularly important by a range of academics (Hall, 2009; Payne-James et al., 2014; Terrill and Paoline, 2012) and practitioners, with the Independent Police Complaints Commission highlighting this as an important issue (IPCC, 2014). Third, a focus on the silences around TASER is particularly valuable given the silences around weapon technologies both within the agnotology literature, which is interested mainly in environmental and public health issues (Hoffman, 2013; Michaels, 2008; Proctor, 2008; Schiebinger, 2008; Smith, 2005), and within the STS literature (Woodhouse et al., 2002). I now turn to this latter literature in order to demonstrate why and how we should study silences.

Silences in the STS literature

There is an increasing recognition by many in STS that silences and absences of knowledge are important areas for study and that they have been understudied to date. Even those who are unconvinced that silences and absences constitute a sub-field of study in their own right still point to the need for more attention to be given to such topics (Hilgartner, 2014).

Whilst absences have existed in many areas of science and technology, and academics have responded to these in various ways, debates have tended to fracture around realist and relativistic positions (Rappert, 2001; see also Pickering, 2005). One approach, aligned with the common treatment of ignorance as a spur to knowledge (Proctor, 2008), is to try to fill the silences. The intention is to establish truth claims about particular technologies and their effects – claims dependent, to a greater or lesser extent, on the inherent characteristics of the technologies themselves. Such studies can see silence as a way of obscuring or downplaying unwelcome truths. This approach, which Hilgartner (2014) terms a 'sociology of errors', features strongly in the TASER debate and is discussed later on in this chapter.

A second approach has been to analyse how such truth claims and communicative acts are made and accepted, and to look at the functions they fill. Within this approach, those writing from a post-essentialist perspective strongly reject the notion that there are inherent characteristics of technologies. Rather, they frame the technical capacities and effects of technology as social constructions (Grint and Woolgar, 1992). Attention is therefore given to deconstructing the claims made about different technologies and their so-called technical features.

Although these accounts differ in many regards, they are united in the insufficient attention they give to silences. Those writing from a realist approach tend to see silences as gaps to be filled, in the process closing down space for a consideration of the silences themselves and their effects. For post-essentialist writers, silences and uncertainties around different technologies are inherent features of socio-technical life. Silences cannot be filled objectively, once and for all, or with reference to technical characteristics, and thus always have the potential to be differently constituted (Grint and Woolgar, 1992). However, such approaches have been criticised for being unable to pay sufficient attention to the creation and management of silence as part of power dynamics, broader institutional factors and the politics of knowledge production – issues brought to the fore by the 'undone science' literature (Frickel et al., 2010).

The distribution and management of ambiguities

There is thus room for a third way, one which builds on the strengths of post-essentialist approaches, whilst acknowledging the concerns of critics. Such an approach – based on considering how ambiguities, and by extension silences, are distributed (Rappert, 2001) – is rich, complex and contains many elements of interest. In this brief summary I am unable to do justice to all of these elements. Instead, I will pull out four stages that I consider important in this type of analysis – identifying ambiguities, querying resolution, documenting management techniques and charting effects – and identify key questions to be asked at each of these stages.

Identifying ambiguities

In keeping with the post-essentialist perspective, I start by acknowledging that differing accounts of technologies are available and that debates around the effect and impact of particular technologies are complex, unlikely to be settled solely with reference to their technical characteristics, and will vary from case to case. Due to a wide range of factors, then, some areas of ambiguity are inevitable and must be acknowledged as such; from ambiguities around, for instance, the ability of children to speak for themselves (Lee, 1999) to the health effects of CS spray (Rappert, 2001). Rappert usefully identifies multiple ambiguities associated with less lethal technologies, including their effects and medical implications, how they are used in practice, who they are used against, their effectiveness and deterrent value, and how, and by whom, they should be assessed (Rappert, 2001).

Thus, it could be argued, one possible first step in such an analysis might be to document the nature, location and extent of such ambiguities and tensions, and to make them a focus of analysis. Yet if this approach is to avoid the charges made against post-essentialist critics, it must also consider how such ambiguities are created, how they come to be recognised as such, and how they are bound up with issues of control over knowledge (Rappert, 2001). The ambiguities around less lethal weapons, for example, cannot be understood without looking at what knowledge has been produced and considered significant in the first place, and what information has subsequently been released into the public domain.

Querying resolution

Moreover, it is not enough simply to acknowledge and identify such ambiguities without also examining whether they require resolution – or whether they can remain open-ended. Lee (1999) notes that the ambiguities around the ability of children to speak for themselves can sometimes be left open-ended, with no final resolution required. Mol (2003) shows that ambiguities and tensions in the diagnosis of atherosclerosis can often co-exist in the same medical site. Similarly, Singleton and Michael (1993) caution against assuming that ambivalences have to be resolved or defeated and shows how ambiguities can help strengthen particular networks, in their case the cervical cancer screening programme in the UK.

In other cases, however, ambiguities may have to be resolved and difficult questions may have to be answered. Lee notes, for example, that ambiguities around the ability of children to speak for themselves – as well as key discrepancies between how different institutions in the UK saw the role of children and assessed their ability to provide credible evidence – were able to be left unaddressed and unanswered for years, without raising particular issues. Yet, following a crisis involving high levels of suspected child abuse that were diagnosed in one specific hospital but questioned by the local police force, ambiguities around the ability of affected children to provide credible evidence and personal testimony took centre-stage. Under these conditions, such questions could no longer be ignored and came to require an urgent answer (Lee, 1999). The extent to which ambiguities can be 'kept in circulation', rather than demanding resolution, however provisional or temporary this resolution may be, is thus an interesting second question for empirical analysis.

Documenting management techniques

If ambiguities do require resolution, at least temporarily, a key question then becomes where, how, and by whom, such ambiguities are resolved. Faced with hard questions, 'wicked problems' to which there may not be an easy solution (Rittel and Webber, 1973), and contested and partial information over which even experts may disagree, the question then arises as to who, if anyone, has to adjudicate over such issues. Whilst one might start with institutions such as the UN, national courts, or the Association of Chief Police Officers (ACPO, replaced by the National Police Chief's Council, the NPCC, in 2015), given the difficulties detailed above, it will come as no surprise that they often face difficulties in resolving ambiguities and providing definitive statements around their use. If such statements are too prescriptive, they risk being seen as inflexible and impractical by those charged with applying them, thus undermining the legitimacy and credibility of the institutions that provide them. If they are too broad and general, they risk being dismissed altogether – again, with potential impacts on institutional legitimacy.

In such situations, there is more than a mere possibility that the strategies and management techniques that institutions develop to handle ambiguities may end up displacing difficult issues and making them someone else's problem. Thus Lee notes that the drafters of the UN Convention on the Rights of the Child, by stressing that children should be assessed individually, on a case by case basis, *defer* ambiguities to a later point in the proceedings when such specifics will be known. Rappert has argued that, by stressing that decisions around the use of CS spray are ones for individual officers to make based on their own assessment of the situation, ACPO (and now the NPCC) *devolves* responsibility for handling the ambiguities and tensions around the spray down to grass roots level.

Yet, as we have seen above, such processes of deferral and devolution cannot always continue indefinitely. Lee (1999: p. 471) therefore proposes that identifying 'where deferral stops and who bears the burden of responsibility' for handling ambiguities should be key questions for analysis. In asking such questions this approach – by tracing the dynamics through which different actors are more or less able to deflect or defer responsibility elsewhere – both builds on post-essentialist approaches whilst acknowledging the concerns of critics, by allowing for a consideration of power and of institutional and structural factors.

Charting effects

This leaves one crucial final question. What are the consequences of the management techniques deployed in the resolution of ambiguity, for the groups of individuals involved, as well as for the legitimacy of particular technologies and the institutions charged with governing them? In asking such questions, the intention is not to imply that ambiguity is negative in itself, nor that a perfect, ideal way of handling such tensions exists. Indeed, as Lee notes, ambiguities can cause problems wherever they come to rest. Instead, the hope is that documenting the effects of these management techniques may result in a more sophisticated understanding of who – or what – they advantage and disadvantage, and this in turn may lead to a more nuanced assessment of various technologies and the issues associated with them.

In summary, then, such an approach asks a series of questions. What are the existing ambiguities around a particular technology and in the socio-technical network surrounding it? Do these ambiguities require resolution? If so, how and where are such resolutions achieved and managed: which actors become responsible for this resolution, and what mechanisms make this possible? Last, what are the consequences of ambiguities and the particular ways in which they are managed?

From ambiguities to silences

While these questions focus on the management of ambiguities, I argue in the remainder of this chapter that studying such questions can also be useful for an analysis of silences and their consequences. First, however, some points of clarification about the term 'silence' and its qualities are necessary. Whilst ambiguity often refers to information or evidence open to multiple interpretation, by contrast silences are often characterised by the absence of such information in the first place. Thus silences are often associated with absence – an absence of knowledge, an absence of communication, or both.

Yet if silences cannot be understood without absences, neither can they be reduced to them. Indeed, my discussion of silences around TASER sometimes focuses on silences that have been explicitly acknowledged by others and sometimes focuses on making my own claims about silences. In both cases, however, silence is being discussed and brought out into the open. In this sense, my use of the term 'silence' has much in common with Gross's use of the term 'ignorance' to refer to limits to knowledge that are explicitly acknowledged as 'known unknowns'. These gaps can be usefully contrasted both to 'unknown unknowns' or 'nescience' (absences in our understanding of which all are totally unaware) and to silences of which some actors may be aware but may not see fit to publicly communicate (i.e. silence as secrecy) (Gross, 2007). These latter understandings are important, but not my focus here.

This focus on communication also highlights the role of the analyst in constructing (certain) silences and obscuring others. Silences are, at least in part, manufactured by the analyst and by the demands of academic (and other forms of) writing. Indeed, as Wynne notes (1992: p. 116), uncertainty or ignorance can only be defined 'by artificially "freezing" a surrounding context which may or may not be this way in real-life situations'. Thus a responsible discussion of silences needs to cast doubt on the notion that such silences are somehow there to be discovered, uncovered and neutrally reported. Such considerations are particularly important in the TASER debate – where one commentator has felt compelled to ask whether 'there is unusual bias amongst the experts who study TASER' (Jauchem, 2015: p. 62) – but are also highly appropriate in any discussion about silences that presumes some kind of (artificial) status on the part of the researcher (Smithson, 2008).

Cognizant of such considerations, I now outline and discuss two main perspectives on TASER, before considering the merits of an alternative approach based on the management and distribution of silences.

The existing literature on TASER

Analysis of less lethal weapons tends to be polarised (Rappert, 2001) and debates around the safety of TASER are no exception. Reactions to the weapon are so strong that one commentator has described them as a matter of 'love' or 'hate' (Ho, 2009: p. 771), with the weapon characterised as posing a risk to the public that either is 'unacceptably high' or 'low enough to accept' (Hall, 2009). Such opposing tendencies could be termed 'Truth about TASERs' (the title of a 2009 journal article by Ho, the medical director of TASER International) and 'Truth not TASERs' (the name of a Canadian blog on the weapon).

Truth about TASERS

For some – often, but not exclusively, those with links to manufacturers and law enforcement agencies – TASER is well researched, well regulated and largely safe. Ho notes that TASER 'is, perhaps, the most intensely studied and vetted law enforcement tool in existence, and the majority of the human work in this area is supportive of its safety profile' (Ho, 2009: p. 771). Kroll argues that one of the most contentious issues around TASER – that of the device contributing to potentially lethal cardiac arrest – has been 'sufficiently studied and … the epidemiological data convincingly show how rare (if even existent) such side-effects are' (Kroll et al., 2014: p. 168), whilst practitioners note that TASER is 'a unique tool that experience has proven is effective and overwhelmingly safe' (the executive director of the Force Science Research Centre, in Jenkinson et al., 2006: p. 239).

From this perspective, analysts such as myself who are interested in silence and limits to knowledge should be asking why the 'truth' is silenced, and so many 'misunderstandings' persist (Jauchem, 2015). The literature advances a range of explanations for why this might be the case; from medical experts giving 'bias(ed) interpretation of the facts' (as opposed to 'complete, balanced, and impartial accounts'), to the current 'tort litigation climate' (Heegaard et al., 2013: p. 260; Kroll et al., 2014: p. 168), to the mysterious workings of electricity (NPCC, 2015). Explanations are also sought in concerns that TASERs 'will end up in the hands of illegal arms dealers and child soldiers in the Third World and will be deployed as instruments of torture' (Jenkinson et al., 2006: p. 240). From such a perspective, a key task for analysis is not to point out areas where communication is missing, but to draw attention to 'misunderstandings' in the literature that does exist, including the 'exaggerated' association between TASER and deaths in custody (Jauchem, 2015: p. 53).

Whatever the merits of such accounts – and, indeed, many of the points raised are worthy of further investigation – they have certain effects, especially if considered in isolation. First, they can serve to shift the terms of the debate, deflecting attention away from silences, from what remains unknown and unsaid, to focus on who is saying what – and why. Second, they can shift the emphasis away from those who manufacture, supply and use the weapons to justify their use, and towards those who would seek to make claims about, or question, less lethal technologies. Third, and relatedly, viewing controversies and 'misperceptions' as stemming from external factors – the broader societal 'climate', deeply held innate beliefs – may detract attention from the important roles that manufacturers and policing agencies can play both in generating, and in tackling, the silences and misperceptions around TASER.

Truth not TASER

For other accounts, far from TASER's lethality being 'exaggerated', the weapon remains 'potentially lethal' (Amnesty International UK, 2015). As such a key question is not why the weapon remains controversial and misunderstood, but why 'relatively few deaths … have been found to be caused by TASERs' (Truth Not TASERs, 2008). Answers are variously found in 'the manufacturer's aggressive approach against coroners and medical examiners' (Truth Not TASERs, 2008; see also Stanbrook, 2008)[3] and in its sponsoring of research into the weapon. O'Brien and Thom (2014: p. 422) state that 'evidence about the safety of TASERs is dominated by studies whose authors have financial interests in the commercial production, sale, and promotion of TASERs by police. … and this has the potential to influence the volume and content of publications.' Indeed a paper written by Azadani et al. (2011), and criticised by Vilke et al. (2012) and Kunz (2012), found that those studies funded by TASER were nearly eighteen times more likely to find the device was safe than studies without such an affiliation. As such, some of these accounts are less concerned with the silences around the weapon, but with the various silencing processes – legal action, aggression, financial interests – that work to downplay the risks associated with TASER use.

Some may also argue that attempts by the manufacturer to engage with those interested in their product might be an example of a silencing trend. For example, my own trip to TASER International, funded by the company, could be seen in different ways: as an attempt to convince independently funded academic researchers of the merits of TASER and to correct 'misunderstandings'; as an attempt to give them access to (certain) 'facts'; or as an attempt by the company to 'squelch any messages that could hurt its bottom line' (Stanbrook, 2008: p. 1402). Some have seen contact by TASER UK employees as an attempt to intimidate, to 'make clear that (the company) monitors what people say and write about TASER and that they are keen to take action against those who put a foot wrong' (Taylor, 2012).

Such engagement could also fill another role: that of offering researchers exposure to what could be characterised as the 'epistemic community' (Haas, 1992: p. 3) around TASER. This community is drawn from a wide-range of stakeholders – from company officials to policy makers, law enforcement agencies to medical experts – who share 'normative and principled beliefs' (for example, the need for law enforcement officials to have a range of weapons at their disposal), 'causal beliefs' (for example, the factors that contribute to death following TASER), 'notions of validity' (based around quantitative evidence) and a 'common policy enterprise' (for example, the need to counteract misunderstandings around the lethality of the weapon). From such a perspective, divergent interpretations merely underscore the need for the responsible researcher to pay more attention to the silencing processes at work and how the truth – this time around the negative effects of the weapon – may be silenced.

Whilst these accounts differ greatly, they also have something in common. As an example of Hilgartner's 'sociology of error', they attempt to 'document the corruption and distortion of truth' (Hilgartner, 2014: p. 84), with the truth presumed to be, respectively, that TASERs are highly safe and 'misunderstood', or that their safety record has been overstated. The role for the analyst is either to correct misunderstandings around the weapon, or to 'de-spin the pro-TASER propaganda' (Excited Delirium, undated).

Yet much of this debate is infected with the 'faulty assumption' that we are able to 'fairly evaluate' TASERs 'without a more thorough understanding of what ... [they] actually do, or of what remains unknown about what they do' (Sussman, 2012: p. 1345). Whilst judgments about the weapon are often based on 'the known benefits of TASER use balanced against the known negatives of TASER use', Sussman argues that what we 'do not know' should be given greater weight in future analysis of the weapon (p. 1345). It is against this backdrop that I argue that an explicit consideration of silences and their consequences can usefully add to the debate.

The distribution and management of silences around TASER

In this final section I apply the four questions detailed earlier to the distribution and management of silences around the comparative safety of TASER. I argue that, whilst responsibility for addressing the silences around the safety of TASER is located at a number of points, processes of deferral and devolution mean that rank-and-file officers bear the primary responsibility for addressing these gaps. Officers are often required to make split-second decisions about whether, and what, force to use and, in this instance, are unable to defer questions and silences about the safety of their weaponry any further. Yet such silences are only temporarily filled and laid to rest. Once the incident in question is over officers are often – but not always – able once again to shift silences and questions about the weapon and its safety elsewhere.

Documenting silences

To begin, it is important to clarify exactly what silences we are talking about. In the UK context, silences exist around the comparative safety of the weapon in several ways. First, with the partial exception of Jenkinson et al. (2006), a paper focusing on an old model of TASER no longer in use in the UK, there have been no peer reviewed studies of the comparative injury rates of TASER and other weapons in England and Wales. Inferences are made on the basis of American data, but the two jurisdictions differ in terms of the nature of police–civilian relations, the training and guidance provided on TASER, and the other force options available for use. American analyses routinely compare TASER to the use of force options not in use in England and Wales, and to some whose use has been explicitly ruled out due to safety concerns. Moreover, the American literature tends not to explicitly compare TASER to the alternative force options available to officers (Terrill and Paoline, 2012: p. 178).

Second, silences in the literature are not surprising given the absence of data on which to base quantitative analysis. Officers in England and Wales are required to fill out a seven-page form every time TASER is used and to give details about injuries sustained. These forms are then collated centrally. However, the medical committee DOMILL (now replaced by SACMILL) has registered its concern that:

> medical audit information is mostly unavailable from incidents in which individuals exposed to TASER discharge have been transferred directly to hospital. Paradoxically, this means that DOMILL is unable to review outcomes in cases that are likely to be at the more serious end of the injury spectrum.
>
> (DOMILL, 2012: p. 14)

Moreover, officers are not currently mandated to fill out similar forms when they use baton, incapacitent spray or empty hand techniques (Payne-James et al., 2014). The result is that data on some injuries inflicted by the police is not available for systematic analysis and it is extremely difficult for academics and practitioners alike to compare the injuries resulting from TASER with the injuries resulting from other force techniques. Such data limitations have, however, been recognised by the Home Secretary, who has launched an 'in-depth review of the publication of Taser data and other use of force by police officers' – with which I am involved – 'to present options for publishing data on how police officers are deploying these sensitive powers, who they are being used on and what the outcome was' (Home Office, 2014a), so this may yet be subject to change.

Third, current data sources systematically privilege the accounts of certain actors and silence others. By definition, under a system of officer recording it is the officer's accounts of injury that hold sway over the alternative accounts that might have been provided by those who were subject to TASER. Moreover, the debate tends to centre around visible injury. Under this definition no attention is given to the long-term psychological effects following TASER exposure (an area that DOMILL specifically noted needed further investigation) or to the pain of being 'Tased' (an area that Sussman argues needs further attention), much less to the pain produced by other force options (DOMILL, 2012; Sussman, 2012).

Fourth, as the most recent DOMILL (2012: p. 12) report notes, whilst the 'overall risk of serious injury associated with UK use of ... [the weapon] is low', silences still remain around TASER safety, with important issues – such as the 'risk of cardiac capture' associated with certain shots to the chest and concerns over prolonged exposure and prolonged discharge of the weapon – remaining 'unknown' or 'unexplored'. Indeed Sheridan, a doctor with the Defence Science and Technological Laboratory tasked with providing advice on the physiological effects of TASER to the UK Home Office and to SACMILL, has called for more studies on the risk of cardiac capture posed by the weapon, in particular newer TASER models. He notes a 'major knowledge gap in our understanding of the cardiac effects of the TASER X2 and X26P' and calls for more research into the association (or lack thereof) between the position of TASER barbs and 'the induction of arrhythmias' (Sheridan, 2014: p. 167).

In contrast to accounts that start by identifying 'truths' around TASER, this account has started by identifying that which is absent and silent. These are issues which are inseparable from questions around the politics of knowledge production, the power that institutions and actants may have to create silences and the limits to such power.

Querying resolution

The fact that silences can be identified, however, doesn't mean that they necessarily have to be filled. Rather than just presuming that an answer is required, it is important to ask whether, and to what extent, silences have to be negated. In the case of TASER, an analysis of official Home Office TASER statistics for England and Wales would suggest silences around the relative safety of TASER compared to other force options do have to be filled, however temporarily. According to the Home Office statistics for July to December 2013, on average TASER was fired in probe firing mode four times a day in England and Wales[4] and drawn, sparked or used in red-dot mode 24 times a day throughout this period (Home Office, 2014b). In all these instances, decisions have been made about the relative safety of the device. Moreover, in countless dangerous situations, TASER-trained officers, far from being paralysed, will deliberately decide not to use TASER at all and will use other options instead. Yet refraining from using a weapon still requires judgments to be made about its relative safety and whether such a use of force constitutes a proportionate response to the circumstances faced.

Moreover, if the weapon is fired then decisions need to be taken concerning where to aim the weapon, from what distance and for how long the discharge should continue. In such situations, then, silences around the safety of the weapon, and of particular methods of use, are having to be addressed and resolved, however provisionally. As such, a key question becomes how these silences are managed and where – and by whom – they are negated.

Analysing management techniques

Far from being completely silent on the issue of TASER, many bodies in England and Wales do discuss its suitability and safety, in spite of the gaps mentioned above. Testing of less lethal weapons, including TASER, is conducted by the Home Office prior to use and the medical implications reviewed, formerly by DOMILL and now by SACMILL, with these reports available to officers and the public (e.g. DOMILL, 2012). Weapons must have ministerial approval before being authorised for use, although chief constables are able to use unauthorised weapons should they see fit (see Dymond and Rappert, 2014) and some forces require officers to seek authorisation from senior officials before deploying TASER.

Officers receive training and guidance on the use of a range of weapons and empty hand techniques and have at their disposal a decision-making tool – the National Decision Model – to assist them in their deliberations.[5] Moreover, the College of Policing, ACPO and, latterly, the NPCC, have produced guidance on the use of TASER and additional documents in response to questions frequently asked by the public and the media. One such entry on the NPCC's blog notes that 'no use of force is risk free' but that 'physical restraint, batons, police dog … can have a much more long-term impact on someone compared with a TASER' (NPCC, 2015).

At first glance, then, it could be argued that, far from being a silence, there is an abundance of communication on the topic. Yet, if we look at some of these documents in more detail, we can see that the communication that does exist contains notable silences about the weapon's comparative safety. To illustrate this I will look briefly at two key sets of texts: DOMILL's statements on the use of less lethal weapons (DOMILL, 2012) and the College of Policing TASER Guidance, developed in partnership with ACPO and contained within a broader document on Authorised Professional Practice in Armed Policing (College of Policing, 2014a).

DOMILL statements

The most recent DOMILL statement on TASER finds that the 'risk of serious adverse medical outcome from exposure to the TASER is low, provided that the system is employed by trained users in accordance with ACPO policy and guidance' (2012: point 73) and usefully provides advice on the risks associated with TASER deployment on certain vulnerable populations (such as individuals with heart disease, asthma, epilepsy, or who have taken certain drugs). However the document is unable to provide a clear guide as to the relative safety of different force options. Indeed it states that 'there are many ... (conditions) that may render those affected more likely to experience an adverse response as a consequence of use of the TASER (*or other form of force*)' (2012: point 70; emphasis added). In so doing, it, quite literally, brackets issues around the comparative safety of less lethal weapons and how the safety record of TASER fares with respect to other uses of force – issues that are often of key concern to officers.

Such statements are not, perhaps, surprising, as the Home Office Code of Practice on Police use of Firearms and Less Lethal Weapons explicitly excludes 'weapons routinely issued to patrol officers for self-defence purposes' from medical assessment (Home Office, 2003: point 1.3.2). As a result the Committee has not conducted similar assessments of batons, empty hand techniques or chemical irritant sprays[6] – an omission recognised by the Triennial Review of SACMILL which recommended that SACMILL should be able to review 'all current systems' in use (Ministry of Defence, 2014: p. 4). Thus the statements of the Committee, whilst useful in many ways, are unable to provide clear guidance to officers as to whether the use of TASER is preferable to, or more risky than, other forms of force.

College of Policing TASER guidance

The College of Policing's guidance on TASER also refrains from making definitive statements as to the comparative safety of the weapon. Whilst the NPCC's blog, quoted above, notes that other weapons 'can have a much more long-term impact on someone compared with a TASER', similar statements are not found in official guidance that is issued to officers. Instead the guidance is silent on the relative safety of the weapon or when it should be used, noting that 'it is not practicable or possible to provide a definitive list of circumstances where TASER would be appropriate'. Once again, such silences are not surprising, given the lack of UK specific statistics on the safety of TASER when compared to other weapons, which complicates efforts to make such assessments. Moreover whilst the document lists 'risk factors' for the weapon's use, these are only factors that 'may influence the operational use of TASER', not necessarily situations in which use should be prohibited or use of other forms of force preferred. Instead, the document simply notes that TASER is 'one of a number of tactical options available when dealing with an incident with the potential for conflict' (College of Policing, 2014a).

The DOMILL statement quoted above noted that 'the duration of application of Taser discharge should be limited to that necessary to achieve the desired operational effect' and explained that 'although DOMILL does not provide operational advice on Taser point-of-aim, the Committee notes that any risk [of cardiac capture] that does exist would be mitigated by avoiding, where tactically feasible, the firing of barbs into the frontal chest overlying the heart' (DOMILL, 2012: p. 12). However, the guidance given to officers does not elaborate on such statements. It does not provide any specific direction as to the recommended maximum levels of exposure, save to note that: 'the five-second cycle can be repeated if the incapacitation results does not appear to take effect. Officers should review other options as there may be technical or physiological reasons why the device is not working as expected'; and: 'Any medical risk may be increased the longer or more often the device is discharged' (College of Policing, 2014a).

Nor does the guidance provide detailed instructions on the distance between the officer and subject or the point of aim, save to note that undefined 'sensitive areas' should be avoided and that 'the maximum range of the device is ... 21 feet or 6.4 metres' and the 'effective range ... may be a lesser distance' (College of Policing, 2014a). This contrasts markedly to the guidance given on the less lethal Attenuating Energy Projectile, which details a specific point of aim for officers (in this case, the belt-buckle area) and sets minimum distances, prohibiting the use of the weapon at distances of 'less than one metre' unless there is a 'serious and immediate threat to life' (College of Policing, 2014b).

As such, whilst the publicly available guidelines on TASER have much to offer, they could hardly be said to offer detailed guidance as to when the use of TASER is proportionate and necessary, how the risk of weapon compares to other options available to officers, or how it should be deployed if and when its use is deemed appropriate. Indeed, the UN Committee Against Torture, in its 2013 visit to the UK, called for more specificity on TASER guidance, noting that: 'the State party should revise the regulations governing the use of such weapons, with a view to establishing a high threshold for their use, and expressly prohibiting their use on children and pregnant women. [...] The Committee urges the State party to provide detailed instructions and adequate training to law enforcement personnel entitled to use electric discharge weapons and to strictly monitor and supervise their use' (UN Committee Against Torture, 2013: pp. 9–10). Thus, whilst a large amount of communication and documentation exists around the weapon, paradoxically the guidance remains silent on certain key decisions that officers will have to face.

As a result, decisions around the safety of different force options are deferred to a later point in the proceedings. Due to the difficulties in providing prescriptive advice, the silences around TASER's safety are to be filled not when the guidance is being drafted, or when medical statements are being written and approved, but at some future point in time. The responsibility for addressing the silences around TASER is therefore devolved to individual officers equipped with TASER. The guidance states that it is the officer in possession of TASER who is 'legally and organisationally accountable' for decisions around its use. Officers are 'individually responsible and accountable for their decisions and actions ... [including] decisions to refrain from using force as well as any decisive action taken, including ... the use of a less lethal weapon' (College of Policing, 2013). Officers may thus find themselves in a situation in which they urgently need to choose between a variety of force options – amongst them TASER – but have been given few definitive statements as to which options may be least injurious.

These silences are perhaps inevitable given the difficulties inherent in providing clear, credible policy statements that can be applied to a range of situations and under circumstances in which knowledge may be silent or silenced. Those faced with having to make statements in such conditions, be they inside or outside of academia, have to find some way of balancing the general and the specific. A common way of doing this, both within policing environments (Rappert, 2001) and in other contexts including hospital and court settings (Lee, 1999), is to be silent at certain points and to fall back on the 'general presumption' that specifics matter (Lee, 1999: p. 466). However, the widespread nature of this approach makes it all the more important to look at its effects and consequences, the subject of the next section.

Charting the effects of silence and its management

The silences around the comparative safety of TASER and the processes of deferral and devolution that have been used to manage them, have a range of consequences for officers, individuals subject to TASER, and for the network more broadly. Giving officers individual responsibility for deciding when to use different types of force gives them flexibility to decide how to handle situations, whilst maximising the amount of discretion available to them. Yet in giving officers the *opportunity* to make decisions, this approach also gives them the *obligation* to do so.

Officers are only in a position to be able to make decisions around the use of TASER as a result of a much broader socio-technical network that exists around the weapon and as a result of collective decisions that have been made elsewhere. Yet, at least on paper, it is the officers who are individually accountable. They are thus tasked with having to make complex determinations over the likely effects of TASER vis-à-vis other weapons – effects which we know to be highly contested, marked by noticeable gaps and silences, and subject to considerable disagreement even within the medical profession. This can put considerable pressure on officers and also risks masking the importance of previous decisions made by other actants. Such decisions include, for example, the length of shock automatically delivered by the weapon and the nature of the shock generated, which may also impact on how force is used and with what consequences.

However, the pressures and obligations to fill the silences around the safety of TASER may only be experienced by officers temporarily. When faced with situations in which force needs to be used, the responsibility for filling the (awkward) silences around TASER may indeed be said to stop at the officer's feet – decisions need to be made and they often need to be made there and then. Yet given the difficulties faced in making decisions about the appropriate use of force, there may be multiple opportunities for officers afterwards to reopen the issues that were temporarily answered and to reinstate the silences that were provisionally filled. Responsibility for responding to silences and unknowns could be further diffused and located in, for example, the inherent variability of individuals, the circumstances in which police officers are placed, or in 'the inherent difficulty associated with the classification and detection of disease entities' in individuals and especially in those whose behaviour is said to necessitate the application of police force (Anais, 2014: p. 62).

In such ways the silences around TASER and its comparative safety – silences which concentrate pressure on officers in advance of use-of-force decisions – may help to partially alleviate pressure on officers once decisions have been made. Faced with knowledge gaps, it may be difficult for individuals subject to TASER, or those representing them, to evaluate whether decisions were reasonable – or to categorically prove that they were not – as there are few firm yardsticks against which to measure such behaviour. Thus, whilst deferring decisions down to officers and giving them responsibility for filling the silences around TASER use may well be intended to enhance accountability, it may risk weakening it.

At the same time, the silences around the weapon may also make it difficult for officers to demonstrate that they have used reasonable force, given the lack of consensus around which kinds of weapons and which patterns of use are less injurious than others. The data gaps may also mean that senior police representatives are denied the information necessary to make a strong case for TASER based on evidence drawn from England and Wales.

This, in turn, has implications for public confidence. The IPCC (2014: p. 25) has identified an:

> obvious mismatch between the public perception that TASER is a high level use of force that should only be considered when faced with the most serious threats of violence, and the police's most frequent rationale for use, that TASER presents a lower risk than other equipment such as CS spray, physical restraint or a baton.

It is hard to see how such a 'mismatch' can currently be resolved given the silences around the comparative safety of the weapon, the lack of qualitative research into the topic and the lack of large scale quantitative analysis of injury rates from different use of force options in England and Wales. It is to be hoped that the review of use-of-force reporting, ongoing at the time of writing, may be of some assistance in this regard.

At the same time, however, the rich STS literature discussed in the first half of this chapter cautions against assuming that the 'facts' of the matter can be easily ascertained and used to explain why certain perspectives win out. It also provides us with a useful reminder of the work, effort and difficulty involved in resolving exactly these kinds of mismatches. My intention here is not to argue that additional evidence will somehow 'objectively' resolve such a mismatch. Instead, I wish to note that additional evidence, particularly that of the quantitative variety, is, as David Thacher puts it, 'often the only sort of knowledge that has any hope of overcoming the ingrained ideological positions that characterize the most polarized policy debates' (Thacher 2001: 388), and that the current silences and knowledge gaps around TASER are to the benefit of few of the actors involved. Moreover in showing that the silences around TASER and the way in which they are managed defy polarised readings, I hope to have moved us beyond simple characterisations of the weapon either as 'a valuable tool' or as a 'potentially lethal weapon' (Stanbrook, 2008: p. 1401).

Conclusion

In this chapter I have argued that the STS literature has tended to give insufficient attention to silence and that an approach originally developed by Lee and Rappert to advance our understanding of the management and distribution of ambiguities, may also help us in our understanding of silence. I argued that four stages of this approach – identifying ambiguities and silences, querying whether these are resolved, querying how and by whom they are resolved, and charting their effects – can be particularly helpful in this respect. I illustrated the added value of this approach by advancing one possible account of the silences around the comparative safety of TASER in England and Wales – an important issue that deserves more attention than I have been able to give it here – and I contrasted this approach with two dominant perspectives on the weapon ('Truth about TASERs' and 'Truth not TASERs'). An approach centred on the management and distribution of silences can usefully complement such accounts by drawing our attention to how silences are resolved, by whom and with what effects. As such, it can help us recognise that the effects of particular technologies can be complex and widespread and are contingent on, and inseparable from, the silences around these technologies and how they are managed. At the same time, however, an analysis of the debates around TASER also contributes towards an understanding of silence as not *just* an absence of communication, but as an absence within the communication patterns that do exist. Silence is not the absence of communication, but can be intertwined with it.

In highlighting the effects of silences, the point here is not that silences are bad *per se*, nor that they could, let alone should, be eradicated. As a product of the tension between particular and generalisable statements, some silences are essential and inevitable – particularly in debates around policing, where discretion is widely celebrated, and in debates around less lethal weapons (Rappert, 2001).

At the same time, however, as the announcement of the review of use-of-force reporting by the Home Secretary indicates, action can be taken to break certain silences. As such it is all the more crucial to document the effects and consequences of the current configurations and to explore novel possibilities for new, alternative resolutions. This will require looking in more detail at how and why certain silences are generated and come to be managed in particular ways, and the effects (intended and unintended) that filling silences may have. It is unlikely that we shall find a way of managing ambiguities and silences that is entirely unproblematic, but paying attention to such issues can help us look at the voices silenced in particular configurations and can inform a critical assessment of alternative approaches.

Notes

1 As part of the research contributing to this chapter, the author received travel costs from TASER International between 2 and 8 November 2014 to attend the Annual Conference of the Institute for the Prevention of In Custody Deaths and to visit the company's headquarters in Scottsdale, Arizona between 2 and 8 November 2014. She has also been commissioned by the NPCC to undertake analysis of the use-of-force data captured across UK police forces and to identify best practice.
2 TASER®, TASER X26, TASER X2 and TASER X26P are registered trademarks of TASER International, Inc., registered in the US. All rights reserved.
3 Members of the TASER International Scientific and Medical Advisory Board have noted, in response to Stanbrook (2008), that such accusations relate to an 'extreme situation in which the medical examiner consistently (in 3 cases) ruled that the use of an electronic control device constituted homicide ... [but resulted in] striking of the electronic control device as the cause of death' (Kroll et al., 2008: p. 343).
4 According to the July 2013–December 2013 Home Office statistics TASER was fired across England and Wales 867 times and used in one of the non-firing modes 4,400 times in this 184-day period. However, these incidents may not be evenly spread out and, indeed, many may occur at particular times (e.g. on Saturday nights). Thus the figures given are averages only.
5 For further details of such systems see: College of Policing (2014a, b); Dymond (2014); Dymond and Rappert (2014).
6 Chemical irritant sprays were reviewed by an alternative committee. For further details of their testing and introduction, see Rappert (2003).

References

Amnesty International UK, 2015. TASERs 'not a panacea' – concerns at Police Federation vote. Press release, 10 Feb. Available at: www.amnesty.org.uk/press-releases/TASERs-not-panacea-concerns-police-federation-vote. Accessed 11 April 2015.

Anais, S., 2014. Making up excited delirium. *Canadian Journal of Sociology*, 39(1), pp. 45–64.

Azadani, P., Tseng, Z., Ermakov, S., Marcus, G. and Lee, B., 2011. Funding source and author affiliation in TASER research are strongly associated with a conclusion of device safety. *American Heart Journal*, 162(3), pp. 533–537.

College of Policing, 2013. Use of force, firearms and less lethal weapons. Available at: https://www.app.college.police.uk/app-content/armed-policing/use-of-force-firearms-and-less-lethal-weapons. Accessed 10 July 2015.

College of Policing, 2014a. Conducted energy devices (Taser). Available at: https://www.app.college.police.uk/app-content/armed-policing/conducted-energy-devices-taser/. Accessed 22 June 2015.

College of Policing, 2014b. Attenuating energy projectiles. Available at: https://www.app.college.police.uk/app-content/armed-policing/attenuating-energy-projectiles. Accessed 22 June 2015.

DOMILL, 2012. Statement on the medical implications of use of the TASER X26 and M26 less lethal systems on children and vulnerable adults. Available at: http://data.parliament.uk/DepositedPapers/Files/DEP2012-0729/96605%20Library%20Deposit.pdf. Accessed 10 February 2014.

Dymond, A., 2014. The flaw in the TASER debate is the TASER debate: what do we know about TASER in the UK, and how significant are the gaps in our knowledge? *Policing: A Journal of Policy and Practice*, 8(2), pp. 165–173.

Dymond, A. and Rappert, B., 2014. Policing science: the lessons of TASER, *Policing: A Journal of Policy and Practice*, 8(4), pp. 330–338.

Excited Delirium, undated. Excited Delirium.com: mission statement. Available at: http://excited-delirium.blogspot.co.uk. Accessed 11 April 2015.

Frickel, S., Gibbon, S., Howard, J., Kempner, J., Ottinger, G. and Hess, D., 2010. Undone science: charting social movement and civil society challenges to research agenda setting. *Science Technology, and Human Values*, 35(4), pp. 444–473.

Grint, K. and Woolgar, S., 1992. Computers, guns, and roses: what's social about being shot? *Science, Technology, and Human Values*, 17(3), pp. 366–380.

Gross, M., 2007. The unknown in process: dynamic connections of ignorance non-knowledge and related concepts. *Current Sociology*, 55, pp. 742–759.

Haas, P., 1992. Introduction: epistemic communities and international policy coordination. *International Organization*, 46(1), pp. 1–35.

Hall, C., 2009. Public risk from TASERs: unacceptably high or low enough to accept? *Canadian Journal of Emergency Medicine*, 11(1), pp. 84–86.

Heegaard, W.G., Halperin, H. and Luceri, R., 2013. Letter by Heegaard et al regarding article, 'Sudden Cardiac Arrest and Death Following Application of Shocks From a TASER Electronic Control Device', *Circulation*, 127, p. 260.

Hilgartner, S., 2014. Studying absences of knowledge: difficult subfield or basic sensibility? *Social Epistemology Review and Reply Collective*, 3(12), pp. 84–88.

Hoffman, K., 2013. Unheeded science: taking precaution out of toxic water pollutants policy. *Science, Technology, and Human Values*, 38(6), pp. 829–850.

Ho, J., 2009. Can there be truth about TASERs? *Academic Emergency Medicine*, 16(8), pp. 771–773.

Home Office, 2014a. Home Secretary at the Policing and Mental Health Summit: Theresa May: The police must treat people with mental health problems with respect and compassion. Speech delivered 23 Oct., Central Hall, Westminster. Available at: https://

www.gov.uk/government/speeches/home-secretary-at-the-policing-and-mental-health-summit. Accessed 10 July 2015.

Home Office, 2014b. Police use of TASER statistics, England and Wales, 2012 to 2013: data tables. Available at: https://www.gov.uk/government/statistics/police-use-of-TASER-statistics-england-and-wales-2012-to-2013-data-tables. Accessed 17 August 2015.

Home Office, 2003. *Home Office Code of Practice on Police use of Firearms and Less Lethal Weapons.* London: Home Office.

IPCC, 2014. *IPCC Review of TASER Complaints and Incidents 2004–2013.* London: IPCC.

Jauchem, J.R., 2015. TASER® conducted electrical weapons: misconceptions in the scientific/medical and other literature. *Forensic Science, Medicine, and Pathology,* 11(1), pp. 53–64.

Jenkinson, E., Neeson, C. and Bleetman, A., 2006. The relative risk of police use-of-force options: evaluating the potential for deployment of electronic weaponry. *Journal of Clinical Forensic Medicine,* 13, pp. 229–241.

Kroll, M., Calkins, H., Luceri, R., Graham, M. and Heegaard, W., 2008. Electronic control devices. *Canadian Medical Association Journal,* 179(4), pp. 342–343.

Kroll, M., Lakkireddy, D., Stone, J. and Luceri, R., 2014. Response to letter regarding article, 'TASER Electronic Control Devices and Cardiac Arrests: Coincidental or Causal?' *Circulation,* 130, p. 168.

Kunz, S., 2012. Biases in TASER research. *American Heart Journal,* 163(3), pp. 7–8.

Lee, N., 1999. The challenge of childhood: distributions of childhood's ambiguity in adult institutions. *Childhood,* 6(4), pp. 455–474.

Michaels, D., 2008. Manufactured uncertainty: contested science and the protection of the public's health and environment. In: Proctor, R.N. and Schiebinger, L. (eds), *Agnotology: The Making and Unmaking of Ignorance.* Stanford, CA: Stanford University Press, pp. 90–107.

Ministry of Defence, 2014. *Triennial Review Report: Scientific Advisory Committee on the Medical Implications of Less-Lethal weapons.* Available at: https://www.gov.uk/government/uploads/system/uploads/attachment_data/file/311802/20140505-sacmill-triennel-review-report.pdf. Accessed 11 April 2015.

Mol, A., 2003. *The Body Multiple: Ontology in Medical Practice.* Durham, NC: Duke University Press.

NPCC, 2015. NPCC questions and answers on TASER. Available at: www.npcc.police.uk/ThePoliceChiefsBlog/NPCCQuestionsandAnswersonTaser.aspx. Accessed 10 July 2015.

O'Brien, A.J. and Thom, K., 2014. Police use of TASER devices in mental health emergencies: a review. *International Journal of Law and Psychiatry,* 37(4), pp. 420–426.

Payne-James, J., Rivers, E., Green, P. and Johnston, A., 2014. Trends in less-lethal use of force techniques by police services within England and Wales: 2007–2011. *Forensic Science, Medicine, and Pathology,* 10(1), pp. 50–55.

Pickering, A., 2005. Asian eels and global warming: a post-humanist perspective on society and the environment. *Ethics and the Environment,* 10(2), pp. 29–43.

Proctor, R., 2008. Agnotology: a missing term to describe the cultrual production of ignorance (and its study). In: Proctor, R.N. and Schiebinger, L. (eds), *Agnotology: The Making and Unmaking of Ignorance.* Stanford, CA: Stanford University Press, pp. 1–36.

Rappert, B., 2001. The distribution and resolution of the ambiguities of technology or why bobby can't spray. *Social Studies of Science,* 31(4), pp. 557–591.

Rappert, B., 2003. Health and safety in policing: lessons from the regulation of CS sprays in the UK. *Social Science and Medicine*, 5(6), pp. 1269–1278.

Rittel, H. and Webber, M., 1973. Dilemmas in a general theory of planning. *Policy Science*, 155, pp. 160–169.

Schiebinger, L., 2008. West Indian abortifacients and the making of ignorance. In: Proctor, R.N. and Schiebinger, L. (eds), *Agnotology: The Making and Unmaking of Ignorance*. Stanford, CA: Stanford University Press, pp. 149–162.

Sheridan, B., 2014. Letter by Sheridan regarding articles, 'TASER electronic control devices can cause cardiac arrest in humans' and 'TASER electronic control devices and cardiac arrests: coincidental or causal?' *Circulation*, 130, p. 167.

Singleton, V. and Michael, M., 1993. Actor-networks and ambivalence: general practitioners in the UK's cervical screening programme. *Social Studies of Science*, 23, pp. 227–264.

Smith, R., 2005. Medical journals are an extension of the marketing arm of pharmaceutical companies. *PLoS Medicine*, 2(5), p. 138.

Smithson, M., 2008. Social theories of ignorance. In: Proctor, R.N. and Schiebinger, L. (eds), *Agnotology: The Making and Unmaking of Ignorance*. Stanford, CA: Stanford University Press, pp. 209–229.

Stanbrook, M., 2008. TASERs in medicine: an irreverent call for proposals. *Canadian Medical Association Journal*, 78(11), pp. 1401–1402.

Sussman, A., 2012. Shocking the conscience: what police TASERs and weapon technology reveal about excessive force law. *UCLA Law Review*, 59, pp. 1342–1415.

TASER International, 2012. Press kits and info: international statistics. Available at: http://uk.TASER.com/press-kit. Accessed 11 April 2015.

Taylor, R., 2012. Meeting a TASER sales rep. Blog post, 6 Nov. Available at: www.rtaylor.co.uk/meeting-a-TASER-sales-rep.html. Accessed 11 April 2015.

Terrill, W. and Paoline, E., 2012. Conducted energy devices (CEDs) and citizen injuries: the shocking empirical reality. *Justice Quarterly*, 29(2), pp. 153–182.

Thacher, D., 2001. Policing is not a treatment: alternatives to the medical model of police research. *Journal of Research in Crime and Delinquency*, 38(4), pp. 387–415.

Truth Not TASERs, 2008. Submission to the House of Commons Committee on Public Safety and National Security. Available at: http://truthnotTASERs.blogspot.co.uk/2008/05/my-submission-to-house-of-commons.htm. Accessed 11 April 2015.

United Nations Committee Against Torture, 2013. *Concluding Observations on the Fifth Periodic Report of the United Kingdom, Adopted by the Committee at Its Fiftieth Session; 6–31 May*. Geneva: United Nations.

Vilke, G., Sloane, C. and Chan, T., 2012. Funding source and author affiliation in TASER research are strongly associated with a conclusion of device safety. *American Heart Journal*, 163(3), p. 5.

Woodhouse, E., Hess, D., Breyman, S. and Martin, B., 2002. Science studies and activism: possibilities and problems for reconstructivist agendas. *Social Studies of Science*, 32(2), pp. 297–319.

Wynne, B., 1992. Uncertainty and environmental learning: reconcieving science and policy in the preventative paradigm. *Global Environmental Change*, 2(2), pp. 111–127.

13 'An outcry of silences'
Charles Hoy Fort and the uncanny voices of science

Charlotte Sleigh

Few people have heard of Charles Hoy Fort (1874–1932).[1] Sometimes mention of the Fortean Society, however, will ring a bell. 'Oh, isn't he something to do with UFOs?' The answer, from a historical point of view, is 'not really'. The International Fortean Society as it now exists was reinvented in 1961, the UK-based *Fortean Times* in 1973. As such, they were creations of the Cold War, and their UFOs – which were indeed a prominent part of their focus – part of the well-established political and cultural paranoia of that time (Seed, 1999). A pall of weirdness hangs over the Forteans – alien abductions, conspiracy theories – which has served to silence any serious historical investigation of Fort. Ominously, his Wikipedia page (8 May 2015) is flagged as having 'multiple issues'.

Charles Fort was, in his own words, no more a Fortean than he was an elk (Knight, 1971: p. 81). Fort died long before the Second World War and although there are other worlds in his work, there are no UFOs in anything like the Cold War sense (Kripal, 2010: pp. 93–141). The society that was established in his name was largely the work of Tiffany Thayer; it was spawned at a gathering in Fort's apartment in the year before his death. Thayer began publishing the Society's magazine in 1937; it quickly became a vehicle for his own preoccupations, ventriloquised as Fort's. Yet, notwithstanding his subsequent historical silencing, Fort's voice was his most remarkable feature. He possessed considerable and unique talents as a writer, and one aim of this chapter is to bring this voice to historical and literary-critical attention.

Fort's voice, I argue, was above all raised in criticism of contemporary science and its silencing tendency in relation to outsider voices. I focus upon Fort's first book in order to make his argument clear; he attempted to raise a chorus of anomalous data that could not be silenced as individual efforts to critique science usually were. It is my claim that Fort's critical and creative stance on science is a more fruitful way of understanding his oeuvre than the more usual focus on his weird cosmos and strange phenomena. At the chapter's close I reflect, via one of Fort's stories, on why his strategy of raising a chorus of dissent from scientific data was ultimately unsuccessful. In short, it was a thermodynamic failure; just as heat returns to cold, so sound becomes chaotic noise, and finally recedes into silence. Meanwhile, the chapter is shot through with a second-level meditation upon silencing: a consideration of the reasons why Fort is such a problematic figure, having been silenced by historical and literary scholars to date. These reasons are threefold: that he was derivative, an insignificant loner, or just plain mad. None of these, I argue, quite suffices as reason to dismiss Fort.

'So Charles Fort has written a – whatever it is': introducing Fort's life and work

After an unhappy childhood in Albany, New York, during which he nursed ambitions to become a naturalist (Fort, n.d.a: pp. 47–51), Fort began his adult life as a journalist. He was also a published writer of short stories (in newspapers and magazines) and drafted an unpublished autobiography, *Many Parts* (Fort, n.d.a), somewhere in the period 1899–1904. Fort wrote an unknown number of novels of which one remains, published in 1909. In the mid-1910s he received a series of bequests and, liberated from the need to earn (though far from wealthy), he underwent a dramatic change of focus. He began collecting reports of unusual phenomena culled from scientific journals and newspapers, which he wrote up in two manuscripts that have since been lost. Fort named them *X* and *Y*. With the encouragement of the editor and novelist Theodore Dreiser, he reworked the material and added more, to create four books that were published: *The Book of the Damned* (1919), *New Lands* (1923), *Lo!* (1931) and *Wild Talents* (1932).

Fort divided his final fifteen years between New York and London. His wife, Anna Fort (née Filan or Filing; see Bennett, 2009: p. 37), gave detail of three sojourns in the British capital.[2] The first was 'just before Prohibition came in' [1920] and lasted two years. According to Annie, the couple came back to New York for a year before a second, four-year spell in London. They spent a final two years in London after another year in New York. Fort had returned to New York by the time of his death. Whilst in London, Fort made extensive use of the reading room of the British Library, where he found his scientific journals and magazines. The New York Public Library provided him with the equivalent in the USA.

'So Charles Fort has written a – whatever it is'.[3] Fort's report of his completion of *X* to Dreiser went for his published oeuvre too; it is impossible to say exactly what his four final books *are*. They are not fiction, nor are they quite fact. ('A library-myth that irritates me most is the classification of books under "fiction" and "non-fiction"', Fort, 1932: p. 43.) They are not science, nor are they mysticism. They are not personal, nor are they impersonal. They are not even quite prose or poetry. Fort was outraged to discover that the New York Public Library classed *The Book of the Damned* as a 'speculative' work (Steinmeyer, 2008: pp. 190–191); it definitely was not that either. Quentin Skinner's admonition (Skinner, 1969) to beware imposing a myth of consistency upon a written oeuvre applies ten-fold to such a provisional, sly and jokey writer as Fort. Nonetheless, it will be useful to review the content of his tetralogy in the order of publication.

The Book of the Damned consisted of two main interwoven strands. Accounts of strange phenomena are insistently rehearsed as 'data', interspersed with pot-shots at science. The data in question are predominantly sky-falls of one kind or another; as the text progresses they go from inorganic to gelatinous matter and finally to frogs and so forth. There is also discussion of things further out in the sky: of planets and objects in space. Chapter 12 is, for Forteans, the crucial one, in which Fort posits the possibility of previous alien visits to Earth and entertains the famous thought: 'I think we're property' (Fort, 1919: p. 156). It is not, however, at all the central element of this text taken on its own terms. Instead, Fort's strange phenomena serve the purpose of attacking what he sees as the silencing tendency of modern science vis-à-vis problematic results, anomalous data, the achievements of amateurs.

This silencing, according to Fort, was caused by something that he called 'the dominant' in science. The dominant – which it is almost impossible to resist glossing as 'paradigm' (Kuhn, 2012) – is a historically-specific mode of science. It professes to 'explain' natural phenomena but in fact simply re-categorises or 'correlates' them according to accepted knowledge.

> [B]y reasonableness and preposterousness are meant agreement and disagreement with a standard ... Analyze [scientific judgements] and we find that they meant relatively to a standard, such as Newtonism, Daltonism, Darwinism, or Lyellism. But they have written and spoken and thought as if they could mean real reasonableness and real unreasonableness.
>
> (Fort, 1919: p. 246)

Science's ability to define its own reasonableness without external reference is captured in one of Fort's pithy epigrams: 'Science is very much like the Civil War, in the U.S.A. No matter which side won, it would have been an American victory' (Fort, 1931: p. 129). In other words, science was always self-vindicating and the data were always made to speak in its favour. Science 'saved' only the data that were compatible with the dominant. Because the latter data fitted, they could be allowed to speak: written up in scientific journals and promulgated in the popular press. However, the expression of these data was performed by scientists in bad faith, as though it were a proof of the dominant. Logically, this was fallacious: a case of *post hoc ergo propter hoc*. 'We give up trying really to explain, and content ourselves with expressing' (Fort, 1919: p. 294).

In 'saving' the phenomena, says Fort, science may actually make them into something different: re-voice them. One of his instances concerns repeated reports of mysterious tracks. In a medieval 'dominant' these are made by the devil; in the modern one, by animals. The devil has no claws, but animals do, and thus claw prints within the tracks are either present or not present, depending on the dominant. The dominant, in some important sense, actually changes the reality of the phenomena. 'But I shall give reference to two representations of them that can be seen in the New York Public Library',

Fort promises. 'In neither representation is there the faintest suggestion of a claw-mark' (Fort, 1919: p. 294).[4]

New Lands treats the two themes of the *Damned* – data and critique – separately. The first part of the book is a sustained critique of science, astronomy in particular, and its epistemology. This clears the way for the wild possibilities of part 2: that items in space are nearer than we think. Specifically, it may be that another planet or shadow-earth tracks us, and that from it strange items periodically fall in showers. There are also hints that the Earth may be flat or stationary. It would be unfair, I suggest, to take the ordering of the book to mean that the alternative cosmology is the main theme and the epistemological critique only a preparation. For one thing, they are interwoven in the *Damned* and for another, Fort's complex fascination with 'what-if?' modes of reasoning means that either theme could be taken as imaginative foundation of the other. Entertaining thoughts of other planets may be a way of finding faults in science, just as much as the reverse may be possible.

Lo! (1931), written with evidence from the British Museum, more or less continues the themes of the first two books, with a fresh injection of zoological phenomena and arguably even greater panache. Astronomers come in for stronger attack; this is perhaps the book with the greatest number of named scientists in it and the most specific critique.

In his final book, *Wild Talents* (1932), Fort mostly abandoned cosmological themes. Instead the book is filled with tales of human crimes, accidents, deaths and fires – more 'normal' things than the phenomena of the earlier volumes. If one takes the phenomena as genuine, they would appear much easier to explain naturalistically. However, Fort entertains the thought that there are people with 'wild talents', akin to psi phenomena, who produce these effects. There is a great deal of sneering at 'mass psychology', which by his lights requires more by way of complexity and coincidence in its explanations than would witchcraft: a kind of Occam's razor argument. Witchcraft, then, is Fort's final word on the silencing effects of science. 'Religion is belief in a supreme being. Science is belief in a supreme generalisation. Essentially they are the same. Both are the suppressors of witchcraft' (Fort, 1932: p. 249). Fort does not necessarily *believe* in witchcraft, but he *identifies* with it in order to express his experience of being silenced.[5]

Collectively, then, Fort's works attempted to highlight the shortcomings of science, and its arrogance, by presenting alternative data of his own. They were 'damned' because they were inadmissible, inexpressible within the current dominant. Fort challenged the reader:

> Here are the data.
> Make what you will, yourself, of them.
>
> (Fort, 1919: p. 88)

Crucially, Fort's data needed to be manifold – individually silent, they acquired a voice *en masse*.

278 *Silences in the public sphere*

> We shall have an outcry of silences. If a single instance of anything be disregarded by a System – our own attitude is that a single instance is a powerless thing.
>
> (Fort, 1919: p. 274)

Fort's guiding assumption that data, in sufficient quantity, would achieve a collective voice, is evident in his working method – which was a remarkable thing, to say the least. It consisted of an extraordinarily obsessive making and collecting of notes. When Dreiser first went to the Forts' apartment in the early years of the twentieth century, he was astonished to be shown Fort's collection of 'tens of thousands' of metaphors, each one written out on a separate slip of paper (Steinmeyer, 2008: p. 125). There is no record of what happened to them, but the process of noting and filing began again with scientific data in 1912, culminating in a collection of some 60,000 notes, under 1,300 topics: '1300 hell hounds … with 1300 voices', Fort complained (ibid.: p. 135). He destroyed them all before setting sail for England around 1920 (ibid.: p. 192). Once in London the process began again. This time Fort gathered approximately 40,000 hand-written notes on small pieces of card,[6] which he physically categorised by placing them in a pigeon-hole system of his own construction. Additionally, Fort clipped and kept articles from newspapers and kept these too. As his work became known, correspondents wrote to him with their own experiences and observations. These also were filed. It is overwhelming to contemplate the sheer physical presence of Fort's collections – hundreds of kilograms of paper. They cast a shadow of possible mental illness over their creator and, by extension, they silence his books, preventing them from participation in any serious cultural dialogue. More than this, however, Fort's notes test the boundaries of scientific education and research. By their crushing volume they question the assumption that more knowledge, more data, is always better – that there exists an end-point of induction at which one will have gathered enough instances to approach certainty. The appalling excess of Fort's research throws unwelcome light on the impossibility of this epistemology, and in so doing questions the scientific enterprise. Too much research is as suspect as too little – but why?

One reason that might be adduced in favour of not having to take Fort seriously is that his research was derivative: he did not do original research, but only gleaned it from other people's. One might usefully pause, however, and inspect the supposed categorical difference between first-hand and second-hand knowledge. Work on the early modern period in particular has highlighted the complex relationships between these modern typological categories. Richard Yeo's recent study on notebooks (2014) shows how the often copious gathering of apothegms and other verbal forms of knowledge was an intrinsic part of the shaping of the empirical project. Meanwhile, Adrian Johns (1998) and Steven Shapin (1984) have highlighted how the development of writing as a form of virtual witness was essential to creating a stable, transmissible body of knowledge. More recent scientists have also depended substantially on the gathering of vicarious knowledge. Darwin is the most obvious example; although an original and industrious researcher in some of his projects, his books were stuffed with reports from naturalists around the world, from which he built his inductive theories. The published *Correspondence of Charles Darwin* (Burkhardt et al., 1985–) currently runs to 22 volumes, with another eight projected, each of around 1,000 pages. The scale, if not the medium, is comparable with Fort's output; and yet only one seems beyond the pale on account of its voluminousness. Even if one sets the requirement for original research to one side, Fort's activity could be counted as compatible with the work of a scientific populariser, or even a philosopher of science. Finding and digesting other people's research, one might argue, is exactly scientific labour in this respect.

'Are you a follower of the late Charles Fort?': reading communities as judges

A second reason to discount Fort's claims might stem from faith in the mechanisms of science for establishing validity and permanence as symbiotic affirmations of knowledge. The very fact that Fort remained unappreciated in his own day is reason enough to ignore him in the present. Original readers and critics could scent his lunacy and left him well alone – and so should we. We can trust the scientific filters of the past. Although this argument is tautological in form, it is one that a sociological historian might take seriously. Science is whatever counts as science in its given era: if Fort was discounted, then we must accept that. And yet historiography is full of examples which have been silenced by current scientists but re-voiced by historians as vibrant fields of past science. Phrenology is a case in point: clearly bogus by today's standards, it was a genuine area of past scholarship and debate. It has even been identified as 'the most important vehicle for the diffusion of naturalistic and materialistic views in early to mid-nineteenth century Britain' (Shapin, 1983: p. 158). If one wished to create a Whiggish history, as many scientists do, one might say that phrenology paved the way for Darwinian evolution: a vindication of a 'pseudo-science' if ever there was one. Thus the way is open at least to wonder whether Fort might have a place in the history of science after all.

Sociological historians, of course, as their name suggests, search for phenomena on a *social* level. They cannot account for individual psychology. Perhaps the isolated nature of Fort and his oeuvre is a stronger reason to ignore him. Fort is an eccentric – a one-off – ergo his work is insignificant. Again, however, one might usefully probe whether the attribution of eccentricity is accurate or whether it is an artefact of the unpalatable nature of his work.[7] 'Eccentricity', in its etymological roots, implies an absence from the centre, an isolated existence. On a simple factual basis, this is a difficult claim to make. Fort was a bit of a loner, but not a complete recluse. His strong friendship with the novelist and journalist Theodore Dreiser placed him in a network of journalists, realist novelists and sceptics. The membership of the Fortean Society, founded in 1931, yields an immediate list of Fort's literary admirers: the writer and film-maker Ben Hecht, the poet and novelist John Cowper Powys, the critic Alexander Woollcott and Dreiser. Even more significantly, one can look at Fort's work in the context of amateur science journalism and in the context of a readership of 'science fans'.

Recent scholarship has developed an appreciation of the extensive and diverse engagements with science in professionally published periodicals, most especially in the late nineteenth century (Cantor et al., 2004; Henson et al., 2004; Cantor and Shuttleworth, 2004). Fort's story connects with a lesser-known history of amateur journalism at the turn of the twentieth century (Spencer, 1957), an emergent force in the commercial context of US publishing (Zboray and Zboray, 2013). Amateur journalism was a youthful and aspirational phenomenon. Between the ranks of amateur publications and major professional periodicals, there were countless local titles with parochial reach in which would-be writers could try their hand. Authors moved, or attempted to move, through these hierarchies.

Whilst still at his childhood home, Fort began writing for the *Albany Argus*, (alias *The Democrat*). He recounted in *Many Parts* how he had recycled his stepmother's gossip to fill his articles (Fort, n.d.a: p. 218). Upon leaving home for New York City, he became a reporter for the *New York World* (Brooklyn edition) in 1892, progressing to editor of the (unsuccessful) *Woodhaven Independent* in Queens.[8] In 1905 he introduced himself to Theodore Dreiser, then editor of *Smith's Magazine*, and succeeded in publishing short stories for him. Upon Dreiser's moving to another title, he begged Fort for more, but Fort had then moved on to his scientific work (X, n.d.).

In this context, as well as in the context of weird writing, Fort bears comparison with H. P. Lovecraft (1890–1937). Like Fort, Lovecraft combined an interest in science with the writing of strange, other-worldly texts, and both were interested above all in astronomy. Unlike Fort, however, Lovecraft kept a strict generic demarcation between science and fiction, producing copy that was always clearly identifiable as one or the other. Lovecraft came to the attention of the *Pawtuxet Valley Gleaner* as a result of writing to the state-level *Providence Sunday Journal* and the national *Scientific American* – both whilst still at school. He began writing astronomy for the *Gleaner*, simultaneously contributing a monthly column on the same topic to the *Providence Tribune*. He also formed links with the United Amateur Press Association, though this is an unduly grand title for what was a rather youthful and disorganised organisation (Burleson, 1983: p. 6; Fossils, n.d.). Lovecraft's original ambition to write for the *Providence Journal* had been stymied by the fact that a family friend, the professional astronomer Winslow Upton of Brown University, already had a long-running column in it (Lovecraft 2005: p. 100). Writing for the *Tribune* was a second choice, and as a youthful amateur Lovecraft felt the sense of exclusion rather keenly. The nature of professional astronomy and his relationship with it as an amateur was a constant sub-theme to his writing on the science, just as it was for Fort.

Astronomy was well established as an amateur science in this period (Marché, 2005), and during the decades of Fort's activity American amateurs sought to organise themselves into groups for greater sharing of knowledge and methods; the Society for Practical Astronomy, for example, was formed in 1910 (Williams, 2000). Clubs were also founded for the study of special astronomical phenomena, for instance, the American Meteor Society in 1911 (Williams, 2000). Close by Fort, the Amateur Astronomers Association of New York was founded in 1927 (Amateur Astronomers Association of New York, n.d.). It is difficult to ascertain Fort's level of involvement, if any, with practical astronomy. Annie Fort could not remember her husband ever meeting with astronomers, but recalled his pleasure in looking at the night sky for hours on end and his great knowledge about it.[9] Fort's books, especially *Lo!*, reveal a close engagement with recent and historic astronomical science.

Amateur astronomical groups instantiated a confident and occasionally pugnacious faith in amateur science. Lovecraft, for instance, had a particular bee in his bonnet about one-dollar telescopes, which he recommended to all his readers, claiming that as much could be achieved using one of these as using the latest, professional equipment (to Fort, the latter were 'millionaire's memorials', Fort, 1923: p. 139). Fort was particularly critical of the spectroscope as an astronomical tool, pointing out that the same instrument had been used both to 'prove' and 'disprove' Lowell's contention that there was life on Mars (Lowell, 1909):

> The question is not what an instrument determines. The question is – whose instrument? All the astronomers in the world may be against our notions, but most of their superiority is in their more expensive ways of deceiving themselves.
> (Fort, 1931: pp. 250–251)

Fort also had great scorn for astronomers' 'proof' of their theories by finding heavenly bodies where they predicted them. A typical example was the ecstatic reaction to Charles Delaunay's 'brilliant vindication' of the Newtonian system by his discovery of Neptune (see Daston and Galison, 2007: p. 212; cf. Fort, 1923: pp. 12–18). The title of *Lo!* was an ironic evocation of such post hoc announcements. To Fort, this was a patently shoddy method, epistemologically meretricious and, sociologically speaking, nothing more than a confidence trick on the part of scientists:

> My notion of astronomic accuracy:
> Who could not be a prize marksman, if only his hits be recorded?
> (Fort, 1919: p. 134)

Fort gathered astronomical newspaper clippings with particular assiduity. Anything contradicting previous findings was instantly filed away.[10]

Jeremiah Horrocks's 'triumph' in successfully predicting a transit of Venus in 1639, contrary to the predictions of Kepler, was a particular source of inspiration. To Fort, it was a tale of the amateur David and scientific Goliath. 'I suppose this was one of the most agreeable humiliations in the annals of busted inflations', he judged (Fort, 1932: p. 35). Fort made play of the fact that though Horrocks 'was interested in astronomic subjects', he 'had not been heard of by one [professional] astronomer of his time'; he was 'an outsider' (ibid.: p. 34). It does not take a great deal of psychologising to see how Fort related himself to this tale. By extension, he gave succour to the amateur astronomers (see especially Fort, 1931: pp. 390–403). Their observations – without fancy equipment, without fancy theories – were the hope of the science. 'A ... reasonable idea is that if nightwatchmen and policemen and other persons who do stay awake nights, should be given telescopes, something might be found out' (Fort, 1923: p. 118). When a Japanese farm hand in Washington discovered a comet, that went straight into Fort's file.[11] 'If amateur astronomers were as numerous as amateur golf players', Fort suggested, 'we'd realize much more' (Fort, 1931: p. 379).

One can find more legitimate contemporary critics making similar points in the same vein. The British chemist Henry Armstrong, for example, was a powerful opponent of 'dogma' in science and a believer in teaching through experiment (Armstrong, 1903). Closer to Fort, in the sense of being a scientific outsider, was George Bernard Shaw, who mooted a few similarly unusual astronomical notions (Henderson, 1911: pp. 469–470). Fort's countryman and contemporary Henry Adams, was closest of all; his *Education* was publicly published in 1918, just as Fort was writing the *Damned*.[12] In 'The Grammar of Science' (Adams, 1999: pp. 375–384) Adams paints the historic arrogance of science and its failure in the face of recent discoveries – an avalanche of new forces (X-rays, radium) which has exploded its apparently sewn-up universe. This catastrophic disruption has, says Adams, provoked mixed reactions amongst scientists, with some of them scuttling to try and defend the indefensible whilst others attempt to brush off the limits of knowledge as though they were merely temporary. Still others have resorted to deliberate obfuscation, for example in their textbooks:

> Chapter after chapter close[s] with phrases such as one never met in the older literature: 'The cause of this phenomenon is not understood'; 'science no longer ventures to explain causes'; 'the first step towards a causal explanation still remains to be taken'; 'opinions are very much divided'; 'in spite of the contradictions involved'.
>
> (Adams, 1999: p. 414)

Nowhere was a cognitive engagement with the new reality encouraged. Adams expressed this in a most Fortean phrase: there was 'a conspiracy of silence inevitable to all thought which demands new thought machinery' (Adams, 1999: p. 315).

How widely was Fort read in his own day? *The Book of the Damned* sold well, going into a reprint edition in 1920. *New Lands* did not do nearly so well, apparently failing to sell its initial run of 1,000 copies, but *Lo!* was produced in a costly-looking edition, suggesting the publisher's faith in the run. It was also serialised in the science fiction magazine *Astounding Stories* from May to November 1934 and thus came to a large audience, including British fans. The British science fans, as I have described elsewhere (Sleigh, forthcoming), were lower-middle or working class and moderately educated, typically originating in the industrial towns of northern England. They had been brought together by pulp magazines imported from the USA and found in scientifictional writing a way to develop their collective identity. Fort, it seemed, was easily co-opted into this vision.

For British readers at least, Fort was not anti-science but rather firmly in the camp occupied by their own magazines: pro-science, pro-imagination, pro-participation. The official journal of the British Science Fiction Association endorsed the very first issue of the Fortean Society's magazine: 'Not just another fan mag, but something considerably higher in both production and contents' (Carnell, 1937: p. 16).

Writing in the fanzine *Tomorrow*, H. S. W. Chibbett placed Fort firmly in the realm of science as fans understood it:

> It is clear, however, that ... super-normalities occur in Nature. They are not super natural, therefore, and should be diligently studied by Science. For this reason the Group to which I belong makes a practice of collecting and collating data of unusual happenings throughout the world – much in the manner of the late Charles Fort ... [C]lose study of apparent irregularities in Nature will eventually show that they fall into line with generally accepted knowledge. Here scientifiction can play its part, by dwelling upon the data laboriously acquired by the methods of psychic research, and allowing the flame and colour of its imagination to suggest through the media of stories the interpretation and meaning of existence.
> (Chibbett, 1938: p. 8)

In a questionnaire intended to understand the nature of the scientifiction movement, fans were asked: 'are you a follower of the late Charles Fort?' (Hanson, 1938: p. 2). Although the results of this survey are not, apparently, in existence, it is surely revealing that the question was worth asking. A later contributor to the same fanzine defended science fiction against its realist detractors by claiming that some of the things it described could be linked to real events in newspapers – Fort's sources – and in *Lo!* (Birchby, 1939).

'The new American would need to think in contradictions': belief and earnestness

A third perspective on the silencing of Fort might be that, although his methods were in themselves reasonable, his mental framework was not. The problem, one might suppose, was simply that he insisted on relating everything to his ridiculous cosmology, selecting and skewing as he went. Thus, for example, the (near) conclusion to *New Lands*:

> Behind concepts that sometimes seem delirious, I offer – a reasonable certainty –
> That, existing somewhere beyond this earth, perhaps beyond a revolving shell in which the nearby stars are openings, there are stationary regions, from which, upon many occasions, have emanated 'meteors,' ... flaming intimacies of destruction and slaughter and woe.
> (Fort, 1923: p. 249)

Such cosmological claims cannot be entertained seriously (although one might note its curiously medieval air). Yet one need not read Fort's writing as the deluded attempt to 'prove' such madness, for several reasons.

The first set of reasons clusters around issues of style and genre. Fort was an experienced journalist. Many of his short stories reveal a cynical knowledge of how journalism twisted (or invented) facts and created realities through its exploitation of naive readers. His tetralogy completely eschews anything approaching a journalistic style, which he might have employed effectively had he wished to delude and mislead.

It is almost impossible to give a sense of Fort's strange use of language without quoting pages of text. The following, from the middle of *Lo!*, gives a taste:

> A TREK of circumstances that kicks up a dust of details – a vast and dirty movement that is powdered with particulars –
> The gossip of men and women, and the yells of brats – whether dinner is ever going to be ready, or not – young couples in their nightly sneaks – and what the hell has become of the grease for the wheels? – who's got a match?
> It's a wagon train that feels out across a prairie.
> A drink of water – a chaw of tobacco – just where to borrow a cupful of flour – and yet, even though at its time any of these wants comes first, there is something behind all –
> The hope for Californian gold.
> The wagon train feels out across the prairie. It traces a path that other wagon trains make more distinct – and then so rolls a movement that to this day can be seen the ruts of its wheels.
> But behind the visions of gold, and the imagined feel of nuggets, there is something else –
> The gold plays out. A dominant motive turns to something else. Now a social growth feels out. Its material of people, who otherwise would have been stationary, has been moved to the west.
> The first, faint structures in an embryonic organism are of cartilage. They are replaced by bone.

The paths across prairies turn to lines of steel.

Or that once upon a time, purposefully, to stimulate future developments, gold was strewn in California – and that there had been control upon the depositions, so that only enough to stimulate a development, and not enough to destroy a financial system had been strewn –

That in other parts of this earth, in far back times, there had been purposeful plantings of the little, yellow slugs that would – when their time should come – bring about other extensions of social growths.

(Fort, 1931: pp. 266–267)

In this passage many aspects of Fort's writing are illustrated. There are subjects and objects without verbs: 'the yells of brats'; there are objects and actions without subjects: 'The hope for Californian gold'. Such stylistic quirks are developed in Fort's earliest writing, his autobiography and novel. With its frequent line breaks, Fort's style at times approaches a kind of prose poetry. He interleaves themes which, like a leitmotif returning in a piece of music, slowly evolve as they go. Some are running jokes, like the butter that keeps cropping up in the *Damned*. Although the paragraphs are short, they frequently end with dashes rather than full stops. (Poe often did the same, only with ellipses.) They are not complete in themselves, but lead on to the next. One is forced to read provisionally, not knowing whether the next paragraph will confirm or annul the meaning one has ascribed to the present one. This is even more true in sections of the text that are more argumentational in nature. Pages and pages go by, and the reader forgets whether she began on a trail of proof or disproof. 'Or this', 'or that' – it is almost impossible to recall to what these sub-clauses refer and in what respect.

Even within each paragraph the grammar often seems incomplete, provisional, such as the final sentence in the excerpt above. The first is little better: 'A trek of circumstances that kicks up a dust of details' – where is the main clause? – what does this trek *do*? Passive formulations are used to open a sentence and then lead nowhere. They imply a missing agency that is never confirmed, unless by the reader. The people are not, it turns out, the active subjects of the story, but rather the hope, the train, the 'yellow slugs'. Similar devices are used in the more overtly argumentational passages of Fort's books.

One verbal device that recurs perhaps more than any other is Fort's tendency to begin his short paragraphs with the word 'That ...': 'That something, far from this earth, had bled – super-dragon that had rammed a comet – ' (Fort, 1919: p. 287). At the opening of the *Damned*, a series of paragraphs beginning 'that' appears to supply a straightforward articulation of a thesis:

That the quest of all intellection has been for something – a fact, a basis, a generalization, law, formula, a major premise that is positive: that the best that has ever been done has been to say that some things are self-evident – whereas, by evidence we mean the support of something else –

> That this is the quest; but that it has never been attained; but that Science has acted, ruled, pronounced, and condemned as if it had been attained.
>
> (Fort, 1919: p. 9)

The procession of 'that' clauses (and there are many others) appears to be a mischievous echo of legislative formulation:

> We hold these truths to be self-evident, that all men are created equal, that they are endowed by their Creator with certain unalienable Rights, that among these are Life, Liberty and the pursuit of Happiness.
>
> That to secure these rights, Governments are instituted among Men, deriving their just powers from the consent of the governed, That whenever any Form of Government becomes destructive of these ends, it is the Right of the People to alter or to abolish it.

The 'thats' of the Declaration of Independence function differently than Fort's, anchored by an overt statement of belief at the outset. 'We hold these truths to be self-evident'; all the subsequent 'thats' hinge grammatically from this axiom. Fort's 'thats', however, pile up without either an initial anchor or an end-point of completion.

'That' introduces a clause which could be completed 'is deniable', or perhaps more plausibly 'is undeniable'. But it can also be seen in the most natural grammatical sense as the *end* clause of a sentence. 'Because x happens it can be proved *that y is the case.*' Its use thus gestures at the kind of abductive logic that has been ascribed to Poe (Eco and Sebeok, 1983; Sleigh, 2010: pp. 98–99). Abductive logic begins with a conclusion summoned in imagination, then argues backwards to show that this indeed must have been the case. Fort, however, withholds his abductive axiom from the reader, forcing him to reach it for himself. The technique bears obvious comparison with his constant invitations to let the data speak for itself. Indeed, 'that' is often used to introduce the data, as a frame for what a newspaper or journal has said: *that* such-and-such occurred. '*Edinburgh Philosophical Journal*, 2–381: That the earthquake had occurred at the climax of intense darkness and the fall of black rain' (Fort, 1919: p. 33).

According to many accounts – Popper's is perhaps the best known – science is about the maintenance of scepticism, the continuing provisionality of knowledge. Fort's style of non-closure perpetuates this stance to the point of agony. It stretches regular science to create an unbearable fermata (Kaplan, 1993), an ad absurdum critique. Fort's is uncanny science, not just in content but also in form.

At this point the reader may wonder whether Fort is not in fact practising a confidence trick upon his audience. Perhaps his invitation to open-minded contemplation of the data, couched in an ultra-provisional presentation, is all a hustle. Harry Houdini commissioned Lovecraft to write a book debunking pseudoscience (Lovecraft, 2005: p. 11); Fort is perhaps doing the opposite. By encouraging readers to think he is being nothing more than open-minded, Fort is able to sneak his strange ideas past their critical faculties. By making readers identify with Fort as the underdog, he carries them along. Again, one can see strong resemblances to Poe, whose various writings and lectures have never been uniformly and definitively designated as either in earnest or hoaxing (Higginson, 1998; Stott, 2009). Poe's narrators protest, 'Yet, mad am I not', and the reader lends them all the more trust for their admission that the world thinks them insane (Sleigh, 2010: p. 98).

Fort's earliest supporters responded to his writing in a wry sort of way. It seems that the knowing/not knowing tension was a crucial element of their pleasure in his texts. Ben Hecht's review of the *Damned* in the *Chicago Daily News* proclaimed:

> If it has pleased Charles Fort to perpetuate a Gargantuan jest upon unsuspecting readers, all the better. If he has in all seriousness heralded forth the innermost truths of his soul, well and good. I offer him this testament. I believe.
>
> (Fort, 1931: p. 3)

The actor and pulp writer Tiffany Thayer announced: 'But regardless of the absence of anything to believe, I was converted too. I "believed"' (Fort, 1931: p. 3). Thayer's praise is wonderfully contradictory, denying that there is any substance to Fort and placing his own belief in scare quotes. And yet Thayer was Fort's greatest fan. Though ultimately in large part responsible for Fort's unfortunate reputation, at this early stage Thayer affirmed the carefully poised uncertainty of his hero's claims. The considerably more highbrow writer Booth Tarkington also reviewed the *Damned* positively; for him, its literary qualities were foremost. By selective quotation Tarkington suggested that Fort's alarming beasts in the sky were neither fact nor fiction, but incarnations of science itself (Fort, 1923: pp. 1–4). This reading is, as I hope I have already demonstrated, a most plausible one.

Nor did Fort take his own claims too seriously. Reflexivity and humour were never far from his pen, either in his books or his letters. He called his philosophy of intermediatism 'a pseudo-standard', noting that:

> To the intermediatist there is but one answer to all questions:
> Sometimes and sometimes not.
>
> (Fort, 1919: p. 268)

He was not unaware of his developing reputation as a crank (he uses the term of himself in *Wild Talents*), and even the data themselves were subject to doubt. Fort consistently used the word 'yarn' to describe them: 'I go on with my yarns. I no more believe them than I believe that twice two are four' (Fort, 1931: p. 153). Perhaps contrarianism would have been a better word for Fort's philosophy than intermediatism. The point was that he could always argue another way on the basis of his data:

> at any time, let anybody say to me, authoritatively, or with an air of finality, that the stars are trillions of miles away, or ten miles away, and my contrariness stirs, or inflames, and if I can't pick the lock of his pronouncements, I'll have to squirm out some way to save my egotism.
>
> (Fort, 1931: p. 367)

This suggests that his strange cosmology was a ruse, a self-imposed challenge of mounting a counter-case to the descriptions of science. Had science held some other form of nature to be the case, his cosmos would have altered accordingly in reaction. He took explicit pride in his ability to cook up any theory and make it work:

> If I had the time for an extra job, I'd ask readers to think up loony theories, and send them to me. I'd pick out the looniest of all, and engage to find abundant data to make it reasonable to anybody who wanted to think it reasonable.
>
> (Fort, 1931: p. 65)

Perhaps the best source for understanding Fort's quasi-scientific voice (and certainly one of the more compact) is his short story 'The Giant, the Insect, and the Philanthropic-looking Old Gentleman' (Fort, n.d.b). The manuscript is undated but its contents suggest that it was written around the time that the third and final collection of data was finished. It is a tale of the uncanny and also a rehearsal of Fort's own working method vis-à-vis voice. It is a tale of superfluity designed to reveal emptiness: of voices raised to silence a shouter.

The narrator begins by describing how he has been taking notes on science – 48,000 of them – to try and prove his theory that the laws of nature apply also to human beings. He describes how in the course of his researches he is distracted by the sight of another man, Mr Rapp, who is himself watching a house on the corner of the street. It turns out that Rapp is keeping an eye on one Dr Katz, a peddler of quack medicine. For many years Katz has been appearing in public as an advertisement for his nostrum, but now he has fallen ill and has been replaced by another 'Katz'. The narrator and Rapp get talking about the former's notes and by reference to them they come to identify the Katz problem as 'mimicry, aggressive'. They get the idea from records of natural history: 'In India there is a mantis that has taken on the appearance of a flower; by means of its form and pink color, it allures other insects upon which it subsists' (Fort, n.d.b).

How to counter this mimicry and unveil the quackery? The narrator is fortified by his belief that 'for every device of defense there is some weapon of attack in Nature', no less in the human than in any other realm. He combs further through his notes:

And we found the answer soon enough. By its own multiplication this phenomenon is kept in check. We found a hint of this in observations by Mr. Bates and Dr. Wallace that mimicking species are always much rarer than the mimicked.

(Fort, n.d.b)

There was, as the narrator explains, additional data to back this up. It related to the case of the 'Cardiff Giant'. In 1869, a 10-foot petrified man was dug up behind a farmer's barn in Cardiff, NY, and the farmer and his cousin began a brisk trade in exhibiting the hoax. P. T. Barnum offered $50,000 for it and, incensed at being turned down, created a copy which he exhibited as the original. The existence of a second giant, it seems, sowed seeds of doubt in the public mind; *both* became suspect (Tribble, 2008). The narrator exaggerates the account of historical record, stating that 'reproductions of it sprang up all over the country'. However, the result was the same: 'multiplication was the undoing of the Cardiff Giant' (Fort, n.d.b).

Without explaining how, or when, or by whom it is effected, the narrator presents the story's denouement: multiple 'Katzes' appear in the street. One after another, these philanthropic-looking old gentlemen spread through the city. Although it is not stated overtly, the reader is given to understand from what has gone before that these simulacra will undo the power of the original copy. Five are counted off and then the story closes:

Up from an area-way! Upon my word, another of them! Most spiritual-looking and healthiest-looking of all of them: white hair curled; black-specked; blinking up at the tall buildings, so placidly, so exotically, in our wicked city.

(Fort, n.d.b)

Amplification, then, is the key to judgement. Something untrustworthy, when amplified, betrays its untrustworthiness. Multiplying its voice will force it into silence.

Fort's style in his 'literary' writing is itself often an experiment in amplification. He will frequently take a metaphor and extend it until it is stretched beyond aesthetic norms. Eventually it goes so far that it regains its power to work, having passed through a phase of over-extension and out the other side:

Extraordinary nose; made me think of a gargoyle; long and lean and poised recklessly over a heavy underlip – like a precarious gargoyle over a window sill with a red blanket out airing on it. He was nervous, and two white teeth appeared frequently, and bit upon and drew in the lower lip – very much as if he were a dwelling of some tall, tower-like kind – a little butler wearing white gloves, inside, you know – little butler constantly fearing the hovering gargoyle, and forever drawing in the too conspicuous red blanket, with his white-gloved hands, and then putting it out for an airing again.

(Fort, n.d.b)

The first repetition of 'gargoyle' breaks an unspoken aesthetic rule that catachrestic metaphor only works once, and after that becomes clumsy. The introduction of the butler breaks a second rule of applying metaphors too densely. By the time the gargoyle returns as the object of the butler's terror, it has become an amusing familiarity with fresh force. The fact that Fort first collected metaphors as he was later to collect data suggests some affinity between the two in his method. Both are pushed to test the boundaries of narrative, whether in 'fiction' or 'science'.

Fort's narrator in 'The Giant' bases his method – as Fort based his own – upon an amplification of data that constantly risked tipping into insanity. There is a fine line between amplification and superfluity. Fort's friend and editor Theodore Dreiser spent a great deal of time reflecting on this problem (Dreiser, 1974: pp. 184–189), and letters between the two men touch upon it. Dreiser was greatly struck by the enormous wastage of individuals in life; for example, the black widow spider's eggs that were eaten in their thousands. Nor was it just a matter of individuals; entire species were wiped out in the evolutionary process. Superfluity was waste, and waste was death. One could read Fort's collection of data (and his childhood natural history collection) as an attempt to combat death through conservation; and yet he railed at scientists who froze knowledge in meaningless grimaces of ontology. Twice Fort destroyed his collections, perhaps aware that science, in its appetite for data, approaches thermodynamic death:

> Heat is Evil. Final Good is Absolute Frigidity. An Arctic winter is very beautiful, but I think that an interest in monkeys chattering in palm trees accounts for our own Intermediatism.
>
> (Fort, 1919: p. 247)

A superfluity of words approached the heat death of the universe, the final silence. In that death, heat and cold met one another; when all effort, all words, had been expended through infinity there would be absolute cold; absolute silence. One could be silent – but this would be an empty, Arctic beauty. Or one could expend words, collect vertiginous quantities of data – but this would lead back to the frigidity of heat death. Either path led to the same silence. 'Chattering' was unsustainable – ridiculous – but it was the only sane response in the interim.

Fort's practice of silencing through superfluity is clearly inflected with a thermodynamic awareness. In this, Fort again resembles Henry Adams. Adams has been placed by the historians Crosbie Smith and Ian Higginson in a mode of history that is thermodynamically degenerative; unlike the evolutionists' vision of improvement, Adams' prospect was energetic dissipation at work. Heat was evil for him too. The acceleration that Adams perceived in the new century was not a matter of progress, but of 'fragmentation and disintegration' (Smith and Higginson, 2001: p. 103).

Adams, like Fort, entertains a speculative ontology that underpins both human history and the unfolding of the earth within the cosmos. (Recall that in 'The Giant' the narrator has a foundational theory that the laws of nature apply also to human beings.) Adams' chapter 'A Dynamic Theory of History' (Adams, 1999: pp. 395–406) is a sort of ontologically flat account that summons 'forces' in history in ways that sometimes appear metaphorical and other times, finally, naturalistic. 'Man is a force; so is the sun' (ibid.: p. 395).

In 'A Law of Acceleration' (ibid.: pp. 407–414), Adams grapples with the challenge of integrating the new forces into an account of history. It is a challenge because such forces are by definition 'super-sensual', that is, beyond ordinary empirical means of inspection. It is perhaps in this essay that Adams approaches closest to Fort (or vice versa), drawing on cosmic imagery to express the mental leap that must occur.

> The image needed here is that of a new center, or preponderating mass, artificially introduced on earth in the midst of a system of attractive forces that previously made their own equilibrium, and constantly induced to accelerate its motion till it shall establish a new equilibrium.
> … this is probably [a] comet, or meteoric streams, like the Leonids and Perseids; a complex of minute mechanical agencies, reacting within and without, and guided by the sum of forces attracting or deflecting it. … The mind, by analogy, may figure as such a comet, the better because it also defies law.
>
> (ibid., p. 407)

Fort's extra-terrestrial realms, the source of all those sky falls, performed exactly this role. Fort's cosmos, his philosophy, can be both believed and disbelieved. It is both true and not true; it is both meant and not meant. Embodying such contradiction, 'The Giant' flip-flops in meaning like one of those impossible logical statements ('I always tell lies'). It is the science gleaned by the narrator that proves reliable in solving the problem of Katz; and yet, by the same method of amplification, the narrative voice of Fort's tetralogy undermines science. Science is both true and not true. Or as Adams put it, also speaking of science: 'The new American would need to think in contradictions' (ibid.: p. 434). Such nonsense, at least in Fort's mouth, sounds like chatter.

Conclusion

In his *Silence: A Christian History*, Diarmaid MacCulloch (2013) points out that silence can be chosen or imposed: creative or damaging. Fort, it seems, was aware of both kinds. He saw, first, a silencing that he judged had been imposed upon past scientific outsiders and on scientific amateurs of his own day. Thus he concluded *New Lands*: 'Silence that is conspiracy to hide past ignorance; that is imbecility' (Fort, 1923: p. 249).

I am not, of course, arguing that Fort should be treated as a legitimate scientist, but I am trying to demonstrate that none of the prima facie reasons one might have for dismissing him as a serious interlocutor in scientific discourse is in itself sufficient. The very troubling nature of Fort and his work arises because of their similarity to legitimate scientific practices. Fort is alarming because he is *too close* to science. He attempted to create such a babble of voices from science, to invoke such a torrent of 'expert' knowledge-telling, that it would become a kind of white noise and conjure the silence that lay behind it – the silence of true observation.

The silencing of the amateur, was – perhaps as he feared – imposed upon Fort too; and yet it was also what he chose, in his own wry and contrary way. In 1916 he wrote to Dreiser:

> I asked you for advice, and you gave me silence. This is the only sound philosophy. Hereafter I am going to publish only silences, myself. Only nothingness can be Truth.[13]

What could he expect, if he was to speak in silences? A stony silence in response.

Notes

1 Two archives of Fort's papers are in known existence. One set is at the University of Pennsylvania, catalogued as mscoll30. These are referred to in this essay in the format 'Penn [folder number]: [page number]'. The other archive is at the New York Public Library. Reference to these follows the library's system, prefaced 'NYPL'. An additional online resource is at resologist.net. This collection of unpublished and rare writing by Fort is curated by someone going by the name of Mr X. Although the name does not inspire scholarly confidence (it is another symptom of the Fort story), the material on the site gives every impression of being accurately transcribed and meticulously edited. In so far as I have cross-checked it with original materials from the two archives and published primary sources it is completely reliable. I am grateful to Mr X for his answers to my questions during the preparation of this essay, and recommend his website to anyone wishing to begin reading Fort.
2 Anna Fort, interview with Tiffany Thayer, n.d. Penn 12330: 2–3.
3 Fort, letter to Dreiser, 31 March 1916. Penn 2043: 5.
4 This might appear to affirm that the claw marks are and have always been real, irrespective of the dominant in place at any given time. However, Fort rarely means anything straightforwardly.
5 The identification of scientists with witch-hunters has been exploited by more recent conspiracy theorists than the Forteans, namely climate change deniers. See publications of the George C. Marshall Institute, e.g., http://marshall.org/climate-change/climate-skepticism-todays-witch-hunt-and-mccarthyism/.

6 Personal communication from Mr X.
7 Cf. Sleigh (2007: p. 36) on August Forel as 'eccentric'.
8 Letter to Dreiser, 1 December 1919 [1915?]. Penn 2042: 16.
9 Anna Fort interviewed by Tiffany Thayer [transcript]. Penn 12330: 7.
10 NYPL AF-III-456; AF-I-11.
11 NYPL AF-I-336.
12 There is no archival evidence to suggest whether or not Fort read Adams' *Education*, but as I explore below, there are textual hints that he may have done.
13 Fort, letter to Dreiser, 27 August 1916. Penn 2043: 18.

References

Adams, H., 1999 [1907]. *The Education of Henry Adams: An Autobiography*. Oxford: Oxford University Press.
Amateur Astronomers Association of New York, n.d. *The Club*. Available at: www.aaa.org/theclub/ (Accessed 15 May 2015).
Armstrong, H.E., 1903. *The Teaching of Scientific Method and Other Papers on Education*. New York: Macmillan.
Bennett, C., 2009. *Politics of the Imagination: The Life, Work and Ideas of Charles Fort*. New York: Cosimo.
Birchby, S., 1939. Stranger than truth. *Novae Terrae*, 3(5), pp. 7–9.
Burkhardt, F. et al. (eds), 1985–. *The Correspondence of Charles Darwin*. 22 vols. Cambridge: Cambridge University Press.
Burleson, D.R., 1983. *H.P. Lovecraft: A Critical Study*. Westport, CT: Greenwood Press.
Cantor, G., Dawson, G., Gooday, G., Noakes, R., Shuttleworth, S. and Topham, J.R. (eds), 2004. *Science in the Nineteenth-Century Periodical: Reading the Magazine of Nature*. Cambridge: Cambridge University Press.
Cantor, G. and Shuttleworth, S. (eds), 2004. *Science Serialized: Representations of the Sciences in Nineteenth-Century Periodicals*. Cambridge, MA: MIT Press.
Carnell, T., 1937. Initial success. *Novae Terrae*, 2(5), pp. 13–16.
Chibbett, H.S.W., 1938. The supernormal. *Tomorrow*, 2(2), p. 8.
Daston, L. and Galison, P., 2007. *Objectivity*. New York: Zone.
Dreiser, T., 1974. *Notes on Life*. Edited by M. Tjader and J.J. McAleer. University, AL: Alabama University Press.
Eco, U. and Sebeok, T.A. (eds), 1983. *The Sign of Three: Dupin, Holmes, Peirce*. Bloomington: Indiana University Press.
Fort, C., n.d.a. *Many Parts*. Unpublished MS. Edited and expanded upon by X. Unpaginated. Available at: www.resologist.net/parte01.htm
Fort, C., n.d.b. The giant, the insect, and the philanthropic-looking old gentleman. Published version untraceable. Available at: www.resologist.net/story29.htm
Fort, C., 1919. *The Book of the Damned*. New York: Horace Liveright.
Fort, C., 1923. *New Lands*. New York: Boni & Liveright.
Fort, C., 1931. *Lo!* New York: Claude Kendall.
Fort, C., 1932. *Wild Talents*. New York: Claude Kendall.
Fossils, n.d. The United Amateur Press Association. The Fossils: the historians of amateur journalism. Available at: www.thefossils.org/horvat/aj/organizations/uapa.htm (Accessed 15 May 2015).
Hanson, M.K., 1938. Questionnaire no. 3. Insert. *Novae Terrae*, 2(8), pp. 2.

Henderson, A., 1911. *George Bernard Shaw: His Life and Works*. London: Hurst and Blackett.
Henson, L., Cantor, G., Dawson, G., Noakes, R., Shuttleworth, S. and Topham, J.R. (eds), 2004. *Culture and Science in the Nineteenth-Century Media*. Oxford: Ashgate.
Higginson, I., 1998. 'I do know the machinery of the universe': system and individuality in Edgar Allan Poe's Eureka. In: Smith, C., Agar, J. and Schmidt, G. (eds), *Making Space for Science: Territorial Themes in the Shaping of Knowledge*. Basingstoke: Palgrave Macmillan.
Johns, A., 1998. *The Nature of the Book: Print and Knowledge in the Making*. Chicago: University of Chicago Press.
Kaplan, L., 1993. Suspense, para-science and laughter. *SubStance*, 22(2/3), pp. 306–314.
Knight, D., 1971. *Charles Fort: Prophet of the Unexplained*. New York: Gollancz.
Kripal, J.J., 2010. *Authors of the Impossible: The Paranormal and the Sacred*. Chicago: University of Chicago Press.
Kuhn, T., 2012. *The Structure of Scientific Revolutions*. 50th anniversary edition. Chicago: University of Chicago Press.
Lovecraft, H.P., 2005. *Collected Essays: Volume 3: Science*. Edited by S.T. Joshi. New York: Hippocampus Press.
Lowell, P., 1909. *Mars as the Abode of Life*. New York: Macmillan.
Marché, J.D., 2005. 'Popular' journals and community in American astronomy, 1882–1951. *Journal of Astronomical History and Heritage*, 8, pp. 49–64.
MacCulloch, D., 2013. *Silence: A Christian History*. London: Penguin.
Seed, D., 1999. *American Science Fiction and the Cold War: Literature and Film*. London: Taylor & Francis.
Shapin, S., 1983. Edinburgh and the diffusion of science in the 1830s. In: Inkster, I. and Morrell, J. (eds), *Metropolis and Province: Science in British Culture, 1780–1850*. London: Hutchinson, pp. 151–178.
Shapin, S., 1984. Pump and circumstance: Robert Boyle's literary technology. *Social Studies of Science*, 14(4), pp. 481–520.
Skinner, Q., 1969. Meaning and understanding in the history of ideas. *History and Theory*, 8(1), pp. 3–53.
Sleigh, C., 2007. *Six Legs Better: A Cultural History of Myrmecology*. Baltimore, MD: Johns Hopkins University Press.
Sleigh, C., 2010. *Literature and Science*. Basingstoke: Palgrave.
Sleigh, C., forthcoming. Science as heterotopia: the BIS before World War II. In: Leggett, D. and Sleigh, C. (eds), *Scientific Governance in Britain, 1914–79*. Manchester: Manchester University Press.
Smith, C. and Higginson, I., 2001. Consuming energies: Henry Adams and 'the tyranny of thermodynamics'. *Interdisciplinary Science Reviews*, 26(2), pp. 103–111.
Spencer, T.J., 1957. *The History of Amateur Journalism*. New York: The Fossils.
Steinmeyer, J., 2008. *Charles Fort: The Man Who Invented the Supernatural*. London: Heinemann.
Stott, G.S., 2009. Neither genius nor fudge: Edgar Allan Poe and Eureka. *452°F*, 1(1), pp. 52–64.
Tribble, S., 2008. *A Colossal Hoax: The Giant from Cardiff that Fooled America*. Lanham, MD: Rowman & Littlefield.
Williams, T.R., 2000. Getting organized: a history of amateur astronomy in the United States. Ph.D. thesis. Rice University.

X, Mr, n.d. Charles Hoy Fort's short stories: introduction. Available at: www.resolo gist.net/introe2.htm (Accessed 15 May 2015).
Yeo, R., 2014. *Notebooks, English Virtuosi, and Early Modern Science.* Chicago: Chicago University Press.
Zboray, R.J. and Zboray, M.S., 2013. *Literary Dollars and Social Sense: A People's History of the Mass Market Book.* New York: Routledge.

Index

(Figures indexed in italic and Tables indexed in bold)

Aarts, Noelle 18, 89–110
absolute silence 3, 115, 127, 290; *see also* silences
academic institutions 89, 97–8, 104, 109
Acheson, Kris 4
ACPO 257, 264–5
Adam, Ronald 146
Adams, Henry 138, 146, 148, 282, 290–1
agnotology literature 2, 254
Agnotology: The Making and Unmaking of Ignorance 221
agreements 5, 13, 101–2, 237, 276; inter-subjective 212; non-disclosure 22; short-term 107; tacit 6; voluntary 14
Ahani, A. 206
Ahrenfeldt, Robert 146
Akrich, M. 90–1, 97, 106, 108
Alagaddupama Sutra 198
alchemical knowledge 82–3
alchemists 22, 82
alchemy 15, 67, 81–2, 84
Alvarez, Robert 129
Amateur Astronomers Association of New York 281
American Meteor Society 281
Amundson, M. 127
analysts 223, 228–31, 233–5, 237, 259–61; desiring to articulate experiential or tacit knowledge 23; divided 235; public 222, 225, 228–31, 233–4, 236–7
analytical chemistry 234, 237
ancient Egyptians 23, 172–6, 180–1, 184, 187, 189–90
ancient Greeks 174, 189–90
Anderson, John 16, 145
Anderson, Melissa 15

Andrewjeski, Juanita 124
aniline dyes 222–8, 232–3
Antarctic ozone readings 38, 45
Antarctica 18, 31–2, 35, 37–8, 44–5
anthropology, cultural 198
Argentine Islands 32–5, 39
Aristotle 189–90
Armstrong, Henry 282
army 19, 135–8, 144, 148; hierarchy 137; officers 135–6; psychiatrists 136, 141, 149
Aspen Center for Physics 17
Association of Chief Police Officers *see* ACPO
Astounding Stories 283
astronomers 277, 281–2
astronomy 247, 277, 280–1
Atkins, P.W. 228–30
Atomic Energy Commission 119–22
Audubon Society 247

bacteriologists 233
Bailie, Tom 128
Bannister, Richard 234
Basso, Keith 5
BCE 175–6, 190
Beagle 52–4, 57–8, 60
Beck, A. 243–4
Beck Depression Inventory 207; *see also* PANAS scales
Before Common Era *see* BCE
Beller, M. 18
Berlin Papyrus (source of evidence for ancient Egyptian mathematics) 176
Bhagsu Mountain (Himalayas) 201–2
Biagioli, Mario 16
Bingham, R.C. 137

Bitbol, M. 208, 210–11
Bohr, Niels 18
Bok, S. 94, 107
Bondi, Herman 35
bone marrow transplants 156, 160, 163–4
Bonney, R. 247
The Book of the Damned 275
Bowen, Barbara 34–5, 40–1
Boyce, Harold 136, 203
Boyle, Robert 67–9, 71, 73, 78, 81, 83–4
Bradford Observer 225
brain activity 200, 202, 205–6, 208–9
Britain 32, 135–6, 138, 152, 154, 227–8, 234, 236–7; and a fear of national, individual and social degeneration 227; food analysts 228; food manufacturers 228; and the media 222, 226; nineteenth-century disputes involving public analysts 229; supremacy in industry and science 222
Britain, and claims by academics of an erosion of basic academic norms 51
British Antarctic Survey 18, 31, 35, 37
British Army *see* army
British Medical Journal 246
Brock, William 230
Brown, David 14
Brown, Helen Gurley 244
Browne, Janet 52, 56, 58, 199
Bruneau, Thomas 2
Buddhism 197–202, 205, 207, 212
Buddhist 194, 201–2, 204; meditation 23, 193, 195–7, 199, 201, 203, 205, 207, 209, 211; monks 201–2; philosophy 207; traditions 197, 212; vocabulary 200, 204
Bulletin of the Atomic Scientists 123
Burkhardt, F. 48, 52, 54–9, 61–2
Burton, D. 242–3, 249
Butler, D. 13–14

Cage, John 3
cancer 152–4, 166; aggressive prostate 167; high rates of 11, 128; lung 221
cancer research 12, 117, 123–4, 152–5, 157, 159, 161, 163, 165, 167–9
candidates 19, 138–48; blustering 143; enrolling of 140; individual 139; officer 137, 139, 142; public school 144
Cantor, G. 280
"Cardiff Giant" 289

"Cardiff Giant" (petrified man) 289
Carter, C. 90, 93–4
Cell 246
censorship 8, 14
chemical dyes 222, 227, 229, 231, 237
chemists 223–4, 228–31, 233–4, 236–7; academic 236; analytical 229–31, 237; central government excise 228; expert 231; fellow 230; industrial 231; organic 231, 234, 236–7; professional 229–30, 237
chemotherapy drugs 154, 156, 159–60, 163; *see also* drugs
Chernobyl reactor 129
Chibbett, H.S.W. 283
Christianson, G.E. 69–70, 75
Churchill, Winston 135, 144–5
"Citizen Science" 247–8
Claggett, M. 175–6
Cobbold, Carolyn 12, 221–38
cognitive neuroscientific studies 206
cognitive therapy programmes 206
Cold War 116–18, 120, 126, 128, 274
collaborative research 201
Collins, John 69–70, 76
Colombetti, Giovanna 23, 193–212
Columbia River 116, 119–21, 125, 129
Command Psychiatrists (British Army) 136–7
communication 3–5, 7–8, 10–11, 16, 24, 90, 94–5, 107, 162–4, 258–60; and conversations for innovation 92; enhancing 8; policies 24; public 20, 84; scientific 67; suppressing 10, 16; verbal 24, 90, 92, 106, 108
contamination 116, 119–21, 123, 125, 127; ground water 122; nuclear 127; offsite 116; radioactive 126; records of incidents 116
Continuous Plankton Recorder Survey *see* CPRS
Cooper, L. 186–7
Coopmans, Catelijne 23, 193–212
Copperfield, David 224
Cowper, John 280
CPRS 44–5
Crang, J.A. 144, 148
Cripps, S. 144–6
Crocker, Lieut-Gen. Sir John 148
Cronin, B. 49, 51
Crookes, William 230
Crotty, David 245
CS spray 256–7, 268
cultural anthropology 198

Index

Dalai Lama 201–2, 206
Darwin, Charles 18, 48–62, 247, 279; biographies 52; confessed to an ignorance of English geology and of all languages 59, 84; evolution theory 279; and his contemplative sensibility 49; ill health and style of work combined to make quietude a central desire 59; interest in quietude 50; naval experience 55; and *The Origin of the Species* 56; priority is "to write vigorously, briefly and with publication firmly in mind" 61–2; silence 48–9, 51, 53, 55, 57, 59, 61
Darwin, Charles Waring (son of Charles Darwin) 61
Darwin, Emma (wife of Charles Darwin) 48, 50
Davidson, Richard 199, 201, 205
Davies, Sarah 118, 246
de Bruijn, H. 90–1
Defense Scientific Advisory Council Sub-Committee on the Medical Implications of Less-Lethal Weapons in the UK *see* DOMILL
Delaunay, Charles 281
Destructive Emotions and Happiness 202–4, 206, 209
Dexter, Arthur 127
Dickens, Charles 222, 224
Dickson, Bob 45
Dobbs, Betty Jo Teeter 82
Dobbs, B.J. 82
Dobson Spectrophotometers 32
DOMILL 254, 262–6
donor transplants 156–7
Dor-Ziderman, Y. 209–10
Dreiser, Theodore 275, 278–80, 290, 292
drugs 165, 221, 265
Dupré, August 233–5
Dutch Health Council 94
dyes 222–3, 229, 231–4, 236–7; animal 232; coal-tar-derived 221, 225, 227, 229, 237; harmless 227; mineral-based 237; new 12, 222, 224–5, 229, 231, 233; nitro 232; synthetic 12, 221–3, 227–9, 231, 233–4, 236–8
Dymond, Abi 15, 253–70

Edmondson, A. 90, 105, 108
Egyptian culture 172, 174–5, 190
Egyptian history 174–5
Egyptian mathematical 23, 172–7, 179, 181–5, 187, 189; problems 187; techniques 180; texts 173–4, 181–2, 185, 187–90
Egyptian mathematicians 23, 175, 180, 183
Egyptians 172–5, 177, 179–80, 184, 187–90; scribes 183, 188, 190; texts of 182, 186–7
Egyptologists 172–5
England 52, 54, 60, 67, 253–4, 262, 264, 278; eighteenth-century 241; Jesuits 66, 84; northern 283; and Wales 264, 268–9
essays 9, 65–7, 71–2, 199, 291; *New Theory of Light and Colours* 65–6, 69–70, 73–5, 77–9, 81; wartime 141
European Commission 21–2
experimental studies 78, 194, 202
experimental subjects 201–5
Experiments and Considerations Touching Colours 67
Expert Committee on the Work of Psychologists and Psychiatrists in the Services 145–6, 148

Fara, Patricia 71–2, 77
Faraday, Michael 244
Farman, Joseph 18, 31–45; appears on BBC Radio 4's *Today* programme 40; oversees the measurement of ozone at Halley Bay 32; studies ozone in Antarctica 36–7
Findlay, John 116–17, 126
Flexner, Abraham 17
fludarabine 164–6
follicular lymphoma 156, 159, 164–5
Fonseca, J. 90–1, 93, 108
food 12, 93, 221–31, 233–8; additives 222, 237–8; adulteration 12, 222, 224–5, 237; colouring 224–6, 229, 231, 237; foreign 226–7; manufacturers 227, 235, 237; production 221, 229, 234; products 223, 231, 234–5, 237; supply 222, 225, 228, 234, 236
Force Science Research Centre 260
Ford, J.D. 93, 95–6, 106–10
foreign foods 226–7
Fort, Anna 275, 281
Fort, Charles Hoy 23, 274–92; activity 279; activity compatible with the work of a scientific populariser 279; books reveal a close engagement with recent and historic astronomical science 281; and the collection of data 290; and the collections of books 281; and his

"literary" writing 289; and his short story *The Giant, the Insect, and the Philanthropic-looking Old Gentleman* 288; reveals a close engagement with recent and historic astronomical science books 281
Fort, Chrles Hoy: style 289; style approaches prose poetry 285
Fortean Society 280, 283
Fortean Times 274
Fouchier, Ron 13–14
Franke, B. 129, 196
Frankel, M. 14
Franklin, Benjamin 247
Friendly Companion: A Magazine for Youth 227
Funny Folks 226

gamma activity 208
Gardiner, Brian 34, 187
Garfield, S. 222
gargoyles 289–90
Garrison, Kathleen 209
Gascoigne, John 79–80
gender relations 52
Gendlin, E. 195, 210
generic demarcation between science and fiction 280
geneticists 50
Geological Society 59
geologists 34
Gephart, R.E. 116
Gerber, Michele 115–18, 121, 126–8
German chemists 222, 226, 232
German scientists 10, 13, 226, 228, 233
Gillings, R.J. 183–4, 187
Gladwell, M. 91–3
Glazer, P.M. 115, 124, 126, 128–9
Glenn, C. 6
Goblin Market 224
God 42, 128, 173, 196
Goffman, E. 93
gold 82, 284–5
Goldberg, S. 10
Goldstein, B. 90
Goleman, D. 199, 202–3
Golinski, Jan 77
Gordin, M.D. 118, 126
government 115–16, 123, 127, 129, 136, 230, 234, 242, 250, 286; chemists 230, 236–7; committees 235; Customs 230; federal 14, 118, 123, 126; scientists 230, 233
graduate students 50

Grafton, Anthony 49
Grand, Ann 21, 241–50
Grand Coulee Dam 118
Gray, Sean 7
Greater Miami River 127
Greeks 172, 185, 189–90
greenhouses 56
Greenspan 31
Gregory, David 70
Grevsmuhl, Sebastian 45
Grigg, Edward 138, 141
Grint, K. 254–5
grounded theory 209–10
Groves, Gen. Leslie R. 9–11, 117
The Guardian 51
guidance 52, 81, 262, 264–7; clear 265; expert 143; official 265; person's 109; produced 264
Gura, T. 246
Gyngell, C. 14

Haas, P. 261
Haber, L.F. 222
Hall, A.R. 69, 78, 254, 259
Hall, Rupert 78
Halley Bay 32–5, 39
Hamilton, Catriona Gilmour 12, 152–69
Hamlin, C. 229
Hanford 11, 115–30; and the contrasting imperatives of openness and secrecy at 115; and the discharges of nuclear waste into the Columbia River 129; and the disruptive potential of accidents 127; employees 118, 125, 130; employers 126; management 126–7, 129; mismanagement and safety violations 123; and news reporting of silence 117–30; policies and decisions 117
Hanford Atomic Energy Works *see* Hanford
Hanford Education Action League 124
Harrington, A. 199
Harris, N. 36–7, 46, 139
Hassall, Arthur Hill 224
health 55, 58–9, 91, 96, 127, 130, 154, 226, 231, 235; effects 116, 229, 233, 256; human 122; ill 50, 59–60, 227; industrial 119; mental 136; professionals 152, 155; public 224, 229, 235
Health Reformer 226
Hecht, Ben 280, 287
Heegaard, W.G. 260

Heisenberg, Werner 18
Henriksen, K. 90, 105, 108
Henslow, Robert 54, 59
Herodotus 189–90
Hetherington, Kevin 205
Heuvelhof, E. 90–1
Hevly, Bruce 116–17, 126
hexanitrodiphenylamine 232
hieroglyphics 187
hieroglyphs 185
Higginson, Ian 287, 290–1
Hilgartner, Stephen 22, 255, 261
Hill, Alfred 236
Hillaby, John 120
Himalayas 201–2
Hind, Liz 23, 172–91
historians 23, 66, 76, 82, 172, 181, 221–2, 229, 234, 279; natural 247; oral 169
historiography 23, 52, 153, 190, 279
Ho, J. 259
Hodgkin's Disease 154–5, 162
Hodgkin's Society 155
Hoffman, K. 254
Holley, K.A. 97
Home Office 263–5
Home Office Code of Practice on Police use of Firearms and Less Lethal Weapons 265
Hooke, Robert 73–9, 84, 241, 247
Hooker, Joseph 56, 59–62
Hooke's Law 74, 241
Horrocks, Jeremiah 282
hospitals 42, 108, 138, 155, 157–8, 161, 257, 262, 267
Houdini, Harry 286
Houff, William Harper 115
House of Commons 136–8, 141, 144
Houshmand, Z. 201
Hueppe, Ferdinand 233
Huntley, Maxine 119
Huygens, Christiaan 75–9, 84

Iliffe, Rob 66, 73, 78–9, 82
Immhausen, A. 188
immune system 156, 166
imported foods 226–7
In His Silence: A Christian History 291
Independent Police Complaints Commission *see* IPCC
industry 118, 149, 222, 230, 237; chemical 226, 236; dairy 234; nuclear 121; private 244; textile 225; ultra-hazardous 116

informed consent, concept of 12, 153, 159
Ingelfinger, Franz 20–1
innovation processes 90–1, 93, 96, 98, 106–9
Institute for Advanced Study 17
International Fortean Society 23, 274
IPCC 254, 268

al-Jabrw'al-Muqabala, Kitab 189
Jackson, Catherine 234
Jauchem, J.R. 259–60
Jaworski, Adam 4, 6, 90
Jenkinson, E. 260, 262
Johns, Adrian 278
Jones, E. 137
Jones, J.H. 154
Jones, V.C. 11
Jones, W.J. 232
Joseph, G. 180–1
Joseph Farman with a Dobson Spectrophotometer at British Antarctic Survey headquarters, Cambridge, 2011 32
Journal of the American Society for Information Science and Technology 49
journalists 11, 14, 19–21, 115, 118, 120–1, 124–6, 128–30, 279–80, 284; David Proctor 124; Hill Williams 121; local 11, 115, 125, 128, 130; medical 20; science 20; Theodore Dreiser 275, 278–80, 290, 292
journals 11, 13–14, 19–20, 49, 53, 223–4, 226, 241, 245–6, 286; *British Medical Journal* 246; *Bulletin of the Atomic Scientists* 123; *Cell* 246; *Health Reformer* 226; *Journal of the American Society for Information Science and Technology* 49; *Lancet* 224; *Nature* 13, 246; *New England Journal of Medicine* 20; *Philosophical Transactions* 65, 70, 73, 76, 78; *PLoS One* 246; *Poisoned Candies* 226; *Providence Sunday Journal* 280; *Punch* 224; *Science* 13; scientific 19–20, 241, 275–6

Kahun Fragments (source of evidence for ancient Egyptian mathematics) 176, 183, 190
Kaiser, D. 13–14, 119
Kassell, L. 82
Kawaoka, Y. 13
Keane, John 7

Keating, M.J. 164
Kempner, Joanna 14–15
Keynes, R.D. 53–4
al-Khwarizm, Muhammad ibn Musa 189
Kiernan, Vincent 20–1
Kingsley, Charles 224
Krippendorff, K. 93, 104
Kroll, M. 259–60

"Lama Öser" 202
Lancet 224
Lawrence, Peter 50
Laws, Dick 40
Lee, Nick 205, 253, 256–8, 267, 269
lethal weapons 253–4, 256, 259, 264–5, 267, 270; analysis of 259; and firearms 265; and the knowledge gaps around TASER 269; less 253–4, 265; and the regulatory particularities of TASER 253; and the variety routinely issued to patrol officers for self-defence purposes 265
leukaemia 162–3
Levenson, Robert 203, 207
Linnean Society 61–2
Linus, Francis 66, 76, 79–80
Lo ! 275
London 54–5, 59, 81, 136, 162, 233, 275, 278; existence comprises work, noise and haste 54; and Middlesex Hospital in 162; milk 228; and New York 275
Lovecraft, H.P. 280–1, 287
Lucas, Anthony 79–80
Luhmann, N. 95, 108
Lutz, Antoine 203, 205, 208
Lyell, Charles 58, 60–2
lymphoma 154, 166

MacCulloch, Diarmaid 291
MacIntyre, A. 52
MacIntyre, Alasdair 52
MacLure, Maggie 4–5
Macuglia, Daniele 11, 115–30
magazines 226–7, 275, 283; *Astounding Stories* 283; *Friendly Companion: A Magazine for Youth* 227; *Funny Folks* 226; pulp 283; *Smith's Magazine* 280
Manhattan Project 9–12, 17, 116–17
Mathematical Leather Roll (source of evidence for ancient Egyptian mathematics) 176
mathematical texts 172, 175, 181, 186–8, 190

mathematicians 69, 172–4, 183–4, 187
mathematics 65–6, 69, 73–4, 76, 172–6, 178, 181, 185, 187, 189–91
Maturana, Humberto 95
McCluskey, Harold 123
McFadden, K.L. 89
media 11–12, 20, 24, 40–1, 121, 222, 224, 237, 245, 264; coverage 21, 115; embargoes 19–20; international 202; mass 241; national 118; social 246; web-based 250
meditation 23, 193–4, 199–207, 210–12; accounts of 204, 212; effects of 23, 194, 197, 206–8; experience of 197, 205, 209–11; intensive 201; practice of 209; recounting of 197, 205; research 202–3, 207; studies of 208; techniques of 204
meditators 194, 199, 203–5, 207, 209–11; accomplished 210; advanced 201; long-term 208–9; long-term Buddhist 208; long-term Vipassana 209; practising 197; training of 211
Mellor, Felicity 2–24
Merchant, Paul 18, 31–46
Merton, Robert K. 9
Messenger, S. 48, 52, 54, 58
Metzger, Miriam 241
Michaels, D. 254, 256
Middle Kingdom 175–6
Mindful Attention Awareness Scale 208
Mold, A. 152
monks 201–3, 205
Montreal Protocol 44, 46
Mooers, C. 118–19
Moore, D. 52
Moore, M. 249
Morris, Ben 148
Morris, L.V. 89
Moscow Mathematical Papyrus (source of evidence for ancient Egyptian mathematics) 176–7, 182, 185–9
Muter, John 231

National Police Chief's Council *see* NPCC
National Research Council 13
National Science Advisory Board for Biosecurity *see* NSABB
Natural Environment Research Council *see* NERC
natural philosophy 15, 65–6, 69, 74, 81–4
Nature 13, 246
NERC 35, 37, 40–1

networks 90–3, 95, 97–9, 104, 107–10, 267, 279; alchemical 82; communicative 18; conversational 93; dynamic innovation 90; international 45; multi-actor 91; strategic internal 97
Neugebauer, Otto 172, 180
neuroscience 194, 202, 207, 211; affective 206–7; cognitive 206
neuroscientists 206, 208
Neve, M. 48, 52, 54, 58
New England Journal of Medicine 20
New Experiments Touching Cold 67
New Lands 275, 277, 283–4, 292
New Theory of Light and Colours 65–6, 69–70, 73–5, 77–9, 81
New York Times 118, 120, 122
New York World 280
Newman, William 82, 85
newspapers 115, 118–19; *Bradford Observer* 225; *Fortean Times* 274; *The Guardian* 51; *New York Times* 118, 120, 122; *New York World* 280; *Seattle Times* 117–25, 128–9; *The Sheffield and Rotherham Independent* 227; *Washington Post* 14
Newton, Isaac 15, 65–85, 241; communication strategy 67; correspondence output dwindles to a bare minimum 76; correspondence with Collins 76; correspondence with Huygens 66, 76; experiment with a prism 71; and his alchemical interests 82; and his correspondence with Ignace-Gaston Pardies 74–7, 79; and his response to Oldenburg 74; insists that colour was a property of light 71; resigns from the Royal Society 65; and the status of *experimentumcrucis* 74–5
Nicolini, D. 91–2, 96, 108
Nielsen, M. 241
non-Buddhists 209–10
NPCC 257, 260, 264–5
NSABB 13–14
Nuffield Council on Bioethics 51

O'Connor, Lieut-Gen Sir Richard 148
Oldenburg, Henry 65–6, 69–70, 73–6, 78–80, 84
"open scholars" 243
open science 9, 12, 21, 241, 243–5, 247–50; breaks system boundaries 21; environment 241; philosophy 249; practice of 249
Oppenheimer, Robert 10, 17

Opticks 72, 84
oral history 31, 152, 169; individual 158; interviews 18, 31, 45
Oral History of British Science (project) 31
The Origin of the Species 56
Osborne, Harold 120, 125
ozone 32–7, 40–2; in the Antarctic stratospheric 32; concentrations 32, 39, 45; depletion 37, 40, 44; layer 18, 42, 44; measurements 35, 38, 45; recording of 31
The Ozone Crisis 38
"ozone hole" 31, 33, 35–41, 43–5

Paget, Gen. Sir Bernard 144
PANAS scales 207
papyrus 174–6, 183, 186, 190
Pardies, Ignace-Gaston 74–7, 79
Perkin, William 222
Petitmengin, C. 194, 208, 210–11
Pettit, M. 140–1
Philosophical Transactions 65, 70, 73, 76, 78
Pickering, A. 254–5
PLoS One 246
poems: *Goblin Market* 224
Poisoned Candies 226
poisons 223–4
police 253, 257, 260, 262–4, 268; environment 267; and TASER guidance 265; use of firearms and less lethal weapons 265
Pope, Alexander 117, 248
Principia Mathematica 74, 84
Proctor, David 124
Proctor, R.N. 221
programmes 36, 40, 60, 89, 91, 96–107, 109, 121; alchemical 67; cervical cancer screening 256; cognitive therapy 206; educational 89, 101
Providence Sunday Journal 280
psychiatrists 19, 135–49; army 136, 141, 149; enabled 135; role of 19, 142; silent 142; withdrawing 19
public analysts 222, 225, 228–31, 233–4, 236–7; *see also* analysts
publications 10, 123, 201, 203, 243, 260, 280; *Agnotology: The Making and Unmaking of Ignorance* 221; *The Book of the Damned* 275; *Destructive Emotions and Happiness* 202–4, 206, 209; *Experiments and Considerations Touching Colours* 67; *In his Silence: A*

Christian History 291; *In His Silence: A Christian History* 291; *New Experiments Physico-Mechanical: Touching the Spring of the Air and Their Effects* 67; *New Experiments Touching Cold* 67; *New Lands* 275, 277, 283–4, 292; *New Theory of Light and Colours* 65–6, 69–70, 73–5, 77–9, 81; *Opticks* 72, 84; *The Origin of the Species* 56; *The Ozone Crisis* 38; *Principia Mathematica* 74, 84; *Royal Society's Philosophical Transactions* 245; *Silence: A Christian History* 291; *The Twelve Gates* 83; *Wild Talents* 275, 277, 287
Punch 224
Pyle, J. 37, 46

Quist, S.A. 10

radioactive material 120, 123, 125, 127
radioactive waste 116, 121–3, 125–6; high-level 122; leaks of 122; solid 116
radioactivity 117, 119–21, 123, 125–6, 129
Raper, Jack 11
Rappert, Brian 23, 94, 193–212, 253–7, 259, 264, 267, 269–70
Redwood, Theophilus 229
Rees, J.R. 136, 138, 148
Reisner Papyrus (source of evidence for ancient Egyptian mathematics) 176
The relative frequency of multiplication techniques in the Rhind Mathematical Papyrus **181**
research 13–14, 16, 20–1, 24, 50–1, 152–3, 167–9, 241–4, 248–50, 278–9; cancer 12, 117, 123–4, 152–5, 157, 159, 161, 163, 165, 167–9; chemical weapon 11; clinical 159, 168; cognitive-neuroscientific 208; collaborative 201; conducting of 124; culture 50–1, 242; dual use of 14; experimental 81; findings 20–2; funded 249; governance 154, 159, 169; industry-sponsored 22; laboratories 9; nuclear 10, 118; outputs 243, 246–7, 250; practices 153, 168; projects 243, 247, 249–50; psychic 283; qualitative 269; subjects 153, 168; teams 14, 38, 167, 201; unclassified 13
Rhind Mathematical Papyrus (source of evidence for ancient Egyptian mathematics) 176, 178–9, 182

Ricard, Matthieu 201–5, 208; descriptions of changes occuring in his emotional state 203; experience as a meditator 204; and his startle response reflex 203
Rickman, J. 139–44
Ripley, George 83
RMP 176, 178–9, 181–6
Roan, Sharon 38
Rogers, Alan 36, 91–2
Rowland, Sherry 42
Royal College of Chemistry 222
Royal Society 50, 65, 67, 69–70, 73, 75–8, 80, 242, 247, 249
Royal Society's Philosophical Transactions 245
Russell, M. 242–3
Ruud, Casey 124, 128

SACMILL 254, 262–5
Savannah River 128
Schaffer, S. 71, 77
Schiebinger, L. 2, 254
Schilt, Cornelis J. 15, 65–85
Science 13
Scientific Advisory Committee on the Medical Implications of Less-Lethal Weapons *see* SACMILL
scientists 10, 13–18, 20, 23–4, 48–51, 200–4, 237–8, 241–7, 249, 276–9; atomic 123, 126; German 10, 13, 226, 228, 233; nuclear 10; professional 228, 248; Western 201
Scott, R.L. 4–6
Seattle Times 117–25, 128–9
Second World War 19, 115, 117, 274
Shapin, Steven 17, 67, 241, 278–9
Shapiro, A.E. 74
Shaw, George Bernard 282
Shear, J. 208
The Sheffield and Rotherham Independent 227
Shrum, W. 98, 107
Shuttleworth, S. 280
silence 2–24, 31–46, 89–110, 115–30, 135–49, 152–69, 172–91, 193–212, 253–70, 274–92; communicative 4–7; in educational innovation 91, 93, 95, 97, 99, 101, 103, 105, 107, 109; epistemological 9, 22–3; generative 8, 16, 19, 21; managing of 255, 257, 259, 261, 263, 265, 267, 269; non-communicative 4, 23; nuclear 11–12; official 135, 137; an outcry of 274–5,

277, 279, 281, 283, 285, 287, 289, 291; persistent 167–8; in the public sphere of interest 256; scientific 52, 60; strategic 18, 36, 106, 163–4; suppressive 8–9, 11–12, 16, 19; temporary 13–14, 16, 20, 24; textual 16; typology of 7, 9, 22
Silence: A Christian History 291
Singer, Tania 202
Sleigh, Charlotte 23–4, 274–92
Smith's Magazine 280
social media 246
Society of Public Analysts, *see* SPA
SPA 228–30
Stanley, Oliver 136
Starkey, George 83, 85
Steele, Karen Dorn 125
Steele, K.D. 116, 124–5, 129
stem cell transplants 156–7, 159
studies: cognitive-neuroscientific 206; contemplative neuroscience 208; experimental 78, 194, 202
Sussman, A. 254, 261, 263
Suzuki, D.T. 199
synthetic dyes 12, 221–3, 227–9, 231, 233–4, 236–8; in nineteenth-century food 223, 225, 227, 229, 231, 233, 235, 237; uses of 226, 233; *see also* dyes

TASER 253–4, 258–69; and deaths in custody 260; debate 255, 259; deploying 264; deployment 265; evaluation of 261; exposure 263; less-lethal policing weapon 15; models 253–4, 259–64, 266–7; safety of 253–4, 259–60, 262–4, 266–9; technology 253–4; trained officers 264; use of 15, 261, 264–8
TASER International 253, 259, 261
Tegetmeier, W.B. 59
Tenzin, L. 200
Thacher, David 269
Thayer, Tiffany 274, 287
Thomas, James P. 129
Thompson, E. 197, 203, 208, 211
Thomson, Janet 34, 40
Thorne, Lieut-Gen. Andrew 137–8
Tomes, N. 152
transplants 156–9, 163–4; allogeneic 156; bone marrow 156, 160, 163–4; donor 156–7; stem cell 156–7, 159
Travis, A.S. 222
Trist, E. 136, 139–43, 145, 147
Tucker, A. 90, 105, 108

Turnbull, H.W. 65–6, 69–70, 72–6, 78–80, 84
The Twelve Gates 83
typology of silence 7, 9, 22
Typology of Silence **8**

UK 12, 129, 154, 242, 245, 254, 256–7, 262, 265–6, 274; Amnesty International 260; government 50; recipients of MOPP chemotherapy 162; science 51; and the silence existing around the safety of the weapon 262
UK Home Office 263
UK Marine Biological Organisation 44
UN Convention on the Rights of the Child 257
Ungerson, Bernard 148
United Kingdom *see* UK
United States 116, 118–19, 122, 127–8
US 10, 115–17, 119, 125, 130, 200; atomic arsenal 126; dairy lobby 228; government 13; newspapers 119; officials 115; physicists 10, 12; scientists 9
USA *see* US

van der Sanden, Maarten 18, 89–110
Varela, Francisco 95, 206, 208
verbal communication 24, 90, 92, 106, 108
Verouden, Nick 18–19, 89–110
von Hofman, August Wilhelm 222, 226
von Hofmann, August Wilhelm 222
Vostal, Filip 51

Wales 253–4, 262, 264, 268–9
Walker, Samuel 117, 126
Wallace, B.A. 61–2, 289
Walsh, Timothy 195
War Cabinet 145, 148
War Office Committee 148
War Office of Britain 135–6, 148
War Office Selection Board, *see* WOSBs
Washington Post 14
waste 52, 94, 117, 120, 122, 290; by-products of 119; chemical 115, 125; coal-tar 221, 229, 231, 237; hazardous 116; issues 117
waste management 122
water 54, 116, 118–21, 198, 284; contaminated 116, 121; drinking 122
Watson, Bob 42
weapons 11, 15, 36, 253–4, 259–69, 288; development of 11; electric discharge 266; electric-shock 253; nuclear 11,

115–16; technology of 14, 254;
unauthorised use of 264; uses of 266
Weart, S. 13
Webster, Stephen 18, 48–62
Wedgwood, Hensleigh 54
Wellerstein, A. 11
Western scientists 201
Westfall, Richard 66, 71, 76, 79–80, 82
Weyl, Theodore 232
White, Alice 19, 135–49
Whorton, J.C. 224
Wild Talents 275, 277, 287
Williams, Hill 121

Wolff, Eric 34
Woods, John 37–8
Woolgar, Steve 254–5
Words used for geometric dimensions in the Rhind Mathematical Papyrus **186**
World War II 115–16, 118, 135
WOSBs 135, 138–41, 144–5, 147–9

Xu, H. 89, 207–8

Yeo, Richard 278

Zboray, R.J. 280

Taylor & Francis eBooks

Helping you to choose the right eBooks for your Library

Add Routledge titles to your library's digital collection today. Taylor and Francis ebooks contains over 50,000 titles in the Humanities, Social Sciences, Behavioural Sciences, Built Environment and Law.

Choose from a range of subject packages or create your own!

Benefits for you
- Free MARC records
- COUNTER-compliant usage statistics
- Flexible purchase and pricing options
- All titles DRM-free.

Benefits for your user
- Off-site, anytime access via Athens or referring URL
- Print or copy pages or chapters
- Full content search
- Bookmark, highlight and annotate text
- Access to thousands of pages of quality research at the click of a button.

REQUEST YOUR FREE INSTITUTIONAL TRIAL TODAY

Free Trials Available
We offer free trials to qualifying academic, corporate and government customers.

eCollections – Choose from over 30 subject eCollections, including:

Archaeology	Language Learning
Architecture	Law
Asian Studies	Literature
Business & Management	Media & Communication
Classical Studies	Middle East Studies
Construction	Music
Creative & Media Arts	Philosophy
Criminology & Criminal Justice	Planning
Economics	Politics
Education	Psychology & Mental Health
Energy	Religion
Engineering	Security
English Language & Linguistics	Social Work
Environment & Sustainability	Sociology
Geography	Sport
Health Studies	Theatre & Performance
History	Tourism, Hospitality & Events

For more information, pricing enquiries or to order a free trial, please contact your local sales team:
www.tandfebooks.com/page/sales

The home of Routledge books

www.tandfebooks.com